Science, Technology and the Cultural Cold War in Asia

Tsuchiya presents a new insight into the political roles of science and technology during the Cold War era in Asia.

The Cold War was not only a battle of conflicting ideologies and economic systems, but also a competition of cultures and lifestyles, and a battle to win the hearts and minds of people in developing countries. Tsuchiya argues that science and technology were an integral part of how culture was deployed strategically. She discusses the 1950s and early 1960s: the Eisenhower and Kennedy presidencies in the U.S., and the decolonization and nation-building efforts in Japan, South Vietnam, Burma, and Indonesia. She also sheds light on the way U.S. technological aid programs such as Foreign Atoms for Peace, and the overseas information program were received by Asian leaders, technocrats, and scientists.

Provides valuable insight for scholars of Cold War History in Asia and US Foreign Policy.

Yuka Moriguchi Tsuchiya is Professor of History and American Studies in the Graduate School of Human and Environmental Studies at Kyoto University, Japan.

The Cold War in Asia
Series Editor: Professor Malcolm H. Murfett

A series of books that both explores and addresses some of the more important questions raised by the Cold War in Asia. This series isn't confined to single country studies alone, but welcomes contributions from research scholars who are tackling comparative issues within Asia during the time of the Cold War. Quality is our goal and this series reflects this objective by catering for work drawn from a number of disciplines.

If you work in the broad field of Cold War studies don't hesitate to get in touch with the series editor Professor Malcolm Murfett at King's College London (malcolm.murfett@kcl.ac.uk). Books, both single authored and edited manuscripts, should preferably be within the 60,000–100,000-word range, although we are also interested in shorter studies (25,000–50,000 words) that focus on elements of the Cold War struggle in Asia.

If you are working on a project that seems to fit these guidelines, please send a detailed proposal to the series editor. Every proposal will, of course, be subject to strict peer review. If the proposal is supported by experts in the field, it will be our aim to begin publishing the next volumes of this series within a year to eighteen months of the issuing of a contract to the author.

We look forward to hearing from you.

The Southeast Asia Treaty Organisation
Ang Cheng Guan

Defectors from the PRC to Taiwan, 1960-1989
Anti-Communist Righteous Warriors
Andrew D. Morris

Science, Technology and the Cultural Cold War in Asia
From Atoms for Peace to Space Flight
Yuka Moriguchi Tsuchiya

For the full list of titles in the series, visit https://www.routledge.com/The-Cold-War-in-Asia/book-series/CWA

Science, Technology and the Cultural Cold War in Asia

From Atoms for Peace to Space Flight

Yuka Moriguchi Tsuchiya

Routledge
Taylor & Francis Group

LONDON AND NEW YORK

First published 2022
by Routledge
4 Park Square, Milton Park, Abingdon, Oxon OX14 4RN

and by Routledge
605 Third Avenue, New York, NY 10158

Routledge is an imprint of the Taylor & Francis Group, an Informa business

British Library Cataloguing-in-Publication Data
A catalogue record for this book is available from the British Library

Library of Congress Cataloguing-in-Publication Data
A catalog record has been requested for this book

ISBN: 978-1-032-15328-5 (hbk)
ISBN: 978-1-032-15329-2 (pbk)
ISBN: 978-1-003-24364-9 (ebk)

DOI: 10.4324/9781003243649

Typeset in Galliard
by Deanta Global Publishing Services, Chennai, India

For my two mentors, Professor Elaine Tyler May and Professor Marlene Mayo, who taught me both the meaning of, and the joy to be found in, historical research.

Contents

Figures

Acknowledgment

The author would like to express her heartiest gratitude to the individuals and organizations that offered generous help and advice throughout this work. Firstly, the author would like to thank Japan Society for the Promotion of Science (JSPS) Grant-in-Aid from 2017 to 2020 (No. 17H02238) for their generous funding in promoting this research. Secondly, this book project would not have been possible without the precious help of many archivists and librarians both in the U.S. and in Japan. Special thanks go to the archivists and librarians of the U.S. National Archives at College Park, Maryland; the U.S. National Archives at Chicago, Illinois; Dwight D. Eisenhower Presidential Library; Hanna Holborn Gray Special Collections Research Center, University of Chicago Library; Bentley Historical Library, University of Michigan; the Diplomatic Archives of the Ministry of Foreign Affairs of Japan; the National Diet Library of Japan; and Kyoto University Library. Thirdly, the author's appreciation goes to her colleagues and friends who offered generous help and friendly advice in preparing the English manuscript of the book. The original version of this book was published in the Japanese language by Kyoto University Press in 2021, but converting such a volume into the English version is by no means a simple matter of word-to-word translation. Sometimes the concepts, nuances, and examples had to be replaced to make the book accessible to English-language readers. In this respect, the author would like to thank Dr. Daniel Pearce for his amazing linguistic ability and his perseverance in translating part of the book, and in proofreading the whole manuscript. The author would also like to thank Ms. Mayuko Morita, a dedicated office assistant, who corrected typing errors and adjusted the manuscript to fit the style manual. The author is also grateful to her colleagues at Kyoto University, especially Professors Yoshiomi Saito, Izuru Ohta, and Masaki Fujioka, who have constantly provided helpful advice and keen opinions on my research and writing. Graduate students at Kyoto University brought the author much joy and energy through lively discussions in graduate seminars and other meetings. Especially in the reading seminar of spring 2021, Shotaro Shindo, Takanori Yagi, Kento Morie, Toshiaki Kato, Sayaka Fukada, and Naoki Magata engaged with the manuscript and provided thoughtful comments and discussions. The author is also greatly indebted to the professional and friendly advice of many scholars outside of Kyoto University. While it is unfortunately impossible to name all of them, Professor Hiroshi Ichikawa of Hiroshima University, for example, has read the entire Japanese manuscript and provided expert comments as a science historian. Professors Shin Kawashima, Somei Kobayashi, Shinsuke Tomotsugu, Yuko Sato, Hiromi Mizuno, Cha Jae-Yong, Heo Eun, Moon Manyong, Lan Shichi, Yang Zhang, Nick Cullather, and Miriam Kingsberg have been wonderful collaborators on another project, but their insights helped the author with the writing of this book, too. The author is also ever thankful to Professor Elaine Tyler May, academic advisor in the doctoral program at University of Minnesota, and Professor Marlene Mayo, academic advisor in the master's program at University of Maryland. Professors and colleagues at the University of Minnesota were a deciding factor in the forming of the author as a historian and scholar, as were the author's former colleagues at Ehime University in Japan. The author also appreciates the numerous scholars and students with whom she has interacted through countless opportunities in academic conferences and small research gatherings, either on-line or in person. Although COVID-19 has significantly changed the world of academic research, and the author also had difficulty in the last-minute collection of data, she still feels fortunate to have so many good colleagues across national boundaries. One thing the Cultural Cold War teaches us is that, even amid political and military confrontations, and even within the confinement of government-sponsored information programs, sometimes genuine comradeship between scientists took shape. In the fragmented and confrontational world of today, there is still much to be learned from the age of the Cultural Cold War.

Last but not least, I would like to express my heartiest appreciation to Mr. Simon Bates, editor at Routledge (Taylor & Francis Asia Pacific), and Professor Malcolm Murfett of Kings College, London, for their thoughtful advice on my book project, two anonymous reviewers whose comments greatly help me improve my manuscripts, and Mr. Shubhayan Chakrabarti, Editorial Assistant, BSE (Taylor & Francis Books India).

Kyoto, November 2021
Yuka Moriguchi Tsuchiya

Abbreviations

ABCC	Atomic Bomb Casualty Commission
AEC	Atomic Energy Commission
AID	Agency for International Development (USAID)
BAS	Bulletin of the Atomic Scientists
CCF	Congress for Cultural Freedom
CIA	Central Intelligence Agency
CINCPACFLT	Commander in Chief, Pacific Fleet
COSPAR	Committee on Space Research
DMPH	Division of Medicine and Public Health (Rockefeller Foundation)
DOD	Department of Defense
EBWR	Experimental Boiling Water Reactor
FBI	Federal Bureau of Investigation
FDA	Food and Drug Administration
FOA	Foreign Operations Administration
HEW	Department of Health, Education, and Welfare
IACF	International Association for Cultural Freedom
IAEA	International Atomic Energy Agency
ICA	International Cooperation Administration
ICBM	Intercontinental Ballistic Missile
IGY	International Geophysical Year
IHD	International Health Division (Rockefeller Foundation)
IINSE	International Institute of Nuclear Science and Engineering
IRBM	intermediate-range ballistic missile
ISNSE	International School of Nuclear Science and Engineering
JCG	Japan Coast Guard
LNHO	League of Nations Health Organization
MATS	Military Air Transport Service
MIT	Massachusetts Institute of Technology
MMPP	Michigan Memorial Phoenix Project
MOFA	Ministry of Foreign Affairs (Japan)
MSA	Mutual Security Agency
NACA	National Advisory Committee for Aeronautics

NARA	National Archives and Records Administration
NASA	National Aeronautics and Space Administration
OPI	Office of Public Information
NOAA	National Oceanic and Atmospheric Administration
NPT	Treaty on the Non-Proliferation of Nuclear Weapons (Non-Proliferation Treaty)
NSC	National Security Council
NSF	National Science Foundation
OCB	Operations Coordinating Board
OIAA	Office of Inter-American Affairs
ONR	Office of Naval Research
ORSORT	Oak Ridge School of Reactor Technology
OSS	Office of Strategic Services
OWI	Office of War Information
PHS	Public Health Service
PKI	Partai Komunis Indonesia (Communist Party of Indonesia)
PLAF	The People's Liberation Armed Force of South Vietnam
PSAC	President's Science Advisory Committee
PTBT	Partial Test Ban Treaty
PTPI	People to People International
SEATO	Southeast Asia Treaty Organization
TVA	Tennessee Valley Authority
UNESCO	United Nations Educational, Scientific and Cultural Organization
UNRRA	United Nations Relief and Rehabilitation Administration
USAID	U.S. Agency for International Development
USIA	U.S. Information Agency
IBS	International Broadcasting Section, USIA
IMPS	International Motion Picture Service, USIA
IMS	International Motion Pictures Section, USIA
IOP	Policy Planning Section, USIA
IPS	Publication Section, USIA
ITV	International Television Section, USIA
USIS	U.S. Information Service
USOM	U.S. Operation Mission
UTR	University Training Reactor
VOA	Voice of America
WHO	World Health Organization

1 Introduction

1.1 Why Cultural Cold War and S&T?

Cultural Cold War and Science & Technology is a book on the history of diplomatic relations between the U.S. and Asia, but not one on conventional state-to-state relations. Chapters of this book involve not only presidents and high-ranking foreign service officers but also scientists, engineers, students, physicians, and businessmen. These actors were deeply involved in Cold War foreign relations, despite the fact that, in most cases, they were not directly concerned with the policy-making process. Rather, they participated in the arena of international politics as non-governmental or quasi-governmental actors. Moreover, many simply practiced their own business: scientists practiced science and physicians practiced medicine. Nonetheless, their roles were all too important in cultivating favorable state images, enhancing national prestige, winning friends, and warding off foes. This book illuminates how non-governmental specialists were important political actors of the Cold War, and especially in the *Cultural* Cold War.

The U.S. and the Soviet Union attempted to gain the leading edge not only through nuclear armament and economic systems, but also through the superiority of their national cultures. Films, music, literature, and art became another field of their competition. However, the Cultural Cold War was fought not only in so-called "high culture," but also in education systems, infrastructure, modern lifestyles, and science and technology (S&T). S&T was an especially important arena of the Cultural Cold War because it could exert long-lasting and pervasive influences on the economic, political, and social systems of the target countries. S&T generated not only tangible products from rockets to kitchen appliances, but also the intangible: authority, respect, and glamor—which governments could employ as "soft power." S&T contributed to the "massive retaliation" or "mutual assured destruction" policies, but it also served as a barometer of modernity, prosperity, and even the morality of countries. People all over the world looked to the U.S. and the Soviet Union and wondered—or at least so the leaders of the two superpowers thought—which country could bring solutions to poverty, disease, and hunger; which country would sincerely strive to maintain world peace; which country would better assist with the decolonization and modernization of the Third World? The answer seemed to depend on the type and level

DOI: 10.4324/9781003243649-1

of S&T the two superpowers developed, and thus both the U.S. and the Soviet Union paraded their S&T in attempts to win the allegiance of foreign leaders and the favor of foreign citizens.

S&T is closely related with culture. Take the example of a tomato harvesting machine cited by science historian Langdon Winner. From the late 1940s, researchers at the University of California developed a mechanical tomato harvester which was "able to harvest tomatoes in a single pass through a row, cutting the plants from the ground, shaking the fruit loose, and (in the newest models) sorting the tomatoes electronically into large plastic gondolas … headed for canning factories." In response to the new machine-harvesting, agriculturalists cultivated new breeds of tomatoes that were "hardier, sturdier, and less tasty than previously grown," and better able to withstand the harvesting process. Winner cited this example to show that technologies were "ways of building order in our world." Technologies influence "how people are going to work, communicate, travel, consume, and so forth over a very long time"; in such a process, Winner argues that "different people are situated differently and possess unequal degrees of power as well as unequal levels of awareness."[1] Winner's example shows that new technologies influence not only scientists and engineers but also society and culture as a whole. The hardier, but less tasty, tomatoes might even affect people's sense of taste and food culture, leaving irreversible changes in society. The tomato harvester in this example can be substituted with automobiles, kitchen appliances, or nuclear power plants in other contexts.

Cold War scholars and science historians have made sporadic reference to the important role that S&T played in the Cultural Cold War. For example, Kenneth Osgood pointed out that nuclear energy and space technology were important themes of U.S. public diplomacy during the 1950s. Ruth Oldenziel and Karin Zachmann dealt with American kitchens and appliances propagated as symbols of American modernity in European countries.[2] John Krige discussed the United Nations International Conference on the Peaceful Uses of Atomic Energy which stirred the desire of developing countries for modern technologies. In Japan, Hiroshi Ichikawa explored the Soviet Science Academy's efforts to win international recognition for Soviet nuclear energy through international conferences.[3] Moreover, the relationship between S&T and the Cultural Cold War is demonstrated by the U.S. Information Service (USIS) films U.S. government-sponsored films shown in some 80 countries all over the world. For example, of a total of 456 titles in the 1959 Japanese-language *USIS Film Catalog* compiled by USIS Tokyo, 52 films directly concerned S&T. These films included clear-cut messages of how American S&T changed culture and society. For example, one USIS film portrayed a team of American scientists who developed streptomycin, an antibiotic to combat tuberculosis; another introduced the Museum of Natural Science in New York, where ordinary citizens could learn scientific knowledge; still another portrayed how U.S. nuclear technology would bring about modern medicine and good living.[4] These films inspired political leaders, scientists, and ordinary citizens in decolonizing and developing countries striving to build modern societies. All these examples show that S&T and its appeal became a weapon of the Cultural Cold War. However, very few existing scholarly works deal directly with the relationship between the Cultural Cold War and S&T. Audra Wolfe's seminal work of 2018, which will be introduced more in detail later, is likely the only exception.[5]

This book deals with the complicated relationship between American S&T and the Cultural Cold War in Asia, with a particular focus on Japan due to the author's ease with Japanese-language primary and secondary sources but also on Vietnam, Burma (present-day Myanmar), and, to a lesser degree, on Korea and Taiwan. The book consists of English translations of the authors' Japanese-language articles and conference presentations of the past seven years. Some have been substantially revised and expanded with the integration of new evidence and recent scholarship in the field. It covers the early Cold War era (roughly from the mid-1950s to the early 1960s), during which time the U.S. government carried out "overseas information programs" on S&T. The notion of the "overseas information program" was similar to today's "public diplomacy," as the U.S. government attempted to promote national prestige and cultivate likability through the dissemination of favorable information about the country. This book uses "overseas information programs" instead of "public diplomacy" because it was the terminology used by the U.S. government in the early Cold War era. S&T was one of the most important themes of the U.S. overseas information program. Not only actual technologies but also *images* and *information* on U.S. technologies were exported in the effort to win the Cultural Cold War. Much of the specific S&T products in the Cold War were "dual-use technologies" applied both to military and civilian purposes. For example, rocket engines could be used either for launching non-military weather satellites or ICBMs, and nuclear reactors could be used either for generating power or producing nuclear bombs. This book establishes that S&T in the early Cold War era involved not only dual-, but *triple*-use technologies because they simultaneously served military, civilian, and overseas information purposes.

Scholarly works on the U.S. overseas information programs—although this term was not necessarily used by those authors—emerged in both English and Japanese from around the early 2000s. Through these works, the Cold War was redefined as not only ideological, economic, and military confrontations but also a cultural phenomenon involving literature, films, music, art, and other cultural aspects. Early Cultural Cold War scholars such as Frances Stonor Saunders tended to criticize the state power for meddling in culture and hindering the free marketplace of ideas.[6] By contrast, scholars such as Penny Von Eschen and Yoshiomi Saito demonstrated that music and musicians crossed cultural and national boundaries regardless of the intentions of state authorities. Fumiko Fujita discussed the U.S. Information Agency (USIA) cultural programs for Japan, focusing on personal exchanges not always defined by state policies. (The USIA was a government agency for overseas information programs established in Washington DC in August 1953. It oversaw the USIS U.S. Information Service located in more than 80 countries. More information on the USIA will follow later.) Research dealing with non-governmental actors has increased over time. For example,

Takeshi Matsuda explored the Rockefeller Foundation's influences on Japanese intellectuals, and Jiyoung Kim discussed the Rockefeller Foundation's aid to Japanese writers as one aspect of the Cultural Cold War. Giles Scott-Smith, et al., dealt with journals published by the Congress for Cultural Freedom (CCF, more details will be given in Chapter 1). Further, many scholarly works have focused on films, radio, and exhibitions as the media through which the Cultural Cold War unfolded. For example, Somei Kobayashi focused on the East Asian theater of Voice of America (VOA) radio broadcasting, and Mitsuo Ikawa examined the Japanese reactions to the Atoms for Peace exhibitions. The present author has also published books and articles on the USIS films and VOA radio.[7]

The study of the Cultural Cold War is inevitably interdisciplinary because it deals with both "culture" and the "Cold War." Until recently, studies of culture were the province of cultural and social historians, while the Cold War was the territory of political and diplomatic historians. This is probably truer in Japan than in the U.S. because traditional disciplinary boundaries have historically been maintained to a greater degree in the former. Yet even in the Western world, cultural historian Peter Burke remarked that cultural history had once been "a Cinderella among the disciplines, neglected by its more successful sisters," i.e., political, military, and economic histories.[8] Even in U.S. diplomatic history, it was not until the 2000s that an increasing number of scholars focused on cultural and transnational aspects of foreign relations. Former president of the Society for the Historians of American Foreign Relations (SHAFR) Anne L. Foster, looking back at the history of *Diplomatic History*, the most prominent journal of the field, explored how the concept of diplomatic history had been broadened as more women and people of color joined the field.[9] "Culture" and the "Cold War" are not regarded as intertwined concepts in some academic quarters, and the author was often faced with questions from diplomatic historians such as: "Does the focus on culture make any change in the overall history of the Cold War?" "But you cannot evaluate the influence of culture, can you?" "Cultural diplomacy is a subsidiary measure in the course of hard power diplomacy, is it not?"

These questions are partly valid because it is indeed difficult to evaluate the effect of cultural diplomacy. Since literature, films, music, and art are freely appropriated and interpreted by recipients, they do not necessarily have the effect that the Cultural Cold Warriors had intended. It does not mean, however, that culture did not have any political effect. Foreign citizens were fascinated by American government-sponsored films and music, and foreign writers and artists genuinely appreciated U.S. grants and scholarships. Such emotional impacts certainly have long-term influences in their societies even if they cannot be quantitatively evaluated.

S&T is both similar to, and different from literature, music, and art. It is *similar* because S&T provokes emotional reactions such as fascination, awe, respect, yearning, or desire. The U.S. government employed nuclear power or space flight as a theme of films, exhibitions, and television programs, and invited foreign scientists, engineers, and business leaders to U.S. scientific laboratories as part of their overseas information programs. These programs probably left a long-lasting emotional impact that cannot be quantitatively evaluated, just as literature, music, and art would have. Chapter 4 of this book, focusing on the reactions of foreign participants in the U.S. overseas programs, endorses such emotional influences. S&T is also *different* to so-called "high culture" because it leaves not only emotional but also material marks of the overseas information programs. As Chapters 2 and 3 will show, the U.S. provided research reactors, lab equipment, and training programs for foreign scientists and engineers. These activities fall into the intersection of overseas information and foreign aid programs—in fact, they were both at the same time. Government agencies such as the Department of State, USIA, International Cooperation Agency (ICA, later USAID), and the Operation Coordinating Board (OCB) were discussing and coordinating these programs both as overseas information programs and foreign aid programs, as chapters of this book will show. Nuclear reactors, lab equipment, and U.S.-trained engineers amounted to the tangible evidence of how U.S. information programs on S&T influenced the target countries. Just like a tomato harvester, a nuclear reactor would have long-lasting influences on the target society. In a similar way, elite engineers who learned the U.S. methods of lab operation might base the structures of their own national laboratories or higher education system upon this knowledge. Compared to literature, films, music, and art, the U.S. S&T information programs left more visible marks in the target society.

As the Cultural Cold War and S&T were closely related, individual scientists and engineers also became entangled in politics. Science philosopher Bruno Latour portrayed the process in which scientists and politicians, while pursuing different goals, increasingly shared interests. The example he cited was Jean Frédéric Joliot-Curie, assistant to Marie Curie, who married her daughter, Irene Joliot-Curie, and received a Nobel Prize together with his wife. In May 1939, about ten teams of scientists in several countries were competing to succeed in the world's first nuclear chain reaction. The experiment required precious uranium and heavy water. Joliot entered into a legal agreement with a Belgian company, the Union Miniére du Haut-Katanga, by which the company would provide technical assistance, funds, and five tons of uranium oxide "left lying about at its waste sites." In return, Joliot promised to patent all his scientific discoveries and distribute part of the profits to the Union Miniére. Joliot further promised to the French Minister of Armaments Raoul Dautry, who "did not share Joliot's leftist political opinions" but "had the same confidence in the progress of knowledge and the same passion for national independence," the development of an experimental reactor for civilian use which might eventually lead to the construction of nuclear weapons. In return, Dautry offered Joliot generous support "while asking him to change his priorities: if the bomb was practicable, it must be developed first and very quickly." By citing this anecdote, Latour argued that science and politics could not be separated as "on the one hand pure science, on the other pure politics," but they were intertwined in a "seamless web." Dautry wanted "to ensure France's military strength and the self-sufficiency of its energy production," and Joliot aspired "to be the first in the world to produce controlled artificial nuclear fission in the laboratory." Latour pointed out that it did not make sense to "call the first ambition 'purely political' and the second 'purely scientific'" because it was precisely the "impurity" that could "allow both goals to be attained."[10] When we shift the time setting to the Cold War era, scientists and policy-makers were entangled with each other in a web called the Cultural Cold War. Scientists' behaviors were orientated by international politics, and policy-makers counted on scientists' cooperation in their pursuit of national interests.

Science historian Audra Wolfe wrote that "the idea that American science ever operated in a 'free zone' outside of politics is itself a legacy of the ideological Cold War."[11] As we shall later see, the mainstream U.S. intellectual norms of the day valued the "freedom" of culture and science from politics. Its antithesis was the Communist world, in which culture and science served politics. However, it is simply a myth that the culture and science of the "free world" stood independently from policies. As the chapters of this book will attest, massive investments in the research and development of nuclear reactors or generous grants for foreign engineers were ruled by rivalries with the Soviet Union or the U.K., as well as strategies toward the Third World countries. Cold War ideology rendered the union of science and politics invisible, and created a myth of "freedom" from politics. This book will make visible the "web" of the Cultural Cold War, through which S&T and politics were inter-connected.

The term Cultural Cold War comprises two multi-faceted, and often controversial concepts, "culture" and "Cold War." In 2009, when the author published the edited volume *Decentering the Cultural Cold War: The U.S. and Asia*, the term Cultural Cold War was not so widely recognized outside the English-speaking world, but by now it is fairly commonly used not only in U.S. but also in Japanese academia. Yet "culture" and "Cold War" both remain loaded notions, and therefore the author will take a little detour to delineate various approaches and usages to put things in order.[12]

It is probably within the field of cultural anthropology that the definition of culture has been most enthusiastically debated. When the Franz Boaz school in the 1920s began advocating cultural relativism, the notion of culture experienced a dramatic paradigm shift. The dichotomized worldview of the civilized West and uncivilized Others was replaced—albeit only gradually—by the understanding that culture manifests in the environments in which people grow up and are socialized, and that all ethnic groups have meaningful cultures. Ruth Benedict, former student of Boaz, proposed "patterns" of culture, and her wartime Office of War Information (OWI)-sponsored study of the "pattern" of Japanese culture was later published as *The Chrysanthemum and the Sword*, an international bestseller.[13] Clifford Geertz later revised Benedict's thesis by arguing that culture was not concrete patterns of behaviors such as customs and traditions, but rather "control mechanisms" to govern human behavior—in other words, it was something like "what computer engineers call 'programs.'"[14] In the 1980s, reflecting new waves in academia such as postmodernism and feminism, James Clifford and others pointed out the asymmetrical power relations between the subject and the object of anthropological studies, and argued that anthropological writing about culture represented only "partial" truths.[15]

Raymond Williams, widely recognized as the founder of cultural studies, acknowledged that "culture" was "one of the two or three most complicated words in the English language." When first introduced from French to English, its primary meaning was "the tending of natural growth" of animals and plants, although from the 16th to the 19th century, the meaning of "process of human development" was added. On the other hand, the term "civilization" emerged in the 18th century. In 19th-century Germany, culture (Kultur) was used as a synonym for civilization, i.e., becoming "civilized" or "cultivated," and also the "secular process of human development." Johann Gottfried von Herder (1744–1803) brought about "a decisive change of use" in the terminology when he criticized the concept of the "unilinear process" of cultural development as judged from the viewpoint of the European standard. Instead, he proposed "cultures in the plural," that is, diverse cultures depending on nations, periods, or social and economic groups. Reflecting on the complex genealogy of the term, Williams summed up the definitions of culture in three ways: (1) "a general process of intellectual, spiritual and aesthetic development"; (2) "a particular way of life, whether of a people, a period, a group, or humanity in general"; and (3) "the works and practices of intellectual and especially artistic activity." He pointed out that the third definition was most widely used in his time, as in the common understanding of what is now referred to as "high culture," i.e., "music, literature, painting and sculpture, theatre and film," and "sometimes with the addition of philosophy, scholarship, history."[16]

By contrast, in international relations (IR) and political science, according to Shingo Tanaka, there were three stages in which the concept of culture was introduced in the discipline. The first was the studies of "national character" in the U.S. during the Second World War. Their purpose was to explain enemy behaviors from cultural aspects; it is well known that Ruth Benedict and other anthropologists worked for the OWI, as already mentioned. The second stage was the comparative study of the U.S. and the Soviet Union, including work such as Jack Snyder's study of the Soviet "strategic culture" from the late 1970s to the early 1980s. The third stage began around 2000, when constructivist IR theory brought cultures and norms of state and non-governmental actors under scrutiny. Peter J. Katzenstein, one of the most important scholars of this trend, defined culture as "a broad label that denotes collective models of nation-state authority or identity, carried by custom or law."[17] Furthermore, Akia Iriye, who proposed "cultural internationalism" as an alternative principle of international society to replace authoritarian politics, was another important scholar who integrated culture in the discipline. In Japan, Ken'ichiro Hirano opened up a new field of international culture studies.[18]

In the academic discipline of history, which is the most relevant to this book, a new subfield of cultural history emerged in the 1970s. Absorbing influences from cultural anthropology, it developed dramatically and established itself as "New Cultural History" by the end of the 1990s. According to Peter Burke, a pioneer scholar in this field, "New Cultural History" dealt with such diverse themes as sports, table manners, pilgrims, books, music, memory, consumption, fashion, housing, and bodies. Some cultural historians have focused on science, too. Burke cited, among others, *Galileo, Courtier: The Practice of Science in the Culture of Absolutism* (1993) by Mario Biagioli as an example of work dealing with the relationship between culture and science.[19] Consciously or unconsciously, scholars of the Cultural Cold War were most likely influenced by the development of New Cultural History. The author's interest in the Cultural Cold War also germinated during her doctoral study at the University of Minnesota, when she was exposed to the flood of cultural history books and articles published in the early 2000s. However, compared to the diverse topics in the New Cultural History, studies of the Cultural Cold War have overwhelmingly concentrated on culture narrowly defined, i.e., literature, films, music, and art. There are two reasons for this limited scope: first, the U.S. government in the early Cold War era defined culture narrowly, and second, scholars of the Cultural Cold War left science and other non-traditional topics outside their definition of culture. When the U.S. government used the term "culture," in those days, it certainly meant literature, films, music, art. However, their use of the term "overseas information program" covered broader concepts, including literature, films, music, and art, but also S&T. Therefore, if a researcher only looks for "culture" in the finding aides of the U.S. national archives, for example, he/she will naturally be led to literature, films, music, and art. However, thanks to the achievement of New Cultural History, we are now aware that diverse topics can fall within the scope of cultural analysis. S&T is one example of the under-explored themes of the Cultural Cold War, to which this book will pay due attention. In particular, very little has been written about how S&T unfolded as part of the Cultural Cold War in Asia, and this book also seeks to fill this geographical gap.

Science history has an entirely different genealogy from cultural history, and it has a very rich reservoir of studies dealing with (even if only in part) the Cold War era. In Japan, for example, science historian Masakatsu Yamasaki has published a history of nuclear power development in Japan through the prewar to the postwar years. Hiroshi Ichikawa wrote extensively on Soviet science history, especially concerning nuclear science. In the U.S., there are a plethora of science history publications, including the edited volume on the history of S&T in Asia by Mizuno, Moore, and DiMoia, another edited volume on Global Cold War and science by Oreskes and Krige, and a study of the military-industrial complex by Leslie, to name just a few.[20] In spite of their valuable contribution to the discipline, most science history studies have not entertained conversation with the studies of the Cultural Cold War. There have been some exceptions, however, such as Jessica Wang's *American Science in an Age of Anxiety: Scientists, Anticommunism, and the Cold War*, and the aforementioned work of Audra Wolfe.[21]

There has also been heated discussion on the other loaded notion included in the term Cultural Cold War: the Cold War. Westad's *Global Cold War* has demonstrated that the Cold War was not just a conflict between superpowers or their proxies, but a phenomenon involving the decolonizing of nations in Asia, Africa, and Latin America.[22] At the same time, a question has been raised about the validity of relating everything that happened during the period from the late 1940s to the late 1980s to the Cold War. Some scholars have tried to tease out what is attributable to the Cold War and what is not, while others emphasized the continuity from prewar to postwar eras. Colonialism, anti-colonialism, and development are some of the key notions connecting the prewar and postwar periods.[23] Furthermore, there is important scholarship focusing on the conflicts between the Allies, or on the contrary, similarities across the East-West camps. Still other studies have explored the transformations of the nature of the Cold War over time, and/or its termination. A recent work by Hajimu Masuda also shed new light on the understanding of the Cold War in Asia by focusing on the relations between policy-making and popular opinion, and between the global and the local.[24]

This book deals with cultural, scientific, technological, and diplomatic aspects of the Cold War in Asia, especially Japan. At the same time, it pays attention to non-Cold War factors such as the legacy of colonialism, domestic politics, business interests, scientists' and politicians' personal aspirations, and so on. By doing so, it will go beyond the state-to-state relationship hitherto defined by the Cold War framework. It will also overcome the framework of U.S.–Japan bilateral relations by situating Japan as one of the many countries which the U.S. overseas information programs attempted to reach. The U.S. overseas information programs formed a global scheme to expand U.S. hegemony, but they were also limited and appropriated by local contexts. This book will illuminate both the structural power[25] that the U.S. built through S&T and its limitations.

1.2 Public Diplomacy, Information Programs, and Government Agencies Involved

What is called "public diplomacy" or "cultural diplomacy" today was inconsistently referred to as "overseas information and education," "foreign information program," or "overseas information program" in the early Cold War U.S. government.[26] This book adopts the term "overseas information program," which has also been used in a title of a volume of *Foreign Relations of the United States*, a collection of primary documents edited by the Department of State. When referring to the policy-making process rather than concrete programs, this book will use the term "overseas information policy." The term "public diplomacy" emerged in the mid-1960s, and therefore does not precisely cover the characteristics of the early Cold War era—the scope of this book. According to the National Security Council (NSC)'s policy paper NSC5509 (December 1954), the purpose of overseas information programs was, in sum, to emphasize "America's devotion to peace," to make "the American way of life better understood," to capitalize on "the series of Free World accomplishments," and to undermine "Soviet prestige and effectiveness."[27] These stated purposes differ quite significantly from the present-day definitions of "public diplomacy." For example, the Japanese Ministry of Foreign Affairs (MOFA) defines public diplomacy as "diplomatic activities to directly access foreign citizens and public opinions through publicity and cultural exchange, and in cooperation with private sectors."[28]

The USIA was primarily responsible for the U.S. overseas information programs after its establishment in August 1953. Especially in its early days, the USIA officers were treated as if they were "second class citizens" in comparison to foreign service officers in the Department of State. However, as the importance of overseas information programs grew over time, the situation improved; the USIA's credibility was also supported by President Eisenhower and his close aides who placed importance on psychological warfare. In fact, the Cold War as a whole had an aspect of psychological and information warfare. Toward the end of the Eisenhower presidency, the USIA had 202 branches in 85 countries, with 3,771 American staff and 6,881 local employees. The Voice of America (VOA) radio station had 50 million listeners per day, and the USIS films were watched by an audience of 500,000 all over the world. USIA-sponsored TV programs were broadcast in 47 countries. The director of USIA sat in the NSC and the cabinet meetings, and he could meet directly with the president once every few weeks.[29]

However, with regard to S&T, the USIA was not the only government agency which practiced overseas information programs. Concerning Atoms for Peace, the Atomic Energy Commission (AEC) cooperated closely with the USIA and the Department of State. The AEC was established in the place of the wartime Manhattan Project research centers such as the Los Alamos nuclear weapons laboratory in New Mexico; R&D laboratories in Oak Ridge, Tennessee, and Hanford, Washington; and two university laboratories, i.e., University of California's physics laboratory led by Earnest Lawrence and University of Chicago's Metallurgical Laboratory. The Metallurgical Laboratory was reorganized as the Argonne National Laboratory after the war and mainly engaged in R&D of nuclear reactors.[30] With regards to space technology, the National Aeronautics and Space Administration (NASA) coordinated with the USIA. Moreover, private companies and non-governmental organizations also cooperated with the overseas information programs from time to time.

The Operation Coordination Board (OCB) oversaw the overseas information programs as a whole to ensure that they would be in line with overall U.S. foreign policies. The OCB was born out of its predecessor, the Psychological Strategy Board (PSB), and was placed directly under the NSC. In the OCB meetings, representatives of the USIA, the Department of State, the Department of Defense, and the CIA discussed the psychological and informational aspects of various national security policies. Many Eisenhower-era policy papers referred to the "psychological factor" (or "P factor" in their jargon) of various national security policies. It was the OCB's responsibility to orchestrate the "P factor" aspects to ensure different governmental and non-governmental parties would act in coordination with, not in contradiction to each other. Within the OCB, ad-hoc working groups on specific themes were established. In 1955, a working group on Atoms for Peace was established, and in 1958, a working group on S&T came into being.

The International Cooperation Administration (ICA) was a subordinate organization of the Department of State in charge of the practical operation of foreign aid programs. However, this book will reveal that the ICA's role was not only concerned with the management of operations, but that it was also an important player in overseas information programs. For example, the director of the ICA and its Japanese office advised the Department of State on the selection of Japanese students and scholars to be sent to U.S. universities and laboratories, and on the selection of businessmen to participate in the business missions to the U.S. The ICA was actively involved in the U.S. S&T policies in competition with not only the Soviet Union but also with allies such as the U.K. and France.

1.3 Foreign Atoms for Peace

In the United Nations General Assembly of December 8, 1953, President Eisenhower delivered his famous "Atoms for Peace" address. He pointed out that the U.S. monopoly of nuclear power had "ceased to exist" and the knowledge of nuclear technology would eventually be shared by all others. To reverse the "fearful trend of atomic military build-up," the President proposed that advanced countries "make joint contributions from their stockpiles" of fissionable materials to an "international atomic energy agency," and the agency should allocate the fissionable materials "to serve the peaceful pursuits of mankind."[31]

The U.S. Atoms for Peace policy was a reaction to the Soviet initiative in the peaceful uses of nuclear energy, and to the situation in which the continued U.S. monopoly of nuclear weapons would no longer be possible. Yet at the same time, it was an overseas information program to disseminate an image of the U.S. as the leader in peaceful uses of nuclear energy. After the president's U.N. address, the USIA commenced an all-out "Atoms for Peace Campaign" through films, TV, radio, exhibitions, lectures, pamphlets, magazines, and so on. In April 1955, Atoms for Peace was designated as one of the USIA's "global themes" (especially important themes in overseas information programs to be disseminated to many parts of the world). Its purpose was "solidly to implant the idea that the U.S. is vitally concerned with constructive use of the atom" in the minds of average citizens in foreign countries. The USIA in particular regarded 1955 as an important year when Atoms for Peace was "moving out of its stage of preparation into the realm of action." The year witnessed many important events, such as the "United Nations Conference on Peaceful Uses of Atomic Energy" and the "conclusion of the first bilateral agreements" on nuclear technological assistance. The USIA especially encouraged its officers to "exploit" the following situations:

1. Isotopes: Publicize any and all uses of U.S. isotopes in local hospitals and industries and on farms.
2. People trained in U.S.: Give maximum appropriate publicity to trainees who have returned from the U.S. or who are about to go. Interviews, arranging speeches, tape recordings, and by-line articles all would be useful.

3. Bilateral agreements: As these begin to be announced, the event should receive maximum exploitation in the country concerned.
4. Research reactors: Most bilateral agreements will provide for construction of research reactors with US assistance. The first announcement and various steps in progress of construction should be exploited as fully as possible.
5. Power reactors: It is probable that power reactors will eventually be built in many countries with U.S. aid. When such is the case, a major campaign could be evolved around the project.[32]

On June 11, 1955, in the address on the centennial commencement of Pennsylvania State University, President Eisenhower proposed to "offer research reactors to the people of free nations who can use them effectively for the acquisition of the skills and understanding essential to peaceful atomic progress." He further promised that the U.S. would "contribute half the cost" of the reactor, as well as furnish the nuclear fuel. Moreover, he continued, for "such friendly nations as are prepared to invest their own funds in power reactors," the U.S. would provide "access to and training in the technological processes of construction and operation" of the reactors.[33] The president's address was part of the U.S. attempt to put into practice his U.N. speech of December 1953, and also an attempt to explore future overseas markets for U.S. nuclear reactors. Small research reactors would be samples for larger, power reactors to be purchased later on.

Within the U.S. government, the foreign aid policies concerning nuclear energy came to be called "Foreign Atoms for Peace." Foreign Atoms for Peace was a broad concept entailing the export of reactors, technological aid, and overseas information programs, of which Chapter 3 will introduce concrete examples. Even though U.S. domestic electric demands were already met by water and thermal power generation, the government was compelled to continue R&D of nuclear reactors because they were competing with the Soviet Union and the U.K. That was why overseas markets were extremely important both for the U.S. government and private companies. They launched overseas information programs even before the power reactors had been completely developed because they hoped to export them as soon as ready. Films and exhibitions geared toward specialists, training programs, government-sponsored tours to the U.S., and so on, were carried out under the banner of overseas information programs.

Concerning the U.S. Foreign Atoms for Peace for Japan, there are already many distinguished scholarly works, mostly in Japanese. For example, Tetsuro Kato has examined the process by which many Japanese embraced the notion of peaceful uses of atomic energy, and successfully deconstructed the myth that Japanese people were essentially anti-nuclear. Mitsuo Ikawa discussed the effects of the Atoms for Peace exhibitions co-sponsored by the USIA and Japanese newspaper companies through analyses of the USIA's survey results. Jun Tateno, Kenzo Okuda, and others explored why Japan, initially prioritizing the domestic development of nuclear technology and import of

U.K. reactors, became increasingly inclined to adopt U.S. light-water reactors. Shun'ya Yoshimi situated nuclear energy within the context of the history of electricity, and discussed how it stirred people's imagination as "dream technology." Tetsuo Arima portrayed the activities of Matsutaro Shoriki, a Japanese media tycoon and politician, and his subordinates concerning the 1955 Atoms for Peace exhibition in Tokyo. Ran Zwigenberg, in a chapter of his book on Hiroshima, focused on a USIS officer Abol Fazl Fotouhi who succeeded in holding an Atoms for Peace exhibition in Hiroshima as a symbol of U.S. friendship with people of the atomic-bombed city. The present author has also discussed the U.S. Foreign Atoms for Peace for Japan as a psychological tool to curb Japanese anti-nuclear and anti-U.S. sentiments.[34] These existing scholarly works collectively revealed the goals of Foreign Atoms for Peace: to water down the Japanese memory of U.S. atomic bombing on the one hand, and on the other, to cultivate markets for American nuclear reactors. However, these scholarly works have also revealed that Atoms for Peace was never a one-way process from the U.S. to Japan, as they exposed the complicated roles of Japanese politicians, businessmen, and scientists who helped introduce the U.S. nuclear technology.

In contrast to these existing studies, this book will contextualize Foreign Atoms for Peace for Japan in three ways: geographically, chronologically, and target-wise. First, although Atoms for Peace in Japan has been overwhelmingly discussed within the framework of U.S.–Japan relations, this book will frame it in a larger picture. The U.S. government embraced a self-image as the center from which knowledge of nuclear technology would radiate toward all corners of the world. Japan was one of the most important countries to which the U.S. exported nuclear technology, but it was not the only one. Japan's case, therefore, should be re-examined within the global development of the U.S. Foreign Atoms for Peace. Secondly, this book will also contextualize nuclear technology within the longer span of U.S. overseas information programs on S&T. Although an important event in the Cultural Cold War, Atoms for Peace was at its peak for at most three years, from 1955 to 1957. On the other hand, S&T continued to be a vital part of the U.S. overseas information programs throughout the late 1950s and 1960s, albeit focusing on different themes over time. By following through the shifting themes, this book will reconsider nuclear technology as part of the larger Cultural Cold War. Thirdly, this book will redefine the target of Foreign Atoms for Peace. Although past scholarship has overwhelmingly focused on the Atoms for Peace exhibitions for the general public, this book will distinguish Atoms for Peace for the general public and that for the so-called "science elite," including emerging scientists, engineers, and technocrats. The U.S. government explicitly mentioned three different target groups: the general public, the science elite, and other elites. The Foreign Atoms for Peace programs for young science elites were especially important in that they would create long-lasting effects on the science policies of the target countries. By including an examination of the science elite, this book will convey a more precise understanding of the U.S. Foreign Atoms for Peace.

1.4 Chapter Organization

This book consists of two parts: Part I (Chapters 2 to 5) deals with nuclear energy. The wartime Manhattan Project yielded not only nuclear bombs but also the ideology behind the promotion of non-military uses of nuclear energy, which was eventually called "the peaceful use of atomic energy." This instantly became the most important theme of the U.S. overseas information program in the mid-1950s, but its momentum did not last long. The negative image of thermonuclear bomb testing in the Pacific reduced the credibility of U.S. nuclear technology advertising, on the one hand, while on the other, the export of nuclear reactors had already been put into practice by the end of the 1950s, thus eliminating the need for dramatic advertisement. As a whole, Part I will portray the wax and wane of Foreign Atoms for Peace.

Chapter 2 provides the foundations on which concrete cases will be examined by observing the process of scientists becoming involved in anti-Communist politics. It will demonstrate why S&T became an arena of the Cultural Cold War. The chapter focuses on the Congress of Cultural Freedom (CCF), an anti-Communist organization funded by the CIA founded in Paris in July 1950. Émigré scientists such as Michael Polanyi and Eugene Rabinowitch, who embraced "freedom of science," resonated with the liberal anti-Communism ideology held by the CCF. This chapter will explore the troubled relationship between science and anti-Communist politics by focusing on these scientists' activities and their writings in the *Bulletin of the Atomic Scientists*. By doing so, this chapter will demonstrate why and how scientists became key actors of the Cultural Cold War.

Chapters 3 and 4 follow the process of nuclear technology's growing importance in the U.S. overseas information programs. In Chapter 3, U.S. exports of research reactors to Japan, South Vietnam (RVN), and Burma are comparatively analyzed. Tangible products such as reactors and lab facilities were employed as symbols of national pride, the authority of the government, or political alliance. At other times, they were used as displays to induce the repatriation of overseas nationals, or as tools to fend off the approaches of the Soviet Union. Consideration of such flexible meanings attached to reactors and facilities will reveal why nuclear technology was regarded as an effective theme of the overseas information programs. Chapter 4 focuses on the International School of Nuclear Science and Engineering (ISNSE) established within the Argonne National Laboratory near Chicago. The U.S. government established the ISNSE to train foreign scientists and engineers in American nuclear technology. It was a strategic study-abroad program implemented to disseminate U.S. nuclear technology through the science elite of foreign countries; however, foreign trainees also had their own agenda and goals. This chapter will illuminate both the achievements and limitations of the ISNSE.

Chapter 5 focuses on the U.S. thermonuclear tests in the Pacific, which undermined the image of U.S. nuclear technology around the world and contributed to the decline of the Foreign Atoms for Peace campaign. The Lucky Dragon incident, in which Japanese tuna fishermen were exposed to a heavy dose of nuclear fallout from a U.S. nuclear test, triggered an anti-nuclear movement internationally. Subsequently, the U.S. government attempted to showcase their "clean bomb," but the initiative ended up as a public relations disaster. The U.S. government, in turn, tried to control and conceal information on nuclear fallout. This chapter will deal specifically with Operation Redwing (1956) and Operation Hardtack (1958), and the irradiation of the Japan Coast Guard's oceanic survey ships which were engaged in the International Geophysical Year (IGY) program, ironically a symbol of scientific internationalism. If the ideals of the "freedom of science" and training programs in the previous chapters were the fancy clothing in which nuclear technology had been dressed, then the U.S. and Japan's collaboration in controlling and concealing information was part of the unshown seams that held together the underside of the overseas information program.

Part II (Chapters 6 to 8) will deal with the emergence of new overseas information programs replacing Atoms for Peace, notably medical aid and space flight. On top of what was discussed in Chapter 5, the successful launch of the Soviet satellite Sputnik in October 1957 triggered the transition of the U.S. overseas information programs from nuclear energy to other areas of S&T. Chapter 6 will overview the organizational and policy-level changes related to the U.S. S&T policy and overseas information programs. President Eisenhower advocated "Science for Peace," building on the experiences of Atoms for Peace, which was intended to generate a more direct impact on the welfare and health of people in developing countries. Various groups both in and out of government drafted concrete proposals for the new overseas information policies, out of which medical aid and space flight emerged as two of the most important themes. Chapter 7 focuses on the overseas information program on medical aid, with special emphasis on a people-to-people friendship program entitled "Project Hope," a project to send a hospital ship to developing countries. Medical aid almost inevitably required the cooperation of private doctors, nurses, and technicians, and, on the surface, Project Hope was operated by private organizations and the good will of American citizens. However, the whole process was in fact carefully supervised and discretely directed by the U.S. government. As this chapter demonstrates, celebrating the *private* voluntarism in the *government-sponsored* information program had inherent tensions and limitations. Chapter 8 explores another theme of the new information program: space flight. Although the Apollo moon landing is undoubtedly the most famous event in the space race, this book sheds light on the early manned space flight program from the late Eisenhower to the early Kennedy administrations. Since space was regarded as a borderless, free place, it became a desirable theme of the overseas information programs to rouse people's imaginations and aspirations, detached from international conflicts on the earth. NASA and USIA closely cooperated with each other, especially in the propagation of Project Mercury. However, to some countries, for example, Japan, American space technology was not of absolute, but only relative importance, as compared to the domestic, European, and Soviet space technologies.

Notes

1 Langdon Winner, *The Whale and the Reactor: A Search for Limits in an Age of High Technology*, 2nd ed. (Chicago, IL and London: University of Chicago Press, 2020), 26–28. First published 1986. Citations refer to the 2020 edition.

2 Kenneth Osgood, *Total Cold War: Eisenhower's Secret Propaganda Battle at Home and Abroad* (Kansas City, MO: University Press of Kansas, 2006); Ruth Oldenziel and Karin Zachmann, eds., *Cold War Kitchen Americanization, Technology, and European Users* (Cambridge, MA and London: MIT Press, 2009).

3 Hiroshi Ichikawa, "Obninsk, 1955: Sekai hatsuno genshiryoku hatsudensho to Soviet kagakusha no 'genshiryoku gaiko'" [Obninsk 1955: The World's First Nuclear Power Plant and the Soviet Scientists' 'Nuclear Diplomacy'], in *Kakukaihatsu jidai no isan: mirai sekinin o tou* [*Legacy of the Nuclear Development Age: Questioning the Responsibilities for the Future*], eds. Yuji Wakao and Eiichi Kido (Tokyo: Showado, 2017), 26–50; John Krige, "Techno-Utopian Dreams, Techno-Political Realities: The Education of Desire for the Peaceful Atom," in *Utopia/Dystopia: Conditions of Historical Possibility*, eds. Michael D. Gordin, et al. (Princeton, NJ: Princeton University Press, 2010), 151–155.

4 *USIS Film Catalog 1959* (USIS Tokyo).

5 Audra Wolfe, *Freedom's Laboratory: The Cold War Struggle for the Soul of Science* (Baltimore, MD: Johns Hopkins University Press, 2018). Broadly speaking, Kazushi Minami's analysis of the impact of technological transfer from the U.S. to China during the 1970s can be also viewed as a study of the Cultural Cold War and S&T. Kazushi Minami, "Oil for the Lamps of America? Sino-American Oil Diplomacy, 1973–1979," *Diplomatic History*, vol. 41, no. 5 (2017): 959–984.

6 Frances Stonor Saunders, *Who Paid the Piper? The CIA and the Cultural Cold War* (London: Granta Books, 1999); *Cultural Cold War: The CIA and the World of Arts and Letters* (New York: The New Press, 1999).

7 Penny Von Eschen, *Satchmo Blows up the World: Jazz Ambassadors Play the Cold War* (Cambridge, MA: Harvard University Press, 2005); Yoshiomi Saito, *Jazz Ambassadors: Amerika no ongaku gaikoshi* [*Jazz Ambassadors: The History of U.S. Music Diplomacy*] (Tokyo: Kodansha, 2017); Fumiko Fujita, *Amerika bunka gaiko to nihon: reisenki no bunka to hito no koryu* [*The U.S. Cultural Diplomacy and Japan: Cultural and Personal Exchange in the Cold War*] (Tokyo: University of Tokyo Press, 2015); Takeshi Matsuda, *Soft Power and Its Perils: U.S. Cultural Policy in Early Postwar Japan and Permanent Dependency* (Stanford, CA: Stanford University Press, 2007); Jiyoung Kim, *Nihon bungaku no <sengo> to henso sareru <Amerika>* [*'Postwar' Japanese Literature and 'America' Played in Variations*] (Kyoto: Minerva Shobo, 2019); Giles Scott-Smith and Charlotte Lerg, eds., *Campaigning Culture and the Global Cold War: The Journals of the Congress for Cultural Freedom* (London: Palgrave MacMillan, 2017); Mitsuo Ikawa, "Genshiryoku heiwariyo hakurankai to shinbunsha" [The Atoms for Peace Exhibitions and Newspaper Companies], in *Sengo nihon no media event: 1945– 1960* [*Media Events in Postwar Japan: 1945–1960*], ed. Toshihiro Tsuganezawa (Kyoto: Sekai Shisosha, 2002), 247–265; Somei Kobayashi, "VOA shisetsu iten o meguru kanbei kosho: 1972–73" [The U.S.-Japan Negotiation Concerning the Transfer of VOA Facilities: 1972–73], *Journal of Mass Communication Studies*, vol. 75 (2009): 129–147; Yuka Tsuchiya and Shunya Yoshimi, eds., *Senryo suru me senryu suru koe: CIE/USIS eiga to VOA radio* [*Occupying Eyes, Occupying Voices: CIE/USIS Films and VOA Radio*] (Tokyo: University of Tokyo Press, 2012); Yuka Tsuchiya, "VOA *Forum* to kagakugijutsu kohogaiko: reisen radio wa Amerika no kagaku o do tsutaetaka" [VOA *Forum* and S&T Public Diplomacy: Cold War Radio Broadcast American S&T], *Amerika Kenkyu* [*The American Review*], vol. 54 (2020): 67–87.

8 Peter Burke, *What Is Cultural History?* 3rd ed. (Cambridge: Polity Press, 2019), 1. First published 2004.
9 Anne L. Foster, "Introduction," *Diplomatic History*, vol. 41, no. 2 (2017): 225–227.
10 Bruno Latour, *Pandora's Hope: Essays on the Reality of Science Studies* (Cambridge, MA and London: Harvard University Press, 1999), 81–88.
11 Audra J. Wolfe, *Competing with the Soviets: Science, Technology, and the State in Cold War America* (Baltimore, MD: Johns Hopkins University Press, 2013), 4.
12 Toshihiko Kishi and Yuka Tsuchiya, eds., *Bunka reisen no jidai: Amerika to ajia* [*De-Centering the Cultural Cold War: U.S. and Asia*] (Tokyo: Kokusai Shoin, 2009).
13 Ruth Benedict, *Patterns of Culture* (London: Routledge, 2020; London: Houghton Mifflin, 1934). Citations refer to the Routledge edition. Ruth Benedict, *The Chrysanthemum and the Sword: Patterns of Japanese Culture* (London: Routledge, 2020; Lodon: Houghton Mifflin, 1946). Citations refer to the Routledge edition.
14 Clifford Geertz, *The Interpretation of Cultures* (New York: Basic Books, 1973), 44.
15 James Clifford and George E. Marcus, eds., *Writing Culture: The Poetics and Politics of Ethnography*, 1st ed. (Berkeley, CA: University California Press, 1986), 7.
16 Raymond Williams, *Key Words: A Vocabulary of Culture and Society*, rev. ed. (New York: Oxford University Press, 1983), 87–92.
17 Shingo Tanaka, "Taigai seisaku ketteiron ni okeru bunka: shuyo model no hyoka to kongo no kadai" [Culture in the Foreign Policy-Making: Evaluation of Major Models and Future Issues], *Kokusai Kokyo Seisaku Kenkyu* [*International Public Policy Studies*], vol. 12, no. 1 (September 2007): 243–257; Naofumi Miyasaka, "Terrorism taisaku ni okeru senryaku bunka: 1990 nendai kohan no nichibei kankei o jirei toshite" [Strategic Culture in Counter-Terrorism: Cases of U.S. and Japan in the Late 1990s], *Kokusai Seiji* [*International Relations*], vol. 129 (February 2002): 61–76; Satoshi Oyane, "Constructivism no shiza to bunseki: kihan no shototsu, chosei no jissho bunseki e" [The Perspective and Analysis of Constructivism: Toward the Positivist Analysis of the Collision and Coordination of Norms], *Kokusai Seiji* [*International Relations*], vol. 143 (November 2005): 124–140; Peter J. Katzenstein, ed., *The Culture of National Security: Norms and Identity in World Politics* (New York: Columbia University Press, 1996), 6.
18 Akira Iriye, *Power and Culture: The Japanese-American War, 1941–1945* (Cambridge, MA and London: Harvard University Press, 1981); *Cultural Internationalism and World Order* (Baltimore, MD: Johns Hopkins University Press, 1997); Kenichiro Hirano, *Kokusai bunkaron* [*International Culture*] (Tokyo: University of Tokyo Press, 2000).
19 Burke, *What Is Cultural History?* 45, 52, 60–73.
20 Masakatsu Yamazaki, *Nihon no kaku kaihatsu, 1939–1955: genbaku kara genshiryoku e* [*The Japanese Nuclear Development, 1939–1955: From Nuclear Bombs to Atomic Power*] (Tokyo: Sekibundo, 2011); Hiroshi Ichikawa, *Soviet Science and Engineering in the Shadow of the Cold War* (London and New York: Routledge, 2018); Hiromi Mizuno, et al., eds., *Engineering Asia: Technology, Colonial Development and the Cold War Order* (London and New York: Bloomsbury Academic, 2018); Naomi Oreskes and John Krige, eds., *Science and Technology in the Global Cold War* (Cambridge, MA and London: MIT Press, 2014); Stuart W. Leslie, *The Cold War and American Science: The Military-Industrial-Academic Complex at MIT and Stanford* (New York: Columbia University Press, 1993). Masaki Fujioka's *Amerika no daigaku ni okeru soren kenkyu no hensei katei* [*Construction of Soviet Studies in U.S. Universities*] (Tokyo: Horitsu Bunka-sha, 2017), dealing with U.S. history of knowledge rather than history of science,

should also be mentioned as part of the emerging body of scholarship focusing on the relationships between knowledge and foreign relations.

21 Jessica Wang, *American Science in an Age of Anxiety: Scientists, Anticommunism, & the Cold War* (Chapel Hill, NC and London: University of North Carolina Press, 1999); Wolfe, *Freedom's Laboratory.*

22 Odd Arne Westad, *The Global Cold War: Third World Interventions and the Making of Our Times,* new ed. (Cambridge: Cambridge University Press, 2011).

23 Minoru Masuda, et al., eds., *Reisenshi o toinaosu: reisen to hireisen no aida* [*Reconsidering the History of Cold War: Border of the Cold War and Non-Cold War*] (Kyoto: Minerva Shobo, 2015); Mizuno, *Engineering Asia.*

24 As for relationships between allies, see for example, Toru Onozawa, *Maboroshi no domei: reisen shoki Amerika no chuto seisaku* [*The Illusion of Alliance: The U.S. Middle Eastern Policies in the Early Cold War Era*] (Nagoya: University of Nagoya Press, 2016); Toshihiko Aono, *"Kiki no toshi" no reisen to domei: Berlin, Cuba, Détente, 1961–63* [*Cold War and Alliance in the Year of Crisis: Berlin, Cuba, Détente, 1961– 63*] (Tokyo: Yuhikaku, 2012). As for similarities across capitalist and Communist blocs, see for example, Jeremi Suri, *Power and Protest Global Revolution and the Rise of Détente* (Cambridge, MA: Harvard University Press, 2005). There are many scholarly works on the transformation and termination of the Cold War, of which major contributions in Japanese include Hideki Kan, *Reisen to domei: reisen shuen no shiten kara* [*The Cold War and Alliance: From the Perspective of the End of the Cold War*] (Kyoto: Shorai-sha, 2014); Hideki Kan, *Reisenshi no saikento: henyo suru chitsujo to reisen no shuen* [*Re-Examining the Cold War: Transforming Order and the End of the Cold War*] (Tokyo: Hosei University Press, 2010). Hajimu Masuda, *Cold War Crucible: The Korean Conflict and Postwar World* (Cambridge, MA: Harvard University Press, 2015) has also been recently published in Japanese.

25 The author is using this term in the sense Susan Strange has discussed in *States and Markets: An Introduction to International Political Economy,* 2nd ed. (London: Printer Publishers, 1994). She defines structural power as the power "to decide how things shall be done, the power to shape frameworks within which states relate to each other, relate to people, or relate to corporate enterprises." In particular, structural power derived from "knowledge structure," according to Strange, is easily overlooked, and therefore under-estimated. The U.S. information programs on S&T are also thought to have contributed to building a knowledge structure from which the U.S. drew power.

26 "Overseas Information and Education" had been frequently used until the early 1950s, when the distinction between overseas "information" policies and "education" policies was ambiguous. This is evident in the naming of the "Civil Information and Education Section" (CIE), a section of the Allied Occupation Forces for Japan dealing with both information dissemination and educational policies. All kinds of U.S. government-supported educational programs for foreign students, including scholarships, study-abroad programs, and cultural exchange, were grouped into the "overseas information and education activities." However, this expression became obsolete in the mid-1950s as "information" and "education" became clearly distinguished partly because of Senator William Fulbright's strong support of such distinction.

27 NSC 5509, United States National Security Program, December 31, 1954, Part 6 USIA Program, Department of State, Office of the Historian, *Foreign Relations of the United States (FRUS),* 1955–1957, Foreign Economic Policy; Foreign Information Program, Vol. IX, Document 185, https://history.state.gov/historicaldocuments/frus1955-57v09/d185.

28 The Japanese Ministry of Foreign Affairs, FAQ, "What Is Public Diplomacy? What Is Soft Power?" https://www.mofa.go.jp/mofaj/comment/faq/culture/gaiko.html #section1, September 27, 2017.

29 Nicholas Cull, *The Cold War and the United States Information Agency* (Cambridge: Cambridge University Press, 2008), 187.

30 Jack M. Holl, *Argonne National Laboratory 1946–96*, (Chicago, IL: University of Illinois Press, 1997), ixx–xx.

31 IAEA website, https://www.iaea.org/about/history/atoms-for-peace-speech.

32 "Global Theme IV," April 29, 1955; "Global Theme III (revised)," January 3, 1956, RG59, box 55, National Archives at College Park. (Hereafter, NACP); Notes on Director's Staff Meeting," April 29, 1955, RG306, Entry P123, 1953-1965, box 1, NACP.

33 Dwight D. Eisenhower, "Address at the Centennial Commencement of Pennsylvania State University," June 11, 1955, The American Presidency Project, https://www.presidency.ucsb.edu/documents/address-the-centennial-commencement-pennsylvania-state-university.

34 Tetsuro Kato, *Nihon no shakaishugi: genbaku hantai, genpatsu suishin no ronri* [*Japanese Socialism: The Logic of Anti-Nuclear Weapons and Pro-Nuclear Power*] (Tokyo: Iwanami Shoten, Publishers, 2013); Ikawa, "The Atoms for Peace Exhibitions and Newspaper Companies"; Jun Tateno, "Keisuiro no donyu to donen dantaisei: hoki sareta jishukaihatsu hoshin" [The Introduction of Light-Water Reactors and the 'Power Reactor and Nuclear Fuel Development Corporation' System: The Abandoned Autonomous Development Plan], in *Fukushima jikoni itaru genshiryoku kaihatsushi* [*The History of Atomic Energy Development Leading to the Fukushima Accident*], ed. Genshiryoku Gijutsushi Kenkyukai (Tokyo: Chuo University Press, 2015); Kenzo Okuda, "Igirisu karano Calder Hall-gata shoyoro donyu" [Introduction of a Commercial Calder Hall Reactor from the U.K.], in *The History of Atomic Energy Development Leading to the Fukushima Accident*; Shunya Yoshimi, *Yume no Genshiryoku: Atoms for Dream* [*The Dream of Atomic Power: Atoms for Dream*] (Tokyo: Chikuma Shobo, 2012); Tetsuo Arima, *Genpatsu to genbaku: nichi bei ei kakubuso no anto* [*Nuclear Power and Nuclear Bombs: Hidden Battles of Japan, U.S. and U.K. for Nuclear Armament*] (Tokyo: Bungeishunju, 2012); Ran Zwigenberg, *Hiroshima: The Origins of Global Memory Culture* (Cambridge: Cambridge University Press, 2014); Yuka Tsuchiya, "Koho bunka gaiko toshiteno genshiryoku heiwariyo kyanpen to 1950 nendai no nichibei kankei" [The Atoms for Peace Campaign as Public and Cultural Diplomacy: U.S.-Japan Relations in the 1950s], in *Nihibei domeiron: rekishi, kino, shuhenshokoku no shiten* [*U.S.-Japan Alliance: Perspectives of History, Function, and Neighboring Countries*], ed. Toshitaka Takeuchi (Kyoto: Minerva Shobo, 2011), 180–209.

Part I

Cultural Cold War and Nuclear Energy

2 The Cultural Cold War and Nuclear Scientists

Congress for Cultural Freedom and the *Bulletin of the Atomic Scientists*

Part I deals with the important roles that nuclear science and scientists, as well as nuclear technological aid played in the Cultural World War. Nuclear scientists and engineers on occasion became the target of the U.S. overseas information programs, but at other times, intentionally or unintentionally, they cooperated directly with such programs. Research reactors, nuclear labs, radioisotope technologies, and nuclear power plants also became diplomatic tools to attract foreign leaders. They represented American modernity in films and exhibitions, or sometimes were actually exported to the target countries. Atoms for Peace became the central theme of the U.S. overseas information programs for several years beginning in the mid-1950s, though its momentum waned toward the end of the decade.

Chapter 2 will provide a pre-history of the Atoms for Peace initiative, focusing on the process by which nuclear scientists became entangled with the Cultural Cold War. To be sure, the scientists introduced in this chapter did not have any direct influence on the USIA's overseas information programs after the mid-1950s. However, their stories will reveal a "web" (to borrow Bruno Latour's term–see Introduction) on which science and scientists smoothly slid through from the world of science to Cold War politics. Some of the Manhattan Project scientists who advocated international control of nuclear energy and "freedom of science" became involved in the anti-Communist Congress for Cultural Freedom (CCF), which was financially supported by the CIA. This chapter will explore how liberal scientists and an anti-Communist organization found each other and what they shared in common. The author was puzzled when she first found such a connection during her research in the University of Chicago archives in 2016–2017, because the two groups appeared to be politically incompatible with each other. Through examination of Department of State records and magazine articles written by some key scientists, the chronological process by which these scientists gradually steered closer toward anti-Communist ideology emerged. Such an analysis demonstrates the proximity of S&T and the Cultural Cold War, and provides a background to the various cases shown in the ensuing chapters. Both scientists themselves and the general public believed that S&T was neutral and apolitical, and this belief increased the value of S&T as a theme of the U.S. overseas information programs. This chapter will demonstrate, however, that the idea of the neutrality and independence of scientists was shaky at best.

DOI: 10.4324/9781003243649-3

In 1945, scientists of the University of Chicago's Metallurgical Laboratory, one of the wartime Manhattan Project laboratories, founded the Atomic Scientists of Chicago, an organization to advocate international control of nuclear energy and nuclear disarmament. They published the *Bulletin of the Atomic Scientists*, a monthly magazine to discuss scientists' responsibilities in the nuclear age, and to educate readers on the problems brought about by nuclear energy. The magazine was established by Eugene Rabinowitch and his colleague Hyman Goldsmith, and the editorial board included notable scientists such as J. R. Oppenheimer, I. I. Rabi, and Leo Szilard.[1] In the immediate postwar era, many magazines were published in the U.S. with the purpose of enlightening the public on nuclear energy issues. The *Bulletin of the Atomic Scientists*, in particular, established an authentic status, attracting readers not only among scientists but also educated citizens interested in nuclear disarmament and international cooperation.

In 1949, the *Bulletin of the Atomic Scientists* became independent of the Atomic Scientists of Chicago. The Atomic Scientists of Chicago thereafter increased political radicalism, but its membership decreased, and it eventually disbanded in 1959.[2] Disagreements among American scientists over thermonuclear tests, cooperation with the Soviet Union, and the World Government movement contributed to the decline of the unified scientists' movement. By contrast, the *Bulletin of the Atomic Scientists* survived these controversies, and even today (2021), remains a symbol of Western intellectuals' conscientiousness surrounding nuclear energy issues; for instance, the cover design of the "Doomsday Clock" adopted in 1947 has become a symbol of nuclear disarmament (Figure 2.1).[3]

The *Bulletin of the Atomic Scientists* and its founding editor Eugene Rabinowitch have often been discussed in the context of scientific internationalism. For example, Patrick David Slaney argued that Rabinowitch "intended the *Bulletin of the Atomic Scientists* to be an institution of scientific internationalism," and his idea could be explained by what Akira Iriye had called "cultural internationalism," the idea that "the best way to order the world so as to avoid war … was to undercut the appeal of nationalism by bringing different cultures into contact in the hopes of producing mutual understanding and the recognition of shared humanity."[4] Slaney's discussion is true regarding the early (around 1950) writings of Rabinowitch. However, cultural internationalism cannot explain the chronological change of his thought, increasingly resonating with anti-Communist ideology. From around 1953, the *Bulletin of the Atomic Scientists* and its key writers gradually developed closer ties with the CCF. The officially announced purpose of the CCF was to defend "freedom" of culture both from Communism and totalitarianism, and many members were unaware of the organization's connection with the CIA. Therefore, while labeling the organization as simply a U.S. government propaganda agency is overly simplistic, there is no denying that a large percentage of its activities were dedicated to nurturing anti-Communist cultural hegemony around the world. The CCF ran many magazines including the famous *Encounter*, but the *Bulletin of the Atomic Scientists* differed from these CCF-sponsored magazines, as it was an independent magazine based in the

Figure 2.1 The "Doomsday Clock" on the front cover of *Bulletin of the Atomic Scientists* (May 1953). Hanna Holborn Gray Special Collections Research Center, University of Chicago Library.

University of Chicago. Yet, key figures on the editorial board of the *Bulletin* also played important roles in some CCF activities. The analysis in this chapter will show the process by which advocacy for the "freedom of science" was gradually blended with Cold War ideology.

There are many existing scholarly works on the CCF. Some pioneering studies came out in the ten years between 1989 and 1999. For example, Peter Coleman defended the CCF as necessary for the fight against Communism. By contrast, Frances Stonor Saunders criticized the organization as having harmed the open exchange of ideas.[5] During the 2000s, more nuanced and complex views of the CCF appeared, including those of Sarah Miller Harris, who partially justified the CCF's activities; Giles Scott-Smith who concluded that the CCF promoted the cultural values commonly accepted in the contemporary political and economic environment; and Hugh Wilford who pointed out that intellectuals took advantage of the CIA funding as opposed to being exploited by the CIA. Among Japanese language works, Masato Karashima's study focused on Japanese economists who were highly regarded as "anti-Communist liberalists" by Edward Shils, an active member of the CCF and one of the key figures in this chapter.[6] In 2017, just when the author was writing the original version of this chapter as an independent article, an important edited volume was published by Giles Scott-Smith and Charlotte A. Lerg, *Campaigning Culture and the Global Cold War: the Journals of the Congress for Cultural Freedom.* Further, in 2018, Audra J. Wolfe's seminal work, *Freedom's Laboratory: The Cold War Struggles for the Soul of Science*, was published. The author owes much to insights gained from both of these fascinating works in revising the original article into the current chapter. The edited volume by Scott-Smith and Lerg focused on several CCF-sponsored magazines, including those published in Asia, Latin America, and the Middle East. Wolfe contributed a chapter on *Science and Freedom*, evaluating the journal as having only limited influence because of its limited circulation and short publication period, mostly stemming from the editor's troubled relationship with the CCF.[7] Wolfe's solo-authored book, published the following year, pointed out that although *Science and Freedom* was not a successful endeavor, science was an important theme of the CCF's activities, and that U.S. governmental agencies such as the CIA and the Department of State regarded science as an important tool for diplomacy and supported scientists' international activities through the CCF and private foundations.[8] As the first version of this chapter was published as an independent article in 2018,[9] it made no reference to the latter work by Wolfe. This chapter is a revised version of the original, which has incorporated Wolfe's arguments.

2.1 Scientists in Chicago and the Publication of the *Bulletin of the Atomic Scientists*

Eugene Rabinowitch, founder and editor-in-chief of the *Bulletin of the Atomic Scientists*, was born into a Russian-Jewish family in Imperial Russia in 1901. After going into exile in Germany via Poland, in 1926 he received a doctoral degree in chemistry from the University of Berlin (present-day Humboldt University). He then became a research assistant to James Franck at the University of Göttingen, but moved to Niels Bohr's laboratory in Copenhagen in 1933 to escape Nazi persecution. He subsequently moved again to University College London, and in 1939, he finally emigrated to the U.S. to join the faculty of the Massachusetts Institute of Technology (MIT). He naturalized and participated in the Manhattan Project in 1944. After the war, he taught at the University of Illinois. In addition to his publishing of the *Bulletin of the Atomic Scientists*, he was also well-known as a leader of the Pugwash Movement, an international scientists' movement for nuclear disarmament launched by Bertrand Russel and Albert Einstein, for which he served as chairman from 1957 until his death in 1973.[10]

During the war, scientists at the University of Chicago's Metallurgical Laboratory became concerned about the impact of nuclear energy on human beings, and advocated for international control. James Franck, who emigrated to the U.S. after Rabinowitch, headed the Chemistry Division, and Rabinowitch was his subordinate. Leo Szilard, a Hungarian-Jewish émigré scientist who received a doctoral degree under Albert Einstein, also joined the Metallurgical Laboratory. Szilard attempted to deliver a letter to President Roosevelt, asking him to postpone the use of nuclear bombs, but the sudden death of the president prevented him from achieving this goal, and he met with Secretary of State Byrnes of the Truman administration instead. However, his action provoked the anger of Leslie R. Groves, director of the Manhattan Project. Director Arthur Compton of the Metallurgical Laboratory organized the Committee on the Social and Political Implications of Atomic Energy to coordinate the scientists' opinions, and appointed Franck as chairman. The Committee submitted the famous "Franck Report" to Secretary of War Henry Stimson on June 11, 1945, protesting the plan to drop nuclear bombs on Japan. It was actually Rabinowitch who drafted the report; in contrast to Franck who was not comfortable in writing English, Rabinowitch was an "adaptable cosmopolitan" who "had an excellent English vocabulary," and Franck's "trusted interpreter."[11] Although the Franck Report was ignored by the U.S. government, which had already reached the decision to employ nuclear weapons against Japan, the scientists' movement continued. In November 1945, other scientists of the Manhattan Project joined the Atomic Scientists of Chicago, and the Federation of Atomic Scientists was established with approximately 3,000 members. It was renamed the Federation of American Scientists (FAS) the following month.

It was out of these scientists' movements that the *Bulletin of the Atomic Scientists* emerged in December 1945. In its early years, the *Bulletin* was funded by the Emergency Committee of Atomic Scientists led by Albert Einstein, and 20,000 prints were distributed free of charge to local branches of the FAS, foreign scientists, and libraries in many countries.[12] During the 1940s, the *Bulletin* devoted many pages to the civilian control of nuclear energy and declassification of scientific information. Just around the time the *Bulletin* was established, the May–Johnson Bill, which endorsed secrecy and the military control of nuclear information, was submitted to the Congress. The Bulletin contributed to the "success in mobilizing public opinion against military control," and to the ultimate failure of the May–Johnson Bill. The *Bulletin* also provided a forum for heated discussions on the international control of nuclear energy when the "Acheson-Lilienthal Report" and its revised version, the "Baruch Plan," were submitted to the United Nations. Rabinowitch believed that civilian control of nuclear energy, the freedom of science, and international control of nuclear weapons were mutually interdependent. For, according to him, civilian control would promote the declassification of information, and open information would bring about free exchange of scientists across national borders, i.e., "freedom of science." Furthermore, if scientists in various countries, including those from the Soviet Union, could build mutual trust, such relationships would enable international control of nuclear energy.[13] In this period, as Slaney has emphasized, Rabinowitch and the *Bulletin* embraced cultural internationalism. However, Rabinowitch's definition of "freedom of science" would be transformed over time.

It is noteworthy that many of the central members of the scientists' movement in Chicago were émigré scientists who had escaped Nazi persecution and other humanitarian crises. When the Nazis expanded their sphere of influence in Europe, U.S. universities helped Jewish scientists emigrate to the U.S., some of whom were nuclear physicists. When the Metallurgical Laboratory was established as a Manhattan Project lab, Director Compton, in his struggle to build the organization from scratch, recruited those émigré scientists. A lively community of émigré scientists flourished on the campus of the University of Chicago, characterized by cosmopolitanism and "hodgepodge democracy." The Chicago scientists' humanitarian concerns regarding nuclear weapons and their interest in international cooperation are at least partly attributable to this background of émigré community.[14] The U.S. government, however, increasingly became alarmed by the political activities of the scientists. Senator McCarthy declared that the FAS was "heavily infiltrated with communist fellow-travelers," and many scientists became the subject of loyalty investigations, often resulting in losing their jobs or even in suicide.[15] The withdrawal of Robert Oppenheimer's security clearance[16] also occurred in this context.

2.2 Encounter with the CCF

As mentioned above, in 1949, the *Bulletin of the Atomic Scientists* became independent from the Atomic Scientists of Chicago, and went under the management of the non-profit Educational Foundation for Nuclear Science, which specialized in the publication business.[17] The reason for this change was partly because of the need for funding, but the move also severed the *Bulletin*'s ties with the politically active scientists. Although Rabinowitch was one of the leading figures of the scientists' movement together with Franck, he was a moderate: when the FAS was established by absorbing smaller groups, Rabinowitch disagreed with more militant scientists who argued that "only a small compact organization could act with the unanimity responsible for their successful stand against the May-Johnson bill." He argued that the scientists' organization "should not acquire partisan color or become an organization to defend professional interests of scientists."[18] Especially after 1949, he insisted that the *Bulletin* should steer a course away from any political movements. For example, in 1950, when he was asked about the *Bulletin*'s attitude toward the "world federalists or the world government groups in general," he answered that its editors had "taken special pains not to become identified with this particular movement" [underline in the original text]. He argued that the *Bulletin* would "lose its raison d'etre if it were to become a partisan organ of any political denomination." He had "tried to steer a course between the two extremes of political partisanship and complete avoidance of politics," he explained, so that the *Bulletin* should be "a source of impartial information, and a forum for open discussion of the political consequences of new scientific discoveries."[19]

Perhaps because of this apparently "apolitical" stance, the *Bulletin* received generous support from the Ford Foundation and expanded its circulation to 25,000 copies. According to the list of donors prepared for the board meeting of the *Bulletin*, the Ford Foundation contributed $25,000 in 1951 and $35,000 in 1953, in addition to the subscription fees, and this became the source of funds for the publication throughout the 1950s. Also, the *Bulletin* during the 1950s frequently carried advertisements for nuclear energy-related private companies such as General Electric, which indicated that advertising also contributed to the revenue. According to the board meeting records, overseas subscribers (including both institutions and individuals) were: 33 in Australia, 128 in Canada, 85 in the U.K., 69 in France, 62 in Germany, 40 in India, 46 in Italy, 78 in Japan, 38 in the Soviet Union, and so on. Many overseas universities and research institutes, especially in the developed countries, subscribed to the *Bulletin*.[20]

When and how did the editors of the *Bulletin* come into contact with the CCF? The CCF was established on June 26, 1950, in Berlin as a "guardian of the open, pragmatic 'market-place of ideas' ideal of post-war liberalism."[21] At the inaugural meeting, notable intellectuals such as Arthur Koestler, Karl Jaspers, John Dewey, Arthur Shlesinger, Jr., and Sidney Hook were present. The CCF soon became a permanent organization based in Paris. At the center of its administration was Michael Josselson, an émigré intellectual from Estonia, who was in constant touch with Frank Wisner, head of the Office of Policy Coordination at the CIA. Other leaders included Melvin Lasky, who organized the inaugurating meeting and edited the flagship journal *Encounter*; Nicolas Nabokov, who served as the General Manager; and Ignazio Silone, an Italian writer and politician. Edward Shils, who was also involved in the establishment of the *Bulletin of the Atomic Scientists*, was the academic backbone of the CCF. Shils was born into a Russian-Jewish family in Philadelphia. After graduating from the University of Pennsylvania, he worked during the day and studied sociology in night classes in New York and Chicago. He joined the faculty of the University of Chicago in 1938, where he remained for the duration of his career, eventually serving as the Dean of the Sociology Department. During the Second World War, Shils was engaged in government propaganda and intelligence activities in the OWI and the OSS, where he built a network of like-minded U.S. and European intellectuals.[22] Such a background likely prepared him for his involvement in the CCF in the postwar years. The officially stated goal of the CCF was to defend culture both from the extreme right and left: this ideology of regarding both Fascism and Communism as threats to intellectual freedom was not new in the U.S. For example, during the 1930s John Dewey led the (similarly named) Committee for Cultural Freedom, which embraced a similar ideology.[23] However, the Congress for Cultural Freedom differentiated itself in that its activities were mostly dedicated to cultivating anti-Communist elements in various countries, and this was probably the reason why the CIA found it a useful conduit. The CIA appointed Cord Meyer, head of the International Organization Division of the CIA, as special liaison officer in charge of the CCF. In the late 1950s, the CCF expanded its activities to Asia, the Middle East, Africa, and Latin America.[24]

The CCF showed interest in the relationship between S&T and political culture from early on, and this interest drew Rabinowitch closer to the CCF. In the CCF's inaugural conference in Berlin, renowned geneticist Hermann Joseph Muller and other notable scientists gave presentations on the importance of freedom of science. These speakers viewed science as a "canary-in-the-coalmine for the more general condition of freedom." According to Wolfe's meticulous research, based on the Muller Papers and other archival sources, the CCF's organizers "attempted to convert the momentum of Berlin into a permanent anti-Communist movement," and one such outcome was the establishment of the American Committee on Cultural Freedom led by Muller, Hook, and Schlesinger. The U.S. group further created a Committee on Intellectual Freedom, with Muller's help, and Rabinowitch was recruited to join this group.[25] Another key individual who influenced Rabinowitch was Michael Polanyi, a Hungary-born chemist who was in charge of science in the CCF. After receiving a doctoral degree at the University of Budapest and working in Karlsruhe, he accepted a position at the Kaiser Wilhelm Institute in Berlin, and further became director of the chemical-kinetics research group in Fritz Haber's Institute for Physical Chemistry and Electrochemistry. He moved to the University of Manchester in the U.K. in 1933, as he and other scientists of Jewish origin were compelled to leave Germany due to the Nazis' rise to power. After the war, he accepted a chair in social philosophy at the University of Chicago in 1951, but was not able to obtain a U.S. visa due to the strict visa policy of the McCarthy era. Polanyi was a pioneering thinker of the relationship between science and society, and argued against "Socialist planning schemes" and "government control of science." During the war, he had helped found a "freedom in science" movement to counter the British scientific Left. It was in the extension of such activities that Polanyi became active in the CCF, chaired the CCF's Special Committee on Science, and helped to organize the Hamburg Conference in June 1953.[26] The Conference played an important role in creating ties between American scholars, including Rabinowitch and Shils, and the CCF.

The 1953 Hamburg Conference turned out to be a conspicuous moment in the Cultural Cold War. U.S. government records reveal that both the CIA and the Department of State showed interest in the conference. On October 20, 1952, Cord Mayer of the CIA visited J. B. Koepfli, Science Advisor of the Department of State (as for the position of Science Advisor, see also Chapter 6), and told him that the CCF was "planning an international conference on the freedom of science to be held in Hamburg in the summer of 1953" and that "there was to be a preparatory meeting of outstanding scientists and philosophers in Brussels in late December or early January." The CIA and the Department of State had "a very great interest in certain aspects of this congress" and were "anxious to have the 'right' people represent the United States." They first approached I. I. Rabi of Columbia University, who was also an editor of the *Bulletin*, but he declined due to health concerns, recommending Theodosius Dobzhansky instead. Asked by Mayer if he could "persuade Rabi to change his mind," Koepfli contacted Rabi, who explained that he "would not be in sympathy with certain of their [other members of the delegation's] points of view," but would reconsider if "the Department of State felt it imperative that he go."[27] Their conversation indicates that the CIA and the Department of State were concerned with the selection of U.S. delegates to the CCF conferences. Although no evidence has been found that the CIA and the Department of State recommended Shils and Rabinowitch, they must nevertheless have been seen as the "right people" in the eyes of the government agencies because apparently their participation provoked neither doubt nor controversy.

The CIA and the Department of State were also advising the Rockefeller Foundation on matters concerning the CCF. In March 1953, Warren Weaver, head of the Natural Science Division of the Rockefeller Foundation, consulted Koepfli on whether it was appropriate to meet the request from the Hamburg Conference organizers for financial support of $10,000. Weaver was concerned about the possibility that "anyone of the American delegation presented a paper in which he attacked, for example, our immigration and naturalization laws as interfering with the freedom of science." Weaver's remark referred to the U.S. denial of entry visas to many European scientists resulting from the xenophobic atmosphere of McCarthyism, including Michael Polanyi, as mentioned above. In fact, back in 1940, Weaver had tried in vain to assist Polanyi's migration to the U.S.: Weaver himself was a vocal supporter of the freedom of science, and had both trusted and admired Polanyi since the 1940s. It was understandable, therefore, that he was particularly concerned about the unfavorable reputation of the U.S. government among scientists for its immigration and naturalization policy.[28] After Koepfli (Department of State), Meyer (CIA), and Outerbridge Horsey (Department of State) discussed the issue, Koepfli replied to Weaver:

1. I think it is safe to say that the United States Government looks with favor on the forthcoming conference.
2. I think all of the people invited to present papers will have been pretty carefully looked at and that the chances of a sleeper are very slight.
3. I think that the conference will have considerable repercussions and that the other side will have to take cognizance of it, since it would hit them pretty hard.
4. To the extent there are political implications, which I assume the Foundation realizes, I suggested that the question of financial support for the congress was a decision they would have to make themselves.[29]

Upon receiving this reply, the Rockefeller Foundation decided to provide funding. Reiko Maekawa, in her study of the Rockefeller Foundation's support of émigré scholars, pointed out that the Foundation funded many of the CCF conferences with the approval of the Department of State. The above evidence shows both the CIA and the Department of State influenced the Rockefeller Foundation's decisions to support the CCF conferences.[30] It is also noteworthy that the CIA and the Department of State perceived the Hamburg Conference as having "considerable repercussions" since it would hit "the other side" (apparently indicating anti-American elements, including Communists, within the scientific circle) hard, implying that international conferences on science were regarded as an arena of the Cultural Cold War in which the U.S. could exhibit its political and moral superiority.

Another episode strongly suggests that Rabinowitch was also aware of his own responsibility as a government-endorsed representative at the Hamburg Conference; about one month before the conference, Rabinowitch contacted Orr Reynolds of the Office of Naval Research (ONR)[31] and asked him if he could use the Military Air Transport Service (MATS) to fly to Germany "on the basis of his ONR contract." It is not surprising that Rabinowitch had a research contract with the ONR because the ONR provided generous grants to many university researchers. It is noteworthy, however, that he considered using a military flight to participate in a CCF-sponsored conference. The discussion within the U.S. government concerning Rabinowitch's request suggests that his participation in the CCF-sponsored conference was perceived as a quasi-official event. Reynolds consulted the Department of State as to whether Rabinowitch's use of MATS should be approved. Neil Carothers of the office of Science Advisor, Department of State, discussed the matter with Outerbridge Horsey, and replied to Reynolds that:

> the justification for MATS should be based on the other scientific conferences but informally I would say that his attendance at the Congress in Hamburg should not be considered a negative factor, but a positive one in favor of providing such transportation.[32]

The conversation indicates that the Department of State viewed the CCF-sponsored Hamburg Conference and Rabinowitch's participation in it as contributing to the national interest. Rabinowitch, unlike more radical and/or dissenting scientists in Chicago, was endorsed as a scientist who could safely represent the U.S. In other words, he was a useful weapon in fighting the Cultural Cold War.

After the Hamburg Conference, the CCF officially established the Committee for Science and Freedom, and *Science and Freedom* was issued the following year. Rabinowitch and Shils also participated in the inauguration meeting of the journal in July 1954. As Wolfe has discussed, the journal's influence was limited because the circulation was small (7,000 prints) and the CCF's support was cut after just a few years. Polanyi and his son, editor of the journal, were not amenable to the CCF headquarters' instruction to combat global Communism, but were rather preoccupied with academic freedom in the West.[33] It should be pointed out, however, that the information dissemination route for the CCF's science division was not only *Science and Freedom*, but also the more widely distributed *Bulletin of the Atomic Scientists*. Rabinowitch, Shils, and Polanyi disseminated the ideals and activities of the CCF's science division through the *Bulletin*.

For example, the details of the Hamburg Conference were introduced in the *Bulletin* even before the first issue of *Science and Freedom* was published. Polanyi's opening address was printed in the November 1953 issue, and Shils' observation of the conference in the May 1954 issue. In his opening address, Polanyi explained that the purpose of the conference was not only to "rouse academic opinion more widely against the treatment of scholars and of scholarship under totalitarianism" but also to pursue "clearer principles of intellectual liberty than we possess today" in the free societies. He proposed to invoke "the power of a different authority" in contrast to the Communist's abusive attempts to control academia through power, and "to protect the free society and the pursuit of free scholarship within it."[34] Shils, on the other hand, celebrated the great success of the conference, at which "119 scientists and scholars from nineteen countries met." He emphasized that one of the greatest contributions of the conference was "the gradually developed notion that scientific activities institutions and individuals formed a kind of social and cultural system with its own powers of self-maintenance and self-regulation and that this system must of necessity be relatively autonomous." Shils' article also showed that speakers in the conference often cited the Soviet Union as an antithesis to the ideal of autonomous science. For example, Shils cited some speakers' criticisms of the Soviet authorities' abuse of power to distort scientific truth, while pointing out that "the love of truth still survived" among some Russian scientists. Shils also introduced a remark of Alexander Weissberg, a noted Jewish émigré scientist exiled from the Soviet Union, that the "real scientists" do not "allow the doctrines of the ruling political regime to intrude into the heart of their studies." Even if they compromise on the superficial level, Weissberg said, they "do not deform their own minds by attempting to believe in dialectical materialism."[35]

The definition of "freedom of science" emerging from the records of the Hamburg Conference, and introduced in the *Bulletin*, differed from Rabinowitch's definition in the early postwar years (up to around 1950), which focused on the free exchanges of scientists across national borders. The Hamburg version of "freedom of science" was the scientists' autonomous community as a privilege of the "free" nations, the polar opposite of the Communist world. As Rabinowitch became closely associated with the CCF, his interpretation of freedom of science began to change.

2.3 Connecting Rabinowitch, Shils, CCF, and the *Bulletin*

Both Rabinowitch and Shils continued their association with the CCF even after the Hamburg Conference, although of the two, Shils was more committed to the CCF's activities. Shils replaced Polanyi as the chair of the Committee for Science and Freedom.[36] After *Science and Freedom* ceased publishing, the Committee continued activities such as distributing anti-Communist pamphlets to foreign countries, and Shils also created a new CCF-sponsored magazine, *Minerva: A Review of Science, Learning and Policy,* for which he served as the chief editor for 33 years. While playing an extremely important role in the CCF on the one hand, Shils continued to be an editor and writer of the *Bulletin of the Atomic Scientists,* on the other hand. Shils' steadfast anti-Marxism and his determination to "'throw the light of reason' on extremist ideologies of both Left and Right"[37] were expressed both in the CCF and in the *Bulletin.*

Rabinowitch also continued his association with the CCF. He proposed organizing an international conference of the Committee of Science and Freedom in Paris in August 1956.[38] In this conference called the "Study Group" not only key figures of the CCF but also representatives from Yugoslavia, Chile, Spain, and others presented papers. Polanyi gave a lecture titled "The Magic of Marxism," based on his paper published in the June 1956 issue of the *Bulletin of the Atomic Scientists.* In his lecture, Polanyi closely examined why Marxism had a strong appeal to Western intellectuals, and explained that part of the reason was that it concealed moral passions with "a scientific disguise against being deprecated as mere emotionalism."[39]

In the Paris Study Group meeting, Rabinowitch presented a paper titled "The Role of Scientists in Public Affairs and the New Meaning of Academic Freedom," which clearly demonstrated a change in his thoughts on the role of scientists. He predicted that scientists, by virtue of their "objective, factual analysis" and "rational solutions based on this analysis," were "heading toward greater recognition and greater influence" in economic, social, and political policies. However, he continued, scientists who would move into positions of national influence would be "a different type from the Bohrs and Oppenheimers, the Fermis and the Comptons of the past generations." The model scientist of this new generation would be less of an "intellectual" type, but be characterized by "capacity for unprejudiced analysis" and "readiness for experimentation." Rabinowitch predicted that scientists all over the world would eventually share this common outlook, and that because of their similarity, "their increased influence on the national policies of the different countries [wa]s bound to increase the case of international communication." The responsibility of scientists was to contribute, "by education, by persuasion, by political action," to the "rationalization of the behavior of nations and of humanity as a whole– rather than to try ... to assure that the results of their research work [are] used only for peaceful and constructive purposes." As for the relationship between academic freedom and the state control, Rabinowitch explained that, since the overwhelming proportion of research budgets came from government agencies, scientists were faced with "a split of loyalties," both to the academic community and to government authorities. Although he admitted a potential threat to academic independence arising from "increased involvement of the science faculties in government-sponsored research," he did not outright reject state scrutiny and supervision because of the government's "increasing contribution to the budgetary needs" of scientists. His solution to protect academic freedom was an arrangement

> by which government research contribution would be distributed not directly to the professor, but through the intermediary of a university body, independent of the government, so that the latter does not acquire direct power of the purse strings over the individual professor or an individual laboratory.[40]

When he established the *Bulletin* in the 1940s, Rabinowitch defined the scientist's role as enlightening citizens and other scientists on the danger of nuclear weapons, and "freedom of science" as international exchange of scientists including those from the Soviet Union. By contrast, in 1956 he labeled notable (and dissenting) scientists such as Oppenheimer as "the past generation," and expected the new generation of scientists to assume a more influential role by "rationalizing" the government and the citizenry. He also defined "freedom of science" as a privilege to be enjoyed inside the protective walls of the university, while accepting a certain degree of surveillance in exchange for government funding.

Shils also presented in the Study Group under the title of "Observations on the Scientists' Movements in the United States." He explained that the "scientists' movement" could touch the minds of many Americans because the movement "did not direct its thought to any radical criticism of American society," and it "accepted the general structure of government," while it criticized "wrong policies" and combated "encroachments on the freedom of science." The movement's leaders also persuaded legislators that they were "well-intentioned collaborators," and as a result, journalists "heed[ed] with respect" and congressmen "turn[ed] a hospitable ear" to what the scientists said. According to Shils, academic freedom could be established only when there was "a sense of affinity between the academic and non-academic worlds." Academic freedom could be guaranteed, he believed, only through "mutual esteem" and the "equal value of diverse activities" not only among scientists but also politicians, administrators, soldiers, and businessmen.[41]

Rabinowitch's and Shils' papers indicate their belief that freedom of science should be founded on the mutual trust of academia and the state authorities, and that scientists should be well-intentioned collaborators with the government rather than dissenters. This is why the Department of State, the CIA, and the CCF placed trust in them, and also why Rabinowitch and Shils distanced themselves from more radical scientists dissenting from government nuclear policies. In 1967 and 1968, a series of reports in *Ramparts*, the *New York Times*, and the *Washington Post* disclosed the CIA's long-term support of various cultural, youth, and labor organizations, including the CCF. The Pugwash Movement, of which Rabinwitch was chairman, had also been receiving financial support from the Kaplan Fund, a CIA conduit. Rabinowitch, however, did not see this as a problem. Since "neither the ONR nor the CIA attempted to steer the direction of the programs they funded," he thought that "scientific freedom, and therefore academic integrity, had been maintained."[42] Rabinowitch's changing definition of the role of scientists culminated in the unquestioned acceptance of clandestine financial aid to intellectual activities. The Pugwash Movement, which initially advocated for complete nuclear disarmament, gradually came to recognize nuclear armament and nuclear deterrence. Such a transition somewhat resembles the process by which Rabinowitch and the *Bulletin* gradually steered closer to the CCF's anti-Communist stance. Both Pugwash and the *Bulletin* endorsed freedom and autonomy of scientists, but both were increasingly involved in Cold War ideology, narrowing the space for dissent and diversity of opinion.

2.4 The *Bulletin* Articles on Scientists' Roles

The changing definitions of scientists' roles and "freedom of science" were also visible in the contents of the *Bulletin*. Although the *Bulletin* had publicly announced an editorial policy to carry diverse opinions on nuclear disarmament, civil defense, radioactive fallout, and scientists' loyalty, and to stand neutral and objective, the *Bulletin* strongly reflected Rabinowitch's and Shils' views. Rabinowitch penned an "editorial" for almost every issue, and Shils contributed his writings frequently. Polanyi was also referred to as a "frequent contributor," although the frequency of his contributions was much less than that of Rabinowitch and Shils. From examination of *Bulletin* articles by Rabinowitch, Shils, and Polanyi from 1953 to 1958 (60 issues in total) emerges a chronological change in the *Bulletin*'s stance toward several key issues: scientists' roles and responsibilities in society, international and civilian control of nuclear energy, and the definition of "freedom of science."

First, on the roles and responsibilities of scientists, Rabinowitch argued throughout the 1940s and 1950s that scientists should enlighten the government and the citizenry. However, the nature of this enlightenment changed over time. In the October 1953 issue, Rabinowitch was still arguing that the "scare propaganda" which was "initiated by atomic scientists in 1945" should be continued until the "horrible vision" of an atomic war would be "etched forever in the minds of people."[43] However, in the January 1955 issue, he admitted that elimination of nuclear weapons was no longer practical, and argued that the only way to avoid nuclear wars rested on the "creation of a supranational world authority capable of enforcing force." If "America and its allies" dedicate their long-range policies to this aim, and if "American people and their leadership ... commit themselves to a world community as their final aim, and to make this commitment clear to the world," he argued, such a "positive ideal" would successfully oppose "the Communist blueprint of a World Federation of Soviet Republics."[44] Further, in the June 1957 issue, he defined the "most important public obligation of scientists" as "the task of educating mankind to the necessity of a world political and moral renewal to match the current revolution in science and technology."[45] This comparison reveals the transformation of his definition of scientists' role from propagating and educating on the terror of nuclear weapons to constructing a world community to match the S&T revolution. Rabinowitch repeatedly used the term "world community" in various articles without clearly defining it, but the January 1958 issue gives a glimpse of what he meant. In the editorial of this issue, Rabinowitch argued that both nuclear deterrence and nuclear test bans were merely temporary measures like "taking aspirins and going earlier to bed" when a major operation was necessary. A more fundamental way to build peace, he suggested, would be the creation of a "world community" involving even the "peoples under Communist rule." He presupposed that "every discovery in science, and every invention in technology, wherever made, [could] be of benefit to mankind" and that "economic progress [was] a good thing, wherever and under whatever economic system it occur[ed]." Upon this presupposition, he argued that the hope of mankind lay in the "creation of a world community" which would fully utilize the "powers of science" to "satisfy the common spiritual and material interests of all nations."[46] From this passage, what he meant by "world community" was not the free international exchanges among scientists that he advocated for in the 1940s, but a world in which beneficial technologies would circulate freely. He also perceived such a liberal capitalist "world community" as a measure to counter the Soviet's World Federation scheme.

Secondly, Rabinowitch's view on the civilian and international control of atomic energy also changed over time. As historian Shiho Nakazawa has pointed out, toward the end of the 1950s, the U.S. government abandoned entirely the idea of international control or total nuclear disarmament. Instead, the government pursued means to maintain nuclear armament while placing the weapons under the control of nuclear-armed countries. Such a change was reflected in the perceptions of the contemporary scientists, too. In the October 1957 editorial, Rabinowitch declared that complete nuclear disarmament was no longer practical. Instead, he deemed it necessary to promote international control by suppressing deterrence to the lowest possible level, on the one hand, and slowing down the arms race on the other.[47] He also wrote frequently about the need for civil defense, based on the thinking that nuclear weapons would be inevitably used if war broke out. He pointed out the vulnerability of urban centers to nuclear attacks, and advocated for the diffusion of the urban population and preparation for massive evacuations.[48]

Finally, concerning the idea of "freedom of science," Rabinowitch's ideal of the free exchange of scientists was not completely lost, but the emphasis of his argument shifted to, for example, "freedom" as the privilege of scientists in the capitalist countries, or to the allegedly inherent nature of science as compatible with liberalism. For example, in the March 1953 issue, Rabinowitch argued that the principal value of science lay in the "promotion of certain moral and ethical principles," such as "respect for fact and denunciation of deceit and concealment," and the "receptiveness to new ideas" and "readiness to acknowledge the validity, or even superiority of the methods used by others." He expressed concern that such values were challenged both by "anti-intellectualism" and "anti-rationalism" outside of scientific circles and "self-denunciation and self-destruction" within scientific circles, as some scientists were accepting "ruling on scientific matters from authorities outside science." He argued that scientists had to defend the "freedom and autonomy of science" from these two challenges.[49] Shils also advocated in the April 1955 issue that "scientific activity [wa]s the activity of free men," and "its community [wa]s the epitome of the free society." According to Shils, "science provided the model of a free society of reasonable men coordinating themselves voluntarily in the light of a transpersonal standard."[50]

When Robert Oppenheimer, former director of Los Alamos National Laboratory in the Manhattan Project, was declared a "security risk" and purged because of his association with Communists when he was an undergraduate student, both Rabinowitch and Shils defended Oppenheimer and criticized the government. Their criticism was based on the presupposition that the U.S. was an inherently "free" society, and thus the Oppenheimer case was a lamentable "deviation" from the American standard of freedom. Shils wrote, for example, "America's reputation as a free country governed by reasonable and courageous men already damaged in these past few years has further been damaged."[51] Rabinowitch also argued that "it would be a grave new departure in American political life," if the "unpopular or dissenting members of the Administration team" should be accused as "security risks."[52] For the same reason, Rabinowitch also reacted fiercely to the AEC's interference with the free academic activities of scientists. Hermann Muller, geneticist at Indiana University, was planning to present a paper on radiation-induced gene mutations at the Second United Nations Conference on the Peaceful Use of Atomic Energy, but his paper was removed from the program by the AEC. Around that time, radioactive contamination caused by thermonuclear tests was attracting attention both inside and outside of the U.S., and Muller and his fellow geneticists were warning that even a small dose of radiation could cause genetical changes in living creatures.[53] The AEC was fearful that the paper might provoke controversy as it contained references to the genetic effects of the Hiroshima and Nagasaki detonations. "Just when Lysenko's grip on Soviet biology [wa]s showing signs of weakening, the imposition of an 'official line' [wa]s attempted in American genetics," Rabinowitch criticized.[54] Common among these arguments was the polar oppositional view of the essentially free American society versus the unfree societies of the Communist countries, and the assertion that deprivation of scientists' "freedom" was a deviation from the normal condition of American society. The 1940s definition of the "freedom of science," as international exchanges inclusive of the Soviet Union, had undergone a dramatic change.

The shifting definitions of the role of scientists and "freedom of science" in the *Bulletin* were probably not unrelated to the editors' close association with the CCF. According to Reiko Maekawa's study of the New York émigré intellectuals during the 1930s, some turned to anti-Communist liberalism after the Second World War—Sydney Hook, who led the American branch of the CCF, was one of them. Maekawa explained that the intellectuals who were disillusioned by the leftist movements found renewed passion in activities to "protect Western freedom and democracy by supporting anti-Communist liberalism."[55] A similar theory might be applied to the first-generation émigré scientists based in Chicago, including Rabinowitch, who gradually became disillusioned with the initial ideals of the elimination of nuclear weapons or the borderless exchanges of scientists, and increasingly identified themselves with the binary worldview in which they found their passion in defending "freedom," either from Communists or from the occasional "deviation" of their own government.

Scott-Smith employed Antonio Gramsci's concept of hegemony to explain that the CCF was spreading the "apolitical" value that the contemporary Western intellectuals commonly held: that culture should be independent from politics. Ironically, however, the CCF did so through political measures: financial support from the CIA.[56] Editors and writers of the *Bulletin of the Atomic Scientists* also held the "apolitical" ideal of autonomous scientific community, but they increasingly came to view such an ideal within the Cold War discursive framework: that freedom of science was a privilege of Western capitalist society, diametrically opposed to the character of Communist society. Such a framework of thought was related to the worldview widely circulated during the 1950s as Eric Foner has discussed, "the idea that the love of freedom was the defining characteristic of American society had become fully incorporated into the popular consciousness" by the end of the 1950s, and people understood the world as sharply divided into opposing camps of "freedom" versus "totalitarianism."[57] In this context, the *Bulletin of the Atomic Scientists*, although born out of nuclear disarmament ideology, came to find the responsibilities of scientists in the protection and expansion of "freedom," rather than in questioning the issue of nuclear parity between the superpowers.

The process by which scientists such as Rabinowitch came to resonate with the CCF was a prime example of the linkage point between science and international politics. The scientists involved constantly advocated for freedom and the autonomy of science, but as long as they did so as a privilege of the Western countries represented by the U.S., their ideals and anti-Communism supported each other, rather than contradicting each other. Scientists such as Rabinowitch, who maintained an international network, and who did not lose trust in state power, became an invaluable asset for the U.S. government in fighting the Cultural Cold War. This was why the CIA, the Department of State, and the CIA-supported CCF treasured Rabinowitch and his friends as the arsenal in the Cultural Cold War. Not all scientists followed the same path, however. More radical scientists, such as Oppenheimer, who steered a course away from Rabinowitch, fell victim to McCarthyism. Scientists could not be completely "free" from the hegemonic values of the time. If they attempted to express such freedom, they risked being penalized by the state power or by their contemporary society.

This chapter has portrayed the process by which scientists became involved in Cold War politics in matters related to the military use of atomic energy. However, the non-military use of atomic energy also became an important theater of the Cultural Cold War. The U.S. government set the "peaceful use" of nuclear energy as the central theme of their overseas information programs, in which science and scientists once again emerged as important actors of the Cultural Cold War. Research reactors and technological training, for instance, were not simply foreign aid programs, but also measures to promote the national image and win the hearts and minds of foreign leaders and citizens. The next chapter will explore how and why U.S. research reactors were provided to foreign countries, with a particular focus on the example of Japan, but also with reference and comparison to the cases of South Vietnam and Burma.

Notes

1   Hanna Holborn Gray Special Collections Research Center, University of Chicago Library, Guide to the *Bulletin of the Atomic Scientists* Records 1945–1984, https://www.lib.uchicago.edu/e/scrc/findingaids/view.php?eadid=ICU.SPCL .BULLETIN; Roy MacLeod, "Consensus, Civility, Community: *Minerva* and the Vision of Edward Shils," in *Campaigning Culture and the Global Cold War: The Journals of the Congress for Cultural Freedom*, eds. Giles Scott-Smith and Charlotte Lerg (London: Palgrave MacMillan, 2017), 48–49 There is a rich depository of studies on atomic scientists in Chicago both in English and Japanese. For example, in English, Alice Smith, *A Peril and a Hope: The Scientists' Movement in America, 1945–1947* (Cambridge, MA: MIT Press, 1971). First published 1965; Lawrence S. Wittner, *Confronting the Bomb: A Short History of the World Nuclear Disarmament Movement* (Stanford, CA: Stanford University Press, 2009); Jessica Wang, *American Science in an Age of Anxiety*. In Japanese, Masakatsu Yamasaki and Shizue Hinokawa, eds., *Genbaku wa koshite kaihatsu sareta* [*This Is How Atomic Bombs Came into Being*] (Tokyo: Aoki Shoten, 1990), and a series of books and articles by American historian Shiho Nakazawa, including *Oppenheimer: genbaku no chichi wa naze suibaku kaihatsu ni hantai shitaka* [*Oppenheimer: Why Did the Father of Atomic Bombs Oppose the Development of Thermonuclear Bombs*] (Tokyo: Chuokoron-Shinsha, 1995); "Eisenhower seiken koki ni okeru kakugun-shuku kosho: kakujikken teishi o meguru mondai o chushin ni" [The Nuclear Disarmament Negotiations in the Late Eisenhower Era: A Focus on the Test Ban Issue," *Bunka Joshi Daigaku Kiyo: Jinbun Shakaikagaku Kenkyu*, vol. 13 (January 2005): 41–53; "Suibaku kaihatsu hantai kankoku to kagakusha no tachiba" [The Recommendation of Anti-Thermonuclear Bombs Development and Scientists], *Kokusai Kankeigaku Kenkyu* [*Studies of International Relations*], vol. 17 (January 1990): 9–30; "Leo Szilard to genshikagakusha undo: genshiryoku no kaihatsu to kanri no shiten kara" [Leo Szilard and Nuclear Scientists' Movement: A Focus on the Development and Control of Nuclear Energy], *Kokusai Kankeigaku Kenkyu* [*Studies of International Relations*], vol. 18 (January 1991): 51–60.
2   Hanna Holborn Gray Special Collections Research Center, University of Chicago Library, Guide to the Atomic Scientists of Chicago Records 1943–1955, https://www.lib.uchicago.edu/e/scrc/findingaids/view.php?eadid=ICU.SPCL .ASCHICAGO.
3   *Bulletin of the Atomic Scientists* website, https://thebulletin.org/.
4   Patrick David Slaney, "Eugene Rabinowitch, the Bulletin of the Atomic Scientists, and the Nature of Scientific Internationalism in the Early Cold War," *Historical Studies in the Natural Sciences*, vol. 42, no. 2 (2012): 114–142.
5   Peter Coleman, *The Liberal Conspiracy: The Congress for Cultural Freedom and the Struggle for the Mind of Postwar Europe* (New York: Free Press, 1989); Saunders, *Who Paid the Piper?*
6   Sarah Miller Harris, *The CIA and the Congress for Cultural Freedom in the Early Cold War: The Limits of Making Common Cause* (London: Routledge, 2016); Giles Scott-Smith, *The Politics of Apolitical Culture: The Congress for Cultural Freedom, the CIA, and Postwar American Hegemony* (London: Routledge, 2002); Hugh Wilford, *The Mighty Wurlitzer: How the CIA Played America* (Cambridge,

MA: Harvard University Press, 2009); Masato Karashima, "Sengo nihon no shakaikagaku to Amerika no philanthropy: 1950–60 nendai niokeru nichibei hankyo liberal no koryu to Rockefeller zaidan" [Postwar Japanese Social Science and U.S. Philanthropy: Association of U.S. and Japanese Liberal Anti-Communists and the Rockefeller Foundation in the 1950s and 60s], *Nihon Kenkyu*, vol. 45 (March 2012), 155–183.

7 Audra J. Wolfe, "*Science and Freedom*: The Forgotten Bulletin," in *Campaigning Culture*, 28–39.

8 Scott-Smith and Lerg, eds., *Campaigning Culture*; Wolfe, *Freedom's Laboratory*.

9 Yuka Tsuchiya, "'Hankyo' to 'hankaku': 1950 nendai ni okeru kagakuzasshi *Genshikagakusha Kaiho* to Bunka jiyu kaigi" ['Anti-Communist' and 'Anti-Nuclear': *Bulletin of the Atomic Scientists* and the Congress for Cultural Freedom in the 1950s], *Americashi Kenkyu* [*Japanese Journal of American History*], no. 41 (September 2018): 36–51.

10 Hanna Holborn Gray Special Collections Research Center, University of Chicago Library, Guide to the Eugene I. Rabinowitch Papers, https://www.lib.uchicago.edu/e/scrc/findingaids/view.php?eadid=ICU.SPCL.RABINOWITCH.

11 Smith, *A Peril and a Hope*, 22–23. Citations refer to the 1965 edition.

12 Wittner, *Confronting the Bomb*, 3–13; Smith, *A Peril and a Hope*, 278–279.

13 Wang, *American Science in an Age of Anxiety*, 21; Smith, *A Peril and a Hope*, 431–432, 451–453, 482–483.

14 Yamasaki and Hinokawa, *This Is How Atomic Bombs Came into Being*, 48–50, 181–185. I would also like to express my appreciation to Dr. Daniel Meyer, former Director of the Hanna Holborn Gray Special Collections Research Center, the University of Chicago Library, who drew my attention to this aspect of the Chicago scientists' community.

15 Wittner, *Confronting the Bomb*, 35.

16 The incident in which noted scientist Robert Oppenheimer, who led Los Alamos Laboratory in the Manhattan Project, was labeled as "a security risk" and deprived of his security clearance when he disagreed with the government policy to develop thermonuclear weapons.

17 Guide to the Atomic Scientists of Chicago Records 1943–1955.

18 Smith, *A Peril and a Hope*, 279.

19 From Rabinowitch to W. W. Waymack, March 30, 1950, Rabinowitch, Eugene I. Papers, box 19, Hanna Holborn Gray Special Collections Research Center, University of Chicago Library. (Hereafter, Rabinowitch Papers.)

20 From Editorial Staff to Board of Directors and Editorial Board, June 25, 1960; "Report to the Board of Directors of the Bulletin of the Atomic Scientists," from Rabinowitch, October 26, 1963, Rabinowitch Papers, box 19. In the 1960s, the *Bulletin of the Atomic Scientists* met financial difficulties, and began to consider applying for other funding aid or falling under the umbrella of some major publishing company.

21 Scott-Smith and Lerg, "Introduction: Journals of Freedom?" in *Campaigning Culture*, 1–4.

22 MacLeod, *Consensus, Civility, Community*, 46–48.

23 Reiko Maekawa, *America chishikijin to radical vision no hokai* [*American Intellectuals and the Collapse of Radical Visions*] (Kyoto: Kyoto University Press, 2003), 160.

24 Scott-Smith and Lerg, "Introduction: Journals of Freedom?" in *Campaigning Culture*, 1–4.

25 Wolfe, *Freedom's Laboratory*, 79–80.

26 Mary Jo Nye, *Michael Polanyi and His Generation: Origins of the Social Construction of Science* (Chicago, IL: University of Chicago Press, 2011), introduction, 210, Kindle; Wolfe, "*Science and Freedom*," 28–29.

27 Office Memorandum from J. B. Koepfli, October 20, 1952, RG59, Entry 1549, box 9, NACP.

28 Nye, *Michael Polanyi and His Generation*, 23–24, 211.

29 Office Memorandum from J. B. Koepfli, March 16, 1953, RG59, Entry 1549, box 9, NACP.

30 Reiko Maekawa, "The Rockefeller Foundation and Refugee Scholars during the Early Years of the Cold War," *Eibungaku Hyoron* [*Critique of English Literature*], vol. 88 (February 2016): 85–113.

31 The ONR was established after the Second World War to continue the wartime Navy's S&T research in peacetime, and to carry out contract studies with scientists in a wide variety of fields. ONR website, https://www.onr.navy.mil/About -ONR/History/History-Research-Guide. Orr Raynolds was a physiologist who worked for the Navy's Bureau of Medicine and Surgery during wartime, and headed the ONR's Biology Division after the war. In 1957, he was appointed as the Director of the Office of Science, the Department of Defense, and later became the Director of NASA's Biology Research Division. Obituaries, *The Baltimore Sun*, April 27, 1991, http://articles.baltimoresun.com/1991-04-27/news/1991117055_1_physiological-society-american-physiological-reynolds.

32 Office Memorandum from Neil Carothers, June 2, 1953, RG59, Entry 1549, box 9, NACP.

33 Wolfe, *Competing with the Soviets*, 28, 31, 39.

34 George Polanyi, "Protests and Problems," *Bulletin of the Atomic Scientists*, (hereafter *BAS*), vol. ix, no. 9 (November 1953): 322, 340. Hereafter, citations from the *BAS* are by the *Bulletin of the Atomic Scientists*. Records, box 18, Hanna Holborn Gray Special Collections Research Center, University of Chicago Library.

35 Edward Shils, "The Scientific Community: Thoughts After Hamburg," *BAS*, vol. x, no. 5 (May 1954): 151–155.

36 MacLeod, *Consensus, Civility, Community*, 46–50; "Committee on Science and Freedom: Report on Activities in the Period February–August 1956," Rabinowitch Papers, box 21.

37 MacLeod, 46, 50–51.

38 From Polanyi to Rabinowitch, October 31, 1954; From Polanyi to Rabinowitch, January 25, 1955; From Schwarzschild to Rabinowitch, May 11, 1955; From Nabokov to Rabinowitch, August 18, 1955, Rabinowitch Papers, box 20.

39 "Committee on Science & Freedom, Study Group, Paris 1956 Agenda," Rabinowitch Papers, box 21; Michael Polanyi, "The Magic of Marxism," *BAS*, vol. xii, no. 6 (June 1956): 211–214, 232.

40 Eugene Rabinowitch, "The Role of Scientists in Public Affairs and the New Meaning of Academic Freedom (Revised Version)," International Association for Cultural Freedom Records, box 402, Hanna Holborn Gray Special Collections Research Center, University of Chicago Library. (Hereafter, IACF Records.) When the CIA's funding aid was revealed in 1967 and caused controversy, the CCF changed its name to the International Association for Cultural Freedom (IACF), and relied on the Ford Foundation funds.

41 Edward Shils, "Observations on the Scientists' Movements in the United States," IACF Records, box 402.

42 Wolfe, *Freedom's Laboratory*, 170.

43 Eugene Rabinowitch, "The Narrowing Way," *BAS*, vol. ix, no. 8 (October 1953): 294–295, 298.

44 Rabinowitch, "Living with H-Bombs," *BAS*, vol. xi, no. 1 (January 1955): 5–8.

45 Rabinowitch, "The Frozen Map," *BAS*, vol. xiii, no. 6 (June 1957): 208–211, 215.

46 Rabinowitch, "New Year's Thoughts," *BAS*, vol. xiv, no. 1 (January 1958): 2–6.

47 Rabinowitch, "About Disarmament," *BAS*, vol. xiii, no. 8 (October 1957): 277–282; Nakazawa, "The Nuclear Disarmament Negotiations," 43.

48 Rabinowitch, "Must Million March?" *BAS*, vol. x, no. 6 (June 1954): 194–195, 238.

49 Rabinowitch, "Science Faces a Double Danger," *BAS*, vol. ix, no. 2 (March 1953): 34–35, 42.

50 Shils, "Security and Science Sacrificed to Loyalty," *BAS*, vol. xi, no. 4 (April 1955): 106–109, 130.

51 Edward Shils, "Scientists Affirm Faith in Oppenheimer," *BAS*, vol. x, no. 5 (May 1954): 188–191; Shils, "The Slippery Slope," *BAS*, vol. x, no. 6 (June 1954): 242, 256.

52 Rabinowitch, "What is a Security Risk?" *BAS*, vol. x, no. 6 (June 1954): 241, 256.

53 Toshio Higuchi, *Political Fallout: Nuclear Weapons Testing and the Making of a Global Environmental Crisis* (Stanford, CA: Stanford University Press, 2020): 63–68.

54 Rabinowitch, "Genetics in Geneva," *BAS*, vol. xi, no. 9 (November 1955): 314–316, 343. "Lysenko's grip on Soviet biology" indicated the incident in which Soviet biologist Trofim Lysenko rejected the Mendelian inheritance theory and asserted that acquired characteristics were inherited across generations. Since his theory was convenient for the Communist Party's official line concerning the improvement of crop yields, Stalin supported Lysenko, and his political influence expanded, while scientists who disagreed with Lysenko were persecuted.

55 Maekawa, *American Intellectuals*, 192.

56 Scott-Smith, *The Politics of Apolitical Culture*.

57 Eric Foner, *The Story of American Freedom* (New York: W. W. Norton & Co., 1998), 260–261.

3 "Foreign Atoms for Peace" and the Export of Research Reactors

On June 11, 1955, President Eisenhower gave an address at the Centennial commencement of Pennsylvania State University, in which he promised to offer research reactors, fissional fuels, financial aid equivalent to half the price of the reactor, and technical information and training to friendly countries.[1] Although the content may sound somewhat strange for a congratulatory speech delivered at a university's commencement, it was not without reason. Pennsylvania State University had pioneered the building of a Triga research reactor on campus that year, and it was expected to go critical in the near future.[2] The president told the audience that he had visited the reactor earlier that morning, and expressed his hope that it would soon yield productive results. He also encouraged the graduating students that "the age of nuclear energy" will bring about not only "boundless opportunities," but also "new and great human problems" that they should tackle with their "broadly informed, wisely sympathetic, spiritually inspired minds."

Behind the dissemination of research reactors among universities was the president's promotion of the Atoms for Peace policy. Pennsylvania State University used its own funds for the construction of the reactor, while the AEC provided the nuclear fuels. The president's congratulatory remark, therefore, was also a self-congratulation on his policy. Pennsylvania State University awarded President Eisenhower with an honorary doctoral degree that year, which symbolized the university's gratitude for the president's generous support for the research of nuclear science.

The president also emphasized that "throughout the free countries there are men and women of great ability who, given the opportunity, can help further to advance the frontiers of knowledge and contribute to the peace and progress of the peoples of all nations." The declaration to offer reactors and fuels, mentioned at the beginning of this section, came right after this emphasis of the need to expand scientific progress to all of the "free countries."

Of course, the president's address was not a simple conveyance of his personal thoughts, but was drafted after careful deliberations within the government. When President Eisenhower gave the renowned "Atoms for Peace" speech at the United Nations General Assembly on December 8, 1953, the international cooperation he proposed did not materialize so smoothly; within

DOI: 10.4324/9781003243649-4

the heated competition of nuclear R&D unfolding not only between the U.S. and the Soviet Union but also involving the U.K. and France, it increasingly became unrealistic that advanced countries would "pool" fissionable materials for other countries to use. Against this backdrop, the U.S. National Security Council (NSC) drafted two important policy papers: the NSC5431 series titled "Statement of Policy by the National Security Council on Cooperation with Other Nations in the Peaceful Uses of Atomic Energy" (August 1954), and the NSC5507 series titled "Peaceful Uses of Atomic Energy" (March 1955). NSC5431/1 proposed offering technical training, information, consulting services, and so on based on bilateral agreements, through which, for the U.S., "Maximum psychological and educational advantage should continue to be taken." The document also regarded the provision of a small research reactor as a "natural step" for the target country toward introducing nuclear power generation in the future. NSC5507 pointed out that building full-scale American power reactors in foreign countries would bring about "great psychological advantages" for the U.S., but implementation would not be easy given the high cost of nuclear power generation and difficulty in choosing construction sites. In this regard, the document argued that research reactors which required cheap construction and operating costs could become "means of securing psychological advantage in international cooperation at a much lower cost."[3] Research reactors, in other words, were a marketing strategy for the future exportation of larger power reactors, and also the small reactors themselves were a source of psychological and educational gain. In the mid-1950s, power reactors were still in the development stage in the U.S., and there were uncertainties about future exportation.[4] Under such conditions, providing research reactors meant offering a sample of unfinished products and cultivating technicians familiar with U.S. nuclear technology. The U.S. government expected such a gesture would yield gratitude, respect, and long-term commitment to the U.S. government and its technology.

In providing a research reactor to a foreign country, the U.S. Atomic Energy Act stipulated that the two governments should first conclude a bilateral agreement. By the end of the 1960s, the U.S. concluded bilateral agreements with 37 countries, providing financial aid of 350,000 dollars, or half the price of the reactor. All reactor parts were imported from U.S. companies, and in developing countries, construction was also carried out by U.S. contractors. In Asia, Japan, Taiwan, and the Philippines concluded bilateral agreements with the U.S. in 1955, followed by Korea (1956), Thailand (1958), South Vietnam (1959), and Indonesia (1960).[5] Burma (present-day Myanmar) did not conclude a bilateral agreement, but received research facilities (not a reactor) on radioisotopes. How did the U.S. government decide to offer nuclear reactors to certain countries and facilities to others? As the NSC document has shown, offering nuclear reactors was thought to bring about a "psychological advantage" in the competition with the Soviet Union. For the recipient countries, on the other hand, the importation of U.S. reactors potentially meant not just a technological transfer but also a long-term, deep-setting impact on their society, including dependence on U.S.

technology. Conclusion of the bilateral agreement, therefore, was an extremely political decision both for the U.S. and the recipient country.

However, contrary to U.S. government expectations, not all of the countries that concluded bilateral agreements later purchased full-scale reactors or cultivated technological ties with the U.S. As the example of South Vietnam shows, sometimes the U.S. decision to offer a research reactor resulted in disaster. The U.S. exported a research reactor in the middle of their deepening military involvement in Vietnam, but the reactor was ultimately confiscated by the People's Army of Vietnam (the North Vietnamese Army) shortly before the fall of Saigon. The U.S. also provided a reactor to South Korea (ROK), but had to refrain from exporting a larger reactor for many years due to fears surrounding Park Chung-hee's ambition for nuclear armament. As a result, Korean nuclear power generation commenced only in 1978. The Philippines began construction of a nuclear power plant in 1976, although a powerful anti-nuclear movement, especially after the Three Mile Island incident of 1979, ultimately prevented the plant from going into operation. As these examples show, the U.S. ambition to cultivate overseas markets for nuclear reactors did not always succeed.[6] On the contrary, Burma requested the U.S. government to offer a research reactor, but the U.S. did not satisfy their request. The U.S. government judged that, due to a lack of basic scientific research and education, it was too early to provide a reactor to Burma.

This chapter will focus on Japan, South Vietnam, and Burma, three contrasting examples, examining the process of the conclusion of bilateral agreements in the cases of Japan and South Vietnam, and in the case of Burma, the process of *not* concluding the agreement but nevertheless resulting in the provision of experimental facilities. For what purposes did these countries appeal to the U.S. for technological aid, and for what reasons did the U.S. decide to—or *not* to—provide a reactor? A close examination of Department of State and other government records reveals that not only tangible reasons such as demand for electricity but also intangible motivations such as prestige and expectations were contributing factors. The three examples demonstrate that it was not only the material and economic gains of S&T that held political importance, but also its psychological effects. This observation also corresponded with the NSC's expectations for "psychological and educational" effects of nuclear technological aid. While there are many examples of "dual-use technologies" such as the internet and GPS, the examples in this chapter demonstrate that nuclear technology was in fact a "triple-use technology," with military and civilian applications, but also "psychological" purposes—in other words, for promoting overseas information programs.

3.1 Foreign Atoms for Peace

The expression "Foreign Atoms for Peace" came to be used frequently in U.S. government documents more or less after President Eisenhower's address at Pennsylvania State University. The notion covered not only technical and financial aid, the exportation of reactors, and the training of engineers, but also overseas information programs such as films, lectures, and exhibitions. Before examining the concrete examples of Japan, South Vietnam, and Burma, a brief discussion of three dimensions of the concept of Foreign Atoms for Peace will be provided.

The first dimension entailed feverish expectations for nuclear energy from the Third World countries and the foreign aid competition between the U.S. and the Soviet Union (and, to a lesser degree, the U.K. and France). A series of notable events such as President Eisenhower's UN address, his Pennsylvania State University address, and the first UN Conference on the Peaceful Use of Atomic Energy held in Geneva in August 1955, contributed to fueling the desire of the Third World leaders and citizenry for modernization through the use of nuclear energy.[7] Expectations surrounding nuclear energy rose in various fields, not only in power generation but also in the application of radioisotopes in the improvement of crops, food preservation, the treatment of cancer, and the eradication of mosquitos. This is the background which motivated the NSC to believe that providing nuclear reactors would continue to draw "psychological and educational benefits" from foreign countries. While providing technology in exchange for allegiance may appear to be traditional hard power diplomacy, there was more to these agreements than just a simple "deal"—for the countries supplied, nuclear technological aid satisfied the desire for modernization, thereby nurturing gratitude for the generous U.S. help, and in the long run, the idea was that the recipient countries would be taken into the fold of U.S. technological networks and structures.

The second dimension of the U.S. Foreign Atoms for Peace was as marketing strategy for the U.S. nuclear industry, in competition with rivals such as the Soviet Union, the U.K., and France. Early in the postwar years, U.S. industry held heightened expectations for nuclear reactor business. In the wartime Manhattan Project, many industrial engineers had been mobilized, and they were keenly aware of the business opportunities nuclear energy might open up in the postwar world. The McMahon Act (Atomic Energy Act) of 1946 was, in contrast to the May–Johnson Bill which had failed shortly before it, denied the military monopoly of nuclear technology and established the AEC as a civilian organization. Even the McMahon Act, however, did not allow researchers and private companies total access to all nuclear technological information. Against this background, strong demand rose from the electronic, chemical, and other industries to open up nuclear technology information. In the presidential election of 1952, the industries supported Dwight Eisenhower with the expectation that he would revise the law to open up the information. The Eisenhower administration, the first Republican government in 12 years, did not betray the expectations of industry parties. In 1954, McMahon Act was revised, and a new Atomic Energy Act was enacted, which declassified some of the key technological information on nuclear energy.

At that time the business world was optimistic about the future prospect of atomic power generation. Soon, however, they realized that there was still a long way to go before nuclear power generation would become commercially viable.

Due to the high cost of power and uncertainties about accidents and liabilities, the momentum in the business world waned. Given the competition with the Soviet Union and the U.K. (and later also France), the U.S. government perceived a need to persuade industry to continue R&D for nuclear reactors. The Soviet Union was rapidly developing their own nuclear reactors, and succeeded in bringing online the world's first civilian nuclear power generation in Obninsk Power Plant. The U.S. government did not simply appeal to the patriotism of industry parties, but also offered financial support. As a result, by the end of 1956, 30 private nuclear reactors had been built, and a further 59 were in construction or under contract. Even though nuclear power could not compete with water or thermal power generation in terms of profitability, 69 electric companies were at the time investing in 14 nuclear power generation projects.[8] Since the domestic power demand was already met by water and thermal power, private companies attempted to cultivate overseas markets, an initiative which the government supported. Their plan was to first export small research reactors, train local engineers in American nuclear technology, and then to later export larger power reactors. However, American power reactors were completed only in 1958, and therefore the companies were promoting products which did not yet exist. In any case, until the cost of nuclear power generation could become competitive with water and thermal power, the U.S. relied on exporting reactors overseas, and bilateral agreements served this purpose.[9]

The third dimension of Foreign Atoms for Peace was the overseas information programs. After President Eisenhower's UN address, Atoms for Peace became the priority theme of the USIA's overseas information programs, and a massive amount of information was disseminated overseas through USIS films, VOA radio, exhibitions, pamphlets, magazines, and other media. In Japan, a series of Atoms for Peace exhibitions co-sponsored by the Yomiuri Newspaper Company and USIS Tokyo were the most famous exhibitions, but similar events were held in many countries.[10] The overseas information program on Atoms for Peace was most intensively promoted in the countries which had concluded (or planned to conclude) bilateral agreements, although it was also carried out in countries without agreements and without any likely prospect of adopting U.S. nuclear reactors in the near future. This resonates with the initiatives in which space flight was promoted even in countries where space technology was not being developed at all, discussed in detail in Chapter 8. In other words, overseas information programs sometimes aligned specifically with concrete economic goals (such as cultivating markets), while at other times pursued more intangible benefits (such as raising U.S. prestige and building long-term relationships through technology, education, and culture).

In sum, Foreign Atoms for Peace satisfied Third World desires for modernization, cultivated foreign markets for American industries, and functioned as an overseas information program to build technological, educational, and cultural ties. These three dimensions were not mutually exclusive but overlapped with each other. In addition, Foreign Atoms for Peace was carried out differently depending on the local situations of the target countries. The following cases will show how Foreign Atoms for Peace unfolded in Japan, South Vietnam, and Burma, and how it was received in each respective context.

3.2 Japan

Japan was perhaps the most successful case of the U.S. Foreign Atoms for Peace program in that it became a very profitable market for American nuclear reactors. After the 1960s, Japan imported many light-water power reactors produced by the General Electric and Westinghouse companies. Japanese nuclear power generation has remained almost entirely dependent on these American reactors to the present day. Japan had already developed scientific knowledge on nuclear physics spanning from the Imperial era, with notable scientists such as Hideki Yukawa of Kyoto Imperial University, Yoshio Nishina of the Institute of Physical and Chemical Research, and their understudies. Yasuhiro Nakasone, a young congressman who would later become prime minister, visited the U.S. in 1954 to observe the nuclear facilities, and he persuaded the Japanese government to allocate a budget of 235 million yen and commence development of nuclear reactors. Japan was among the earliest countries to conclude bilateral agreements with the U.S. (1955), and imported two research reactors and enriched uranium from the U.S. In 1958, the U.S.–Japan agreement was revised to enable Japan to import power reactors.

Previously, the Japanese government had been planning to introduce U.K. nuclear technology, and in fact, Japan's first power reactor was imported from the U.K. However, for various reasons the Japanese government gradually became convinced that U.S. light-water reactors were the best choice for Japan. First, the first reactor imported from the U.K. was of the carbon-graphite type, which was vulnerable to earthquakes. After a serious fire occurred in Windscale Power Plant in the U.K., the Japanese government became concerned that a similar accident might happen in Japan. Second, during the construction of the U.K. reactor, relations between the U.K. and Japanese engineers and industrialists had been troublesome, and the Japanese involved became wary of doing business with the U.K. Third, Japanese electric business had prewar ties with American companies such as General Electric, and they trusted the U.S companies. Fourth—and this is most relevant to the following discussion—the U.S. government and private business conducted an aggressive sales campaign.[11]

Given the Japanese scientific foundation, growing population, expanding economy, and shortage in power, Japan was an ideal target for the U.S. Foreign Atoms for Peace programs. The establishment of the AEC's Tokyo office on November 15, 1957, evidenced the keen interest of the U.S. in Japan. The office was located in the U.S. Embassy in Tokyo, and it was accompanied by a film library dedicated to the theme of Atoms for Peace. The AEC's overseas offices were established in only four cities around the world: London, Brussels, Buenos Aires, and Tokyo. The U.S. viewed Japan as an Asian hub for the U.S. Foreign Atoms for Peace policy. The AEC's Tokyo office was directed by Herbert W. Pennington, an AEC officer who had previously worked for DuPont, Monsanto

Chemical, and Westinghouse. He was concurrently appointed as a science attaché of the embassy, where his responsibilities included overseeing and coordinating the U.S. nuclear policies in other parts of Asia, and building networks of Asian scientists, engineers, and technocrats.[12]

The U.S. government, while regarding Japan as an Asian hub, was reluctant to give Japan a leadership role in nuclear technology in Asia. For instance, the U.S. government did not support the Japanese government's ambition to build the Asian Nuclear Center in Japan. "Perhaps the most appropriate role for Japan in the Center would be as middlemen," said Richard Sneider, a foreign service officer who would later play another important role in the Okinawa reversion. He expected that the U.S. "may be able to use the Japanese to explain and seek support for some of our policies from the other Asian countries" because "the Japanese are fond of envisaging their role in Asia as a middleman or bridge between the more advanced Western countries and the underdeveloped Asian countries."[13] In Sneider's view, Japan was an agent, or a "middleman" between the U.S. and Asia, but not an autonomous actor in dealing with nuclear energy issues. G. E. Meyer, a Department of State information officer, was also reluctant about assigning Japan more important roles. He stated that "care should be taken to avoid the impression that the United States and Japan are working in concert to as to lessen the benefits and authority of other Asian countries."[14] Outerbridge Horsey, director of the ICA in Tokyo, also said that "while Japan's role should not tend to become too dominant, she may well be able to help us put over the project."[15] These remarks reveal that the U.S. government attempted to use Japan as a hub through which to transfer U.S. technology to other parts of Asia, but did not want Japan to behave independently. Historian Shinsuke Tomotsugu has pointed out the gap between the role which the U.S. expected of Japan and the role which Japan envisioned for itself in Asia. The Japanese Ministry of Foreign Affairs, scientists, and politicians hoped to establish an Asian international cooperation system on nuclear technology which was independent from the U.S., and in which Japan could demonstrate its leadership. According to Tomotsugu, such ambitions were manifested in the Japanese attitudes in the Conference of Countries in Asia and the Pacific for the Promotion of Peaceful Uses of Atomic Energy held in March 1963, and the plan for the Asian Radioisotope Center.[16]

In sum, the U.S. Foreign Atoms for Peace for Japan was advanced by the economic motivation to cultivate the Japanese market for nuclear reactors, and by the strategic expectation that Japan would play the role of a U.S. agent in Asia. Towards these ends, Foreign Atoms for Peace for Japan aimed to alleviate anxiety about nuclear energy, and to inspire confidence in the U.S. nuclear technology. The U.S. government was concerned about the strong anti-nuclear sentiment held by the Japanese public because of their past experiences of Hiroshima, Nagasaki, and Bikini Atoll. In order to shift Japanese public opinion away from nuclear weapons by emphasizing the "peaceful use" of atomic energy, the USIA organized many Atoms for Peace exhibitions, USIS film shows, and lectures. Atoms for Peace exhibits were held almost every month somewhere in Japan. Many of them were sponsored by Japanese industrial organizations such as the Japanese Atomic Industrial Forum (JAIF) or Japanese private companies, while the contents of exhibits were often provided by the USIS. According to the *Atomic Industrial News*, a trade paper issued by the JAIF, even during one year from mid-1957 to mid-1958, Atoms for Peace exhibitions were held in at least five cities (Takamatsu, Kumamoto, Moji, Matsuyama, and Chiba)—one million people visited the exhibition in Moji, and 100,000 in Kumamoto and Matsuyama, respectively.[17]

Some of the exhibitions were industrial shows to demonstrate the practical applications of nuclear technologies and stimulate Japanese engineers' and industrialists' ambition to participate in the field. For example, the Atomic Industrial Exhibition held in Tokyo in May 1957 attracted 96,000 visitors in 6 days, many of whom were industrial engineers and scientists. It was held simultaneously with the U.S.–Japan Joint Atomic Industrial Forums conference, and the most popular displays included a model commercial nuclear ship, a model U.S. nuclear power plant, and the construction process of a water boiler reactor (a research reactor imported from the U.S.).[18] Another example, the Tokyo International Trade Fair held in May 1959, included an "Atomic Energy Pavilion" sponsored by the AEC. In the massive AEC pavilion which occupied 11,000 square feet, "a real-size cross section model of the Shippingport power reactor"—the first commercial light-water reactor in the U.S. which commenced operation in late 1957, and was showcased in the second international conference on the peaceful use of atomic energy held in Geneva in 1958—and "a huge panel showing the dynamics of chain reaction" were displayed. Also, a genuine, operational "UTR university training reactor" attracted visitors' attention. Furthermore, models of the "experimental boiling water reactor at Argonne National Laboratory," the "experimental Homogenous reactor II at Oak Ridge National Laboratory," and a model of the "Nuclear ship Savannah" were popular among the Japanese. The AEC also set up a "technological information center" within the pavilion, where they consulted on U.S. technological information and how to access those materials.[19] These exhibitions, unlike the popular Atoms for Peace exhibitions aimed at the general public, targeted those who had a direct influence on the purchase of U.S. reactors, such as business leaders, engineers, politicians, and technocrats. The film library in the AEC's Tokyo office also stored highly technical films for this audience, which differed from the USIS films intended for general audiences.

In addition to exhibitions and films, the AEC also invited influential industrialists, scientists, engineers, and politicians on tours to visit U.S. nuclear research facilities and nuclear reactors under construction. By showing them cutting-edge U.S. technologies, the U.S. government intended both to win their favor, and also to foster an aura of authority in those who had actually seen the U.S. nuclear facilities. The U.S. Embassy and ICA viewed these tours as an efficient overseas information program that could secure the good will of participants with relatively small cost. In fact, those who were invited on the U.S. tours were given respect for their hands-on experiences after their return to Japan. One article in the tenth anniversary publication of the Japan Productivity Center (an industry-sponsored thinktank established in March 1955 by the suggestion and with the support of the U.S. Embassy) celebrated the positive effect of visiting the U.S.:

> The previously unknown world of America was full of energy and the beautiful embodiment of pioneer spirit. Even if the present conditions seem excellent, and even if they *are* actually excellent, they [Americans] try to further improve them, adapt their economic and social life to the changing circumstances, and strive to apply new technologies and methodologies. This "spirit of productivity" has realized a prosperity and high standard of living that none on earth have ever experienced before. A huge spiritual benefit can be gained from getting in touch with such a country and such people. To express it a little dramatically, it is a life-changing experience which provides a new worldview. One may be skeptical about such a dramatic change being brought about by a mere 5–6 weeks of tour. Of course, there is no visible sign of change in one's personality. However, there is a subtle but certain change in the spiritual landscape of the person [who visited the U.S.].[20]

Considered under a modern-day lens, the above passage seems to grossly exaggerate the effect of visiting the U.S. However, in the mid-1950s, Japan was only a few years free of the Allied occupation and stood at the threshold of its economic growth. A select few were able to visit foreign countries, and the U.S. was the most advanced foreign country many aspired to. The above article shows the degree of significance that the U.S. tours had for the Japanese businessmen. It is not important whether each individual visitor actually experienced such a dramatic spiritual change, but rather, how *other people* perceived them. Those who observed the nuclear facilities in the U.S. won a certain respect back home, and their opinions found earnest ears to listen. In this regard, the tour was indeed an effective overseas information program for the U.S. government.

The U.S. government also expected that formerly anti-U.S. individuals might change their attitudes after visiting the U.S. For example, in September 1956, a mission consisting of six Japanese Diet members, two Atomic Energy Bureau staff, one participant from the Japanese Atomic Energy Research Institute, and three participants from electric companies, visited the U.S. for three weeks. Four of the mission members visited the ICA Tokyo to report on their tour. Among them was Saburo Unno, a physicist, engineer, and a congressman of the Japan Socialist Party. Before departure, the attitudes of the four members had been a mixture of "doubt" and "antipathy." Upon returning to Japan, however, they "earnestly reported on the delight in their experiences in the U.S." The ICA Tokyo reported to the ICA headquarters in Washington that they would empower the "elements of the Japanese business community" who were "seeking to persuade the Government to purchase a U.S. reactor."[21]

The U.K. was the main competitor of the U.S. in the Japanese reactor market. As previously mentioned, the Japanese government favored the U.K. reactor technology which was more advanced than that of the U.S. in 1956. Japanese industry, by contrast, had maintained their business ties with U.S. private companies such as General Electric which had been established since the prewar Imperial period. As the Japanese business world maintained a high level of trust in U.S. technologies, the U.S. government moved to have them persuade the Japanese government to adopt U.S. technologies over the U.K.[22] The U.S. government became concerned when the Japanese government sent a mission to the U.K. in January 1957, led by Ichiro Ishikawa, a Japanese Atomic Energy Commissioner. The group had visited the U.K., the U.S., and Canada, but upon acquiring the mission's report, the U.S. discovered that the main focus of the mission was on the U.K., as the report spent only 4 pages on the U.S. and Canada combined, while 31 pages were devoted to the U.K. including details of costs, safety protocols, and concrete conditions for purchase. In the report's conclusion, the mission advised that Japan should import a power reactor from the U.K., where such reactors were already available; it was "too early" to deal with the U.S.[23] Through the report, the U.S. government was painfully reminded of their setbacks in the Japanese market.

In March 1957, Frank A. Waring of ICA Tokyo, in his telegram to ICA in Washington, pointed out that the Japanese government was "now in [the] process [of] making decision [on] whether [to] follow UK or US lead in atomic power development." The Japanese government officials were "presently inclined [to] purchase British Calder-Hall reactor," while the "strongest potential [for] U.S. support lies with private industrialists" who might "prefer U.S. designed reactors" because of their long-term relationships with U.S. companies. "If Japanese industrialists [could] be assisted in appreciation of U.S. superiority," Waring argued, "they would be most effective influence in persuading [the] government [to] conclude comprehensive atomic energy agreement with the U.S." He proposed to the ICA and the Department of State to send a team of 12 Japanese representing the 5 largest electric companies and other companies interested in power reactors. He believed that the Japanese industrialists were "potentially the strongest single factor on the Japanese scene favoring the purchase of U.S. designed reactors," and strengthening their voice would contribute to the "final governmental decision as to the type of power reactor to be purchased by Japan."[24]

Moreover, the ICA attempted to dress up this mission as a spontaneous proposal from the Japanese businessmen. On March 15, Waring met with Taizo Ishizaka, chairman of the Japan Business Federation (*Keidanren*) and former chairman of Japan Productivity Center. There Waring proposed, "as a possible aid in encouraging a closer relationship between Japan and the United States in the atomic energy field," that it might be "worthwhile for the Japan Productivity Center to sponsor a mission to the United States." He recommended that the mission consist of "leading Japanese industrialists interested in atomic energy, including the electric power companies and such machinery manufacturers and shipbuilders as Mitsubishi, Mitsui, Hitachi and Shibaura." If "these policy-making industrialists would visit the leading U.S. manufacturers of atomic energy equipment and observe the latest developments in the peaceful uses of atomic energy field," Waring continued, they would be "capable of speaking with a more authoritative voice regarding the policy-making decisions of the Japanese Government and within their own firms." However, he suggested to Ishizaka "to raise it with the Center as his own proposal" since the suggestion might be "misconstrued if it came from the embassy." Waring's report indicated that Ishizaka responded in the affirmative.[25]

The Japan Productivity Center sent many missions to the U.S. not limited to nuclear energy but concerning various fields, and the ICA supported these tours both in terms of itinerary and funding. The ICA's history traces back to the Mutual Security Agency (MSA). The Mutual Security Act of October 1951 amalgamated various foreign aid programs into one, resulting in the establishment of the MSA. The MSA was reorganized under the Eisenhower administration into the Foreign Operations Administration (FOA), and re-established as the ICA, a subordinate organization of the Department of State. Under the following Kennedy administration, it would be further reorganized into the U.S. Agency for International Development (AID). The ICA's headquarters were in Washington DC, while its branches were established in the U.S. embassies of various countries. The overseas branches of the ICA were also called the U.S. Operation Mission (USOM), and were staffed by employees of the U.S. Embassy of each country.[26] Although the ICA officially assumed the practical operation of the Department of State policies, Waring's activities show that it was not simply a subsidiary of the department—the ICA promoted Foreign Atoms for Peace in Japan quite autonomously, from information gathering activities to policy proposals. ICA Director John Hollister was also a member of the OCB which coordinated the psychological aspects of various government policies, and which might further indicate the importance of ICA as more than just a practical conductor of Department of State policies.[27]

The Atomic Industrial Forum (AIF), established in 1955 by U.S. private companies interested in nuclear energy, and its Japanese counterpart, the Japanese Atomic Industrial Forum (JAIF), also contributed to the introduction of U.S. reactors over those of the U.K. The Joint U.S.-Japan AIF conference held in Tokyo in 1957 was akin to a public relations festival for U.S. nuclear technologies. The U.S. Embassy in Tokyo acknowledged that the conference was successful in:

a) Establishing a closer relationship between the United States and Japanese businessmen interested in the industrial development of the atom.
b) Creating an awareness among Japan's scientists, government officials, and industrialists of United States progress in the field of peaceful development of atomic energy.
c) Encouraging Japanese government officials seriously to re-examine the possibility of purchasing a United States designed reactor in the near future, together with the probable purchase of an improved Calder-Hall model.

The "unflattering comments reportedly made by some United States participants about British atomic energy developments," and "certain remarks contained in the speech of Dr. Kenneth Davis" marred an "otherwise completely harmonious conference," the embassy added, although these remarks did not "adversely affect United States objectives in the atomic energy field in Japan." The *Yomiuri* newspaper, which had previously sponsored the Atoms for Peace Exhibitions at the behest of the newspaper's ambitious owner Matsutaro Shoriki, also reported that the conference doubtlessly brought the Japanese nuclear policy "a large step [closer] toward the U.S." Although the U.K. reactors remained an "explosive issue," the embassy concluded that the Joint AIF conference was successful in "impressing the Japanese with the superiority of U.S. nuclear technologies."[28]

The U.S. Embassy and the Department of State were watching carefully not only the U.K. but also the Soviet approach to Japan. One of the key Japanese informants was Sashichiro Matsui, Chief, Fourth Section, International Cooperation Bureau of the Japanese Ministry of Foreign Affairs. Viewing Matsui as the "number one authority of nuclear energy in the Ministry of Foreign Affairs" and a pro-American bureaucrat, the U.S. Embassy often conducted interviews with him. Entering the Ministry of Foreign Affairs in 1935, Matsui's first contact with the U.S. Embassy probably came through his service in the Central Liaison Office, a contact point between the GHQ/SCAP and the Japanese government during the Allied occupation. He was not a scientist, but had translated an educational book on nuclear energy, *You and the Atom* (1955) authored by Gerald Wendt, a chemist and science communicator working for UNESCO's public relations division. Matsui was keenly interested in the promotion of Atoms for Peace in Japan.[29]

For example, when several Japanese newspapers reported the Socialist Party Diet member Shigeji Shimura's words that "the Soviet Union was ready to offer nuclear reactors and fuels at any time" in November 1956, the U.S. Embassy immediately called Matsui to confirm the information. According to the *Yomiuri*, "reliable sources in Peking" had reported that the Soviet Government was "planning to propose to Japan the conclusion of an agreement on peaceful use of atomic energy around November 10." The same sources had also disclosed that "negotiations on concluding such an agreement were begun in Moscow immediately after the Japanese-Soviet peace negotiations were concluded" and an official memorandum had "already been sent to Japan over the joint signature of Nesmeyanov, President of the Soviet Academy of Sciences, and Slavskii, chief of the Soviet Atoms for Peace Bureau." The contents of the memorandum were very generous:

1. A fundamental agreement will first be concluded in order to promote technical exchanges on atomic energy between the USSR and Japan. Other agreements will be concluded when such exchange gets under way.
2. The USSR will agree to export control equipment for atomic reactors, measuring instruments, enriched uranium, and other necessary materials and equipment, in compliance with the Japanese Government's requests, and also to provide opportunities for Japanese technicians to inspect and study in the USSR.

3. Japan will not be restricted by the USSR if it wants to give such equipment and materials to other nations.
4. Japan is free to dispose of waste from atomic reactors. If Japan needs Soviet assistance, the USSR will always be ready to help her.
5. Japan can use Soviet factories with the Soviet Government's permission, if she desires to dispose of atomic waste there.
6. Japan will be offered the unconditional use of the Joint Nuclear Research Institute in Moscow.
7. In order to promote technical exchange between the two countries, Japanese high school graduates will be allowed to study in either Moscow University or Leningrad University for 5 years at the expense of the Soviet Government.
8. The Japanese Government will be urged to permit the entry of Soviet students into Japan in order to pursue medical studies involving radioactivity.
9. The Soviet Government will invite Japanese technicians to visit 14 state-operated atomic energy research institutes in the USSR, if Japan so desires.[30]

Matsui, asked by the U.S. Embassy about the Soviet government's offer, responded that the Japanese government had not received such an offer, although he believed that it might be possible after the Japan–Soviet joint declaration to be enacted in December. Yet, Matsui expressed that it would be "unlikely that any agreement would be negotiated with the Soviet Union as long as a Conservative Government [was] in power."[31] A few months later, Matsui voluntarily reported to the U.S. Embassy that the Japanese Atomic Energy Commission had asked the Ministry of Foreign Affairs if they could obtain permission to send researchers to the Soviet Union, and that he advised the Ministry not to give such permission as it would be detrimental to the U.S.–Japan relations. Matsui explained that the plan to send researchers to the Soviet Union was originally proposed by Socialist Party members, and that "Japan would never approach the Soviet Union as long as Nobusuke Kishi (LDP [Liberal Democratic Party]) remains in the Prime Minister's seat."[32]

The U.S. Embassy was also collecting information on the opinions of key Japanese individuals. For example, a foreign service despatch dated February 20, 1957, reported that there was "as yet no advocate of the early purchase of a U.S. reactor in an influential position within the Japanese Government." "Japanese AEC Commissioner Ichiro ISHIKAWA, former AEC Chairman Matsutaro SHORIKI and … the new AEC Chairman Koichi UDA," were positive about purchasing a commercial power reactor. However, Ishikawa believed that "Japan should, at the present time, purchase an improved version of the British (Calder-Hall) reactor," because the U.S. did not "have a large size power reactor in operation as do the British." Japanese AEC Commissioners Fujioka and Yukawa believed that it was "premature for Japan to purchase any power reactor at this stage of Japanese atomic energy development." There was also anxiety among the Japanese toward "dependency on any one country," as the "U.S. was the only source for enriched uranium, the fuel of the U.S. designed reactors, while natural uranium, the fuel of the British reactors, [could] be obtained from many sources." The Japanese Government was also concerned about "the probable opposition of the Socialist Party, the scientific community and certain sections of the Government, reportedly led by the Finance Ministry." Certain LDP Diet leaders did "not wish to arouse Socialist parliamentary opposition" and damage the "bipartisan approach with regard to Japan's policies in the 'peaceful uses of atomic energy' field." The scientific community's opposition stemmed from the "Japanese scientists' desires to develop their own reactor designs" while the Finance Ministry was reportedly concerned about "the financial implications of possible governmental purchases of reactors with attendant fuel costs."

By contrast, Japanese industrialists, particularly the utilities, were "being wooed ardently by U.S. reactor manufacturers, such as Westinghouse and General Electric." As a "result of favorable past experience with U.S. suppliers of electric power generation equipment, and the ability of the U.S. Export-Import Bank to offer favorable credit terms," the embassy hoped that "the Japanese industrialists [would], in the not too distant future, persuade the Japanese Government to open negotiations for an atomic energy agreement with the United States."[33]

In sum, the U.S. Foreign Atoms for Peace for Japan was focused on the future sales of power reactors to Japan. The U.S. government, fearing that the advantage of U.S. reactors might be threatened by U.K. or Soviet approaches, promoted various activities from exhibitions and tours to intelligence initiatives. In contrast to the cases of South Vietnam and Burma, discussed below, it was very likely that Japan would later purchase larger power reactors. The U.S. therefore carried out information programs that would maximize the appeal of U.S. nuclear reactors, such as inviting business and political leaders to the U.S. and holding industrial exhibitions where U.S. nuclear technologies were demonstrated to Japanese engineers and businessmen. These types of information programs targeted the business and science elite, rather than the general public. Japan ended up being a very successful case for the U.S., although this success cannot be attributed to the information programs alone: the Japanese industry and government were also enthusiastic about introducing atomic power. The Japanese government ultimately decided to rely heavily on U.S. light-water reactors for nuclear power generation, and Japan became the most important market for the U.S. reactor companies such as General Electric and Westinghouse. Nuclear power became an integral part of the Japanese livelihood—so natural that the Japanese citizens did not even realize that the reactors providing electricity to their homes and industries were imported from the U.S.—until the tragic incident involving Fukushima Nuclear Plant No. 1 on March 11, 2011, when the fuel rod of a General Electric power reactor melted down after it lost power supply as a result of the Tohoku earthquake and ensuing *tsunami*. U.S. power reactors have long sustained Japanese culture heavily dependent on electric power and on the myth that nuclear energy was clean and safe.

3.3 South Vietnam

In contrast to the Japanese case, the case of South Vietnam was a total failure from the perspective of the U.S. Foreign Atoms for Peace policy. The research reactor supplied to the South Vietnamese did not yield any return for the U.S. in the long run; instead, it created a grave security risk, as it was ultimately confiscated by North Vietnam, although the U.S. succeeded in removing and shipping the nuclear fuel rods back home at the last minute before their confiscation.[34] When the U.S. and South Vietnam began talks on a bilateral agreement in 1958, the U.S. still whole-heartedly supported Ngo Dinh Diem—it was several years before his totalitarian attitude went beyond the limits of U.S. tolerance, which led to the CIA-backed coup d'état against him.[35] In the same year, however, Ho Chi Minh was already institutionalizing the Socialist economy in the North, and in the following year, he decided to liberate the South by armed struggle, and began building the Ho Chi Minh trail. Thus, it was in no sense a safe and stable environment in which the U.S. decided to export a reactor to South Vietnam.

Department of State documents indicate that the idea to provide a research reactor to South Vietnam was conceived in the summer of 1958, and that it was the Department of State and one Vietnamese scientist that promoted the idea. On August 23, 1958, President Ngo Dinh Diem disclosed to U.S. Ambassador Durbrow in Saigon his plan to build an Atomic Research Center. Diem requested U.S. assistance in building basic research facilities, but he did not request a nuclear reactor. Diem's plan was to spend eight to ten years nurturing atomic science in Vietnam (in the U.S. government records, the term "Vietnam" was often used to indicate South Vietnam), and in the long run, to build a nuclear power plant to provide electricity. He further expressed his intention to send delegates to the Atoms for Peace international conference to be held in Geneva in the same year. Ambassador Durbrow asked the Department of State to set up meetings between Vietnamese delegates and U.S. representatives in Geneva and Vienna (meaning U.S. representative in the IAEA within the U.S. Embassy in Vienna) to evaluate the capacity, training experiences, and interest of Vietnamese scientists and examine what extent of U.S. assistance would be appropriate.[36]

To the US government's surprise, however, Dr. Bu Hoi (or Buu-Hoi; in Vietnamese, Nguyễn Phúc Bửu Hội), leader of the South Vietnamese delegates to Geneva, said in a media interview that South Vietnam was planning to purchase a research reactor from a private U.S. company. As mentioned above, a prerequisite for a foreign country to receive a research reactor was the conclusion of a bilateral agreement with the U.S. government, but South Vietnam had not even initiated talks about such an agreement. Bu Hoi was an internationally renowned biochemist educated at Sorbonne University in the 1930s, and was also a "prince," related to the Vietnamese royal family. He had originally supported Ho Chi Minh's anti-colonial movement, though he parted from Ho Chi Minh when his movement leaned toward Communism, and offered his support to Ngo Dinh Diem in 1951. Bu Hoi was a moderate, preferring dialogue and co-existence to armed conflicts. He was also a Buddhist, the religion persecuted by Ngo Dinh Diem's government. Despite Diem's rejection of Bu Hoi's moderate line, he nonetheless supported Diem, working as a liaison with the United Nations and with the U.S.[37] Given his fame and influence, it was not surprising that South Vietnamese newspapers reported Hoi's words on the purchase of a nuclear reactor on their front pages. One of the influential newspapers, *Journal d'Extremme-Orient*, cited Hoi's comments that "We possess an important intellectual know-how, thanks in large part to French science." "We could have a surplus of intellectuals," thus, Hoi continued, "it is necessary for us to find for them opportunities in scientific research." According to Hoi, Vietnam could "become the most important Southeast Asiatic country from the intellectual point of view" and "contribute to the solution of problems of the region with the help of the two countries of which we have close relations, France and the United States." Faced with this situation, the U.S. Embassy in Saigon telegrammed the Department of State that it would continue a conversation with Diem about nuclear technological aid, "especially in terms of the plan to establish [the] Asian Nuclear Center."[38]

The Asian Nuclear Center plan was part of the Colombo Plan initiative—originally designed to settle the British war debt (the so-called "sterling balance") to the Commonwealth nations—started in 1955, and aimed at the establishment of a research and training center of nuclear S&T in Asia. When the U.S. joined the Colombo Plan, it proposed a plan to establish a research and training center for nuclear science in Asia, and suggested that member countries be prioritized in receiving research reactors. The Japanese government was eager to invite the Center to Japan, but the U.S. was fearful of giving the impression to Asian leaders that the Center would benefit only the U.S. and Japan. Thus Colombo, Ceylon, was selected as the site of the Center, for Ceylon was part of the British Commonwealth and it would not be difficult to secure the consent of the British government. In early 1956, however, the U.S. moved the site to Manila, Philippines, due to the strategic need to strengthen pro-American President Magsaisai's regime. However, this plan, too, came to a deadlock when Magsaisai died in a tragic airplane accident. The U.S. interest in the Asian Nuclear Center itself subsequently waned, and in March 1959, the U.S. government decided to suspend the plan indefinitely.[39] However, in the summer of 1958, when South Vietnam was about to begin nuclear research programs, the U.S. government had not yet abandoned the plan, and was soliciting Asian countries to join the project. The Department of State, therefore, viewed Diem as a potential partner in the Asian Nuclear Center project.

In early September, the South Vietnamese paper *Time of Vietnam* reported that Bu Hoi, still in Geneva, publicly announced that he would start preliminary negotiations with the General Atomics Company to purchase a 30-kw Triga Mark II research reactor. Alarmed by the news, the U.S. Embassy in Saigon telegraphed the Department of State, saying that "if the purchase is considered premature, it would be best to make this clear to the government of Vietnam now."[40] Secretary Dulles replied, however, that "if a reactor is to be utilized in well-conceived program of nuclear research and training, Department and AEC (Atomic Energy Commission) would encourage its purchase and the Vietnamese government's application for grant under President's 1955 offer of $350,000." Dulles further wrote that "regarding atomic energy equipment other than reactors," the U.S. government was also preparing to provide grants "subject to prior approval of equipment list by the U.S. government."[41]

This favorable reaction of the Secretary of State accelerated the nuclear reactor project in South Vietnam. In early October, Bu Hoi, who was attending the IAEA general assembly, made known to the U.S. delegate his desire to conclude a bilateral agreement. He had already sent sample agreements of Denmark and Guatemala to Saigon. He was planning to purchase a Triga research reactor with U.S. financial aid and build a nuclear research institute in Dalat, a city approximately 180 miles north of Saigon. General Atomics, in cooperation with Keiser Engineers, were already planning to survey the site. The U.S. Embassy in Vienna evaluated Bu Hoi highly as "friendly, communicative, [and] appears [to be a] strong supporter of US" and also "realistic and appreciative of the limitations [of the] Vietnamese economy."[42] The U.S., viewing Bu Hoi as a key contact, disseminated positive images about him through media. *Saigon Daily News Roundup*, a newspaper in South Vietnam sponsored by the U.S. government, praised Bu Hoi for returning from France and contributing to the South Vietnamese government, despite his being used for Communist propaganda for some time in the past. The newspaper article appealed to overseas Vietnamese intellectuals to follow suit and repatriate.[43] A noted scientist, efficient technocrat, and respected aristocrat, Bu Hoi was expected to be a role model for other Vietnamese intellectuals.

During the month of October 1958, a series of meetings were held between Ngo Dinh Diem and the U.S. Embassy in Saigon, during which Ambassador Durbrow handed Diem an "informal paper," indicating that "the authorities in Washington and our staff here believe that it would be in the interest of Viet-Nam to develop a well thought out and well-planned long-range program for activity in the nuclear energy field." Durbrow told Diem that the U.S. government was aware of the fact that Bu Hoi was "discussing with an American firm the question of the purchase of a nuclear reactor" and that he had already "forwarded to the Government of Vietnam a proposed agreement for cooperation with the United States."[44]

On the other hand, the U.S. Embassy in Saigon was collecting information from various circles on whether providing a reactor to South Vietnam served the U.S. national interest. For example, on November 26, 1958, the embassy staff interviewed several Vietnamese scientists, including Lu Tuon Anh, Bu Hoi's subordinate, who would later become deputy director of the Nuclear Research Institute. From what Lu Tuon Anh told the embassy staff, it turned out that General Dynamics (the parent company of General Atomics) and the AEC had been assisting Bu Hoi toward the conclusion of a bilateral agreement and construction of a nuclear reactor. C. A. Rolander of General Dynamics had helped in drawing up the bilateral agreement, and the draft had already been sent to the AEC and received an informal nod. Moreover, Bu Hoi, Lu Tuon Anh, and Rolander had already discussed a bilateral agreement with John A. Hall, AEC's chief of the reactor division, who was leading the U.S. mission to the IAEA conference in Vienna, and AEC Chairman John A. McCone, who was present at the UN conference on nuclear energy at Geneva. Furthermore, "Mr. McReynolds" of General Dynamics had visited Saigon and urged Keiser Engineers, the company in charge of the reactor construction, to accelerate the operation. Lu Tuon Anh also told the U.S. Embassy that the negotiations on the bilateral agreement had been handled directly by the President's office, and that the Foreign Office of South Vietnam was not involved at all. Lu Tuon Anh was quite optimistic about the operation of a Triga reactor, as, according to him, there were "twelve Vietnamese scientists available for work in the reactor center," who were trained in either the U.S. or France. He said that Bu Hoi and General Atomics were anxious to get the reactor into operation as quickly as possible, and President Diem was providing "$400,000 for the first year of the operation of the nuclear center and $300,000 for the second year. In addition, $600,000 is being provided for installation cost." The remainder of the funding, according to Anh, would come from the U.S. However, the embassy was not sure about the source of funding on the Vietnamese side.[45]

The U.S. Embassy, shocked to learn that private U.S. companies, the AEC, and the government of South Vietnam were discussing a bilateral agreement without any consultation with them, and that President Diem and his close circle were promoting it without a well-considered future plan, sent a lengthy telegram to the Department of State, advising them to be cautious about the reactor project. According to the telegram, the embassy and USOM highly doubted if exporting a reactor at this time would be politically advantageous "in light of the other developmental needs of the country." It was important, the embassy pointed out, "to avoid the possibility that the United States might be criticized strongly for assisting the government of Vietnam in a project which appears to have little relationship to Vietnam's real needs." Already many Vietnamese intellectuals had indicated that "President [was] motivated primarily by prestige and political considerations—that is, the desire to bring Vietnamese intellectuals in France back to Vietnam and to confront the Vietnamese people, including North Vietnam, with a powerful symbol of technological progress." In their view, building a reactor had little practical rationale, and would only "consume scarce budget funds greatly needed for developmental activities." Doubts were also expressed about the capability of Bu Hoi to administer a reactor project, as he had little experience in nuclear physics. Among prominent figures expressing such doubts were Dr. Vu Van Thai, Director of Budget, and Nguyen Quang XXXX (illegible), "rector" of Saigon University. The embassy recognized that "political considerations" such as the "prestige which the reactor would afford" and the "drawing back of Vietnamese intellectuals from France" were significant, and also understood that a well-planned program for developing nuclear science in Vietnam should be encouraged. Nevertheless, the embassy doubted the desirability of offering a reactor to South Vietnam with such haste.[46]

The embassy staff also interviewed Bu Hoi a few weeks later, and this interview supported the "previous conclusions that the reasons behind the drive for the erection of a nuclear reactor facility" were "(1) to attract Vietnamese scientists in France back to Vietnam and (2) to acquire prestige, especially vis-à-vis North Vietnam." Bu Hoi said that there were "about 20 Vietnamese scientists in

France" who had experience in nuclear physics and many others in other scientific fields. Although "the French pay them quite well," Bu Hoi continued, the "intellectual possibilities created by the proposed reactor center [would] compensate for the loss of money." According to Bu Hoi, one reason why Dalat was chosen as the site was that "the climate would be attractive to Vietnamese and foreign scientists." He further explained that North Vietnam was publicizing its own nuclear energy program, and that therefore the "propaganda aspect" of a South Vietnamese nuclear project would be important.[47]

Through these interviews, the embassy became convinced that the reactor project was indeed reckless. In a confidential telegram to Secretary of State Dulles, dated December 16, Ambassador Durbrow summarized the communications with the South Vietnamese thus far, and strongly opposed the plan to provide a reactor to South Vietnam. To summarize the lengthy telegram:

1. When the South Vietnamese expressed interest in purchasing a reactor, the embassy requested Department of State guidance. The Department replied that it and the AEC would grant a reactor if it would be used in a well-conceived research and training program.
2. Although many doubts arose within the embassy and USOM as to whether a reactor for South Vietnam was appropriate, the embassy lacked technical competence to make a judgment. The embassy therefore felt it best to reserve judgment until the matter could be fully reviewed with the assistance of a U.S. consultant.
3. The information obtained informally tended to indicate the lack of an effective plan for using a reactor. There was some evidence of opposition from various Vietnamese circles to the reactor project.
4. There was a feeling that the purpose of the reactor project was primarily for local prestige. While the embassy recognized certain validity in such a political consideration, it did not believe it sufficient to override concerns.
5. It appears inappropriate to assist a project which bears little relationship to Vietnam's real needs.
6. The U.S. support of the reactor project may encourage tendencies to seek solutions to South Vietnam's problems through dramatic projects which absorb already scarce funds and administrative talents. It would encourage President Diem to rely on U.S. financial support in all circumstances.
7. The embassy and USOM therefore strongly recommends against consideration of a reactor grant until the government of South Vietnam agrees to discussion of all factors concerning the reactor project.
8. In South Vietnam, people understand that the reactor grant is linked to a bilateral agreement, and therefore the South Vietnamese government might attempt to pressure the U.S. for grant once a bilateral agreement is concluded. The embassy strongly recommends not immediately concluding an agreement.
9. The embassy understands that Bu Hoi intends to visit Washington. The embassy would suggest that he defers.[48]

No copy of Secretary Dulles' reply has been found, but Durbrow's telegram on January 7 indicated that the Department of State had not changed its policy to conclude a bilateral agreement. The telegram reads:

> In view Department's position believe advisable to proceed negotiation bilateral agreement at the same time making it very clear to Vietnamese both in Washington and in Saigon that conclusion of bilateral is merely preliminary step, and does not necessarily commit us to subsequent reactor grant. Embassy wishes to point out that Vietnamese appear already deeply committed both with Kaiser for building construction and General Dynamics for purchase reactor and that favorable impact bilateral signature may be only of fleeting value. Suggest that arrangements be made for prompt AEC review of program following conclusion of bilateral unless review could commence sooner.[49]

According to the telegram, the Department of State intended to tell the South Vietnamese government that a bilateral agreement did not immediately guarantee financial aid (the grant), but the Department was nevertheless willing to conclude an agreement. Moreover, since South Vietnam had already been committed to purchasing a reactor from General Dynamics, discussion toward a bilateral agreement—which should have normally occurred *before* the decision to purchase a reactor—was only catching up with a *fait accompli*. The AEC's review of the safety and usefulness of the reactor program had not been conducted before the agreement, either. Ambassador Durbrow's strong concern did not move the Department at all. Durbrow was a veteran diplomat specializing in the Soviet Union and Europe from the prewar era, and had been appointed as ambassador to South Vietnam in 1957 during the Eisenhower administration. He increasingly became disappointed with the autocratic and corrupt government of Ngo Dinh Diem, and Diem also developed an animosity toward Durbrow, who was critical of his regime. When the Kennedy administration was inaugurated in 1961, the new president adhered to the policy to support Diem. Durbrow was called back to Washington and replaced by Ambassador Frederick Nolting, who had milder attitudes toward Diem (see also Chapter 7).[50] While Durbrow's doubt about the validity of a reactor program might have partly come from his distrust in Diem and his regime, perhaps the U.S. government should have paid his warning more heed, given the destiny of Diem and the reactor in the next few years: Diem was abandoned by the U.S. government and assassinated in a coup d'état, and the reactor was confiscated by North Vietnam. On January 28, 1959, the Department of State, the AEC, and the Vietnamese embassy in Washington held a talk on the bilateral agreement. Lu Tuon Anh was present, although Bu Hoi was absent due to illness. The U.S. government recommended some revisions to the draft agreement, and explained the procedures to be taken before its enactment, such as obtaining President Eisenhower's approval and laying the draft before the Joint Committee on Atomic Energy for 30 days before ratification. The following day, Lu Tuon Anh submitted a $350,000 reactor grant proposal.[51] Ambassador Durbrow was concerned about the possibility that the South Vietnamese government might request additional financial aid because the U.S. had contributed $150,000 to the Philippine government (in addition to the $350,000 grant for the reactor) through special Congressional legislation for establishing the Asian Nuclear Center in Manila.[52]

The process of the bilateral agreement with South Vietnam appeared to be a serious deviation from the initial U.S. thinking that the grant should be offered only when a "well-conceived research and training program" was ready. Nevertheless, the agreement was signed on April 22 and went into effect on July 1. Signing the agreement on behalf of the U.S. were Walter S. Robertson, Assistant Secretary of State for Far Eastern Affairs, and Chairman John A. McCone of AEC. On the South Vietnamese side, Ambassador Tran Van Chuong signed. He had been an ambassador to the U.S. since 1954, and was father of Madame Nhu (for more information about Madame Nhu, see footnote 35.) According to the agreement, the U.S. was to provide "information on design, construction and operation of nuclear research reactors and their use in research, development and engineering projects." Also, U.S. "industrial firms, other organizations and private citizens [were] permitted to supply appropriate nuclear equipment and related services." The AEC was to "lease" to the South Vietnamese government up to six kilograms of uranium-235.[53]

Thus far, this section has examined the path to a bilateral agreement between the U.S. and South Vietnam, based on the records of the Department of State and the ICA. An emerging picture was that the U.S. government offered a reactor, grant, and uranium in spite of the U.S. Embassy's strong opposition based on hands-on knowledge of the local situation. The embassy had argued that the reactor project was premature, that there were strong criticisms of the project among local intellectuals, and that it would not serve the U.S. national interest. However, the AEC, U.S. private companies, Ngo Dinh Diem, and Bu Hoi were enthusiastic about building a reactor, and Secretary Dulles was supportive of their idea. President Diem and Bu Hoi were motivated to construct a reactor for national prestige and propaganda purposes, and for enticing back to Vietnam students and scientists overseas. Even after the enactment of a bilateral agreement, Bu Hoi repeated to a Department of State officer that "frankly, one of the main reasons for constructing the reactor in Vietnam had been psychological," i.e., to encourage overseas Vietnamese students to repatriate.[54] The reactor was a "psychological" apparatus for many actors, though it carried slightly different meanings to each. To Diem, it was a talismanic object to maintain his prestige, and to Bu Hoi, it was a symbol of scientific advancement to attract overseas students and intellectuals. To the Department of State, the reactor was a psychological tool precisely because Diem and Hoi found psychological significance in it.

Bu Hoi's obsession with promoting the repatriation of overseas Vietnamese was related to the fact that many scientists and intellectuals pursued educational and professional opportunities in France, the former colonizer of Vietnam. While many chose to stay in France in order to avoid the chaotic situation of their native country, such personal, academic, and scientific ties with France could become a strength for South Vietnam, if those students would repatriate. However, they could also pose a threat to the U.S. because France was a competitor in nuclear technology. In fact, the South Vietnamese government was secretly negotiating a bilateral agreement with France—the U.S. Embassy in Saigon was shocked to find a Vietnamese newspaper article on January 28, 1961, reporting the conclusion of a bilateral agreement on nuclear technological aid between France and South Vietnam. "The embassy was not previously informed on the agreement," Durbrow telegrammed the U.S. mission to IAEA in Vienna and Pennington in the AEC, Tokyo. According to "Le Tuon Anh, Acting Director for the Office of Atomic Energy claims," the telegram continued, "informal discussions with [the] French [were] initiated prior to [the] U.S. grant," and the "AEC was informed by him and Bu Hoi during their 1959 U.S. visit about the possibility of Franco-Vietnam agreement." Anh further explained that the bilateral agreement stipulated only the outline, that the details of the aid program would be decided later, and that the content would not conflict with the U.S. aid program. The U.S. Embassy demanded that the government of South Vietnam provide a copy of the bilateral agreement, but Bu Hoi replied that it was confidential.[55]

Ngo Dinh Diem, believing the symbolic value of a reactor would strengthen his authority, carried out the ground-breaking ceremony of the construction of the nuclear research center in Dalat on April 9, 1961, the day of a presidential election. He also planned to build a religious monument on the hill behind the research center. Furthermore, he commissioned Ngo Viet Thu, an internationally renowned architect of Vietnamese origin, to change the outer look of the nuclear reactor into a more artistic design. Although General Atomics had already established a contract with Kaiser Engineers on the design and construction of a Triga Mark II reactor, Diem was not fond of the simple, functional design proposed by Kaiser, and asked Ngo Viet Thu to create a "super-modern design" which would "fit the resort atmosphere of Dalat."[56] In short, for Diem, the nuclear reactor was an incarnation of his power, authorized by the U.S., and therefore its functional and research purposes were less important than the psychological effect of its appearance.

As the U.S. Embassy had foreseen, there were mounting problems concerning the construction and management of the reactor. First of all, Bu Hoi, director of the Nuclear Research Institute, and the most reliable contact person for the U.S., was rarely in Vietnam because he also served as the ambassador to African countries and held a position at a French research institute. Nguyen Dinh Hoang, who graduated from Union College in New York and worked for Standard Vacuum Oil Company, returned home to serve as acting director. However, Hoang was not familiar with nuclear technology, and was unable to bear the burden of his heavy responsibilities. He committed suicide in August 1959, only one month after the conclusion of the bilateral agreement. The U.S. Embassy lamented that Hoang's death was a huge blow to the Vietnamese nuclear energy program that had already been suffering from a dire shortage of administrative and technological staff.[57] Furthermore, there was also a shortage of rank-and-file engineers. To solve this problem, the U.S. accepted an increasing number of Vietnamese students in U.S. universities. Herbert W. Pennington, head of the AEC's Tokyo office, even proposed to "let Vietnamese students currently studying in the U.S." receive training in nuclear technology with U.S. financial support. His proposal to teach nuclear science to Vietnamese students, regardless of their specializations, sounds outrageous, but it shows the degree of desperation the U.S. was feeling in their efforts to advance the Vietnamese nuclear program.[58]

The list of Vietnamese engineers prepared by the Department of State shows that the majority of Vietnamese engineers studied in France and other European countries until the mid-1950s, but thereafter, an overwhelming majority pursued studies in the U.S. The trajectory of Le Tuon Anh, who took over as acting director of the Nuclear Research Institute after Hoang's death, was a typical example: after graduating from the Math Department of the University of Ohio in 1954, he acquired a doctoral degree in nuclear science from Saclay University in France in 1958. Thereafter, he received education and training mostly in the U.S. He participated in the first training session of the Oak Ridge School of Reactor Technology (ORSORT) in 1959, received practical training in reactor operation at General Electric, and participated in the U.S. Atoms for Peace exhibition in New Delhi in 1960 to operate the Triga Mark II reactor exhibited there, before returning home to take over Hoang's former position.[59]

The construction of the nuclear reactor in Dalat was completed in 1962, and it went into operation in 1963. By then, however, the political situation in Vietnam was rapidly deteriorating. The OCB "special report" of June 15, 1960, read that "the activities of the Communist guerrillas (Viet Cong) in the South have increased seriously ... and the Communists can be counted on to exploit popular dissatisfaction with Diem's Government." "Diem's Government ha[d] lost some prestige," according to the report, due to "increased popular resentment of official corruption and highhandedness and the shortcomings of its security forces in coping with stepped-up guerrilla activity." Ambassador Durbrow had "recently made extremely strong demarches to President Diem indicating our knowledge of and concern over corruption in the government," the report continued. He "urged on Diem and top officials the importance of strengthening the government's popular support," and made "specific suggestions, such as regular radio addresses by President Diem and better explanations to the people of government programs involved, calculated to give them hope for the future."[60] It is interesting to note that a psychological tool to give "hope" to the citizens was radio, rather than a reactor. Historians have pointed out that ignorance about Vietnamese agrarian society was the greatest enemy for the United States.[61] The U.S. government did not understand that the majority of the Vietnamese farmers had no need for a reactor, and thus the idea that they embraced, that a reactor's appeal would contribute to the preservation of the Ngo Dinh Diem regime, was misguided, and was likely the fundamental reason for the failure of the nuclear reactor project in South Vietnam.

The reactor ended up ceasing operation after only four years because of the increasing offensives of the People's Liberation Armed Forces of South Vietnam (PLAF), or the Viet Cong army. Just months before the fall of Saigon in 1975, as the North Vietnamese Army advanced closer to Dalat, the U.S. government decided to abandon the reactor, but to recover the fuel rods in order to prevent technical information on the fuel from falling into enemy hands. Two American scientists, Wally Hendrickson and John Hollan, were ordered on a secret mission to land in Dalat and remove the fuel rods from the reactor. In the event that removal of the fuel rods failed, they would pour concrete into the core so that the fuel could not be retrieved by the enemy. If concreting also proved impossible, they were instructed to destroy the reactor with dynamite. The two scientists successfully recovered the fuel rods, which were shipped back to the U.S. When the North Vietnamese Army arrived in Dalat, they discovered an empty reactor waiting for them. Later, the Socialist Republic of Vietnam invited Soviet engineers to re-install fuel, and the reactor once again went into operation in 1984 as a uniquely hybrid reactor of a U.S.-made shell and a Soviet-made core.[62]

The case of South Vietnam indicates complex and combined reasons behind the U.S. decision to offer a research reactor. The U.S. private companies wished to market reactors, and the AEC supported them, whereas the Department of State viewed a reactor as a political and psychological tool to support the Ngo Dinh Diem regime. Although the U.S. Embassy, with the hands-on knowledge about the local situation, strongly doubted the validity of such a decision, the Department of State overrode the embassy's opinion. Ultimately, however, the appeal of the reactor was powerless in the face of a corrupt government and fierce attacks by the North Vietnamese Army.

3.4 Burma

The third and last case is that of Burma (present-day Myanmar), which did not conclude a bilateral agreement with the U.S. Burma differed from South Vietnam and Japan in that it was a Socialist country and not friendly to the U.S. Furthermore, it was not a member country of the Asia Nuclear Center plan, to which the U.S. offered nuclear reactors with priority. It is meaningful to focus on Burma because the case demonstrates how and why the U.S. provided nuclear technological aid even to a non-friendly country, and how and why Burma expected to receive that aid. Burma's case reveals that the U.S. employed nuclear technological aid as a psychological measure to attract Burmese political and scientific leaders and to dissuade them from allowing the Soviet Union access to mineral resources.

A key figure in the U.S.–Burma negotiations on nuclear technological aid was U Hla Nyunt, a scientist who had studied nuclear physics at Kyoto Imperial University during the Second World War. Japan had "liberated" Burma from British colonialism and bestowed nominal "independence" upon it during the war, a process through which Japan and Burma nurtured close cultural and political ties, and many talented young Burmese studied abroad in Imperial Japan. Nyunt, who was born in 1924 and studied physics at University College in Rangoon, was one such student. His university education was interrupted by the outbreak of the Pacific War, but from 1944 to 1945, under the auspices of a Burmese government scholarship, he studied at Kyoto Imperial University.

While U.S. government records indicate that Nyunt had "studied physics under Professor Hideki Yukawa, Japanese Nobel Laureate" in Kyoto Imperial University, he was still an undergraduate student at the time, and the author's research at the Kyoto University archives failed to discover any evidence that he received instruction from Yukawa directly. Nyunt returned to Burma at the end of the war, and was temporarily engaged in the war crimes tribunal. He graduated from University of Rangoon in June 1947, and from 1948, he studied at Lehigh University in Pennsylvania as a Burmese government-sponsored scholar. Upon receiving a master's degree in physics, he returned to Burma and became an assistant lecturer of physics at Rangoon University in 1950. He was promoted to lecturer the following year, and in 1953 was elected to the Senate of the university. When the Union of Burma Atomic Energy Center was established in June 1955, Nyunt was appointed as its deputy director, and promoted to director the following year. He participated in scientific conferences in India and Pakistan in 1952 and 1953, the first Indian Atomic Energy Conference in 1954, and the First UN International Conference on the Peaceful Uses of Atomic Energy in Geneva in 1955.[63] From 1955 to 1956, he enrolled in the Argonne International School of Nuclear Science and Engineering (ISNSE: for more information about ISNSE, see Chapter 4). He also led the Burmese delegation to the Asian Atomic Conference in Bombay in 1956, the dedication of India's first nuclear reactor in 1957, and the Second UN International Conference on the Peaceful Uses of Atomic Energy in 1958. Nyunt was the symbol of Burmese nuclear science and the most important liaison with foreign countries in this field. In an interview for the 1960 issue of the Argonne National Laboratory newsletter, Nyunt said that his most important role was to "act as a science missionary, interpreting atomic energy to his government and the people." "He was succeeding," according to the article, as the Prime Minister and the Cabinet of Ministers had become convinced of the importance of nuclear energy. He further explained to the interviewer that the most important benefit he received from his stay at Argonne was "the opportunity for personal contact and exchange of ideas with atomic scientists, not only from the U.S., but from all over the world."[64] While it was true that Nyunt had developed a wide international network, not only with the U.S. but also with various other countries, the U.S. government regarded him as both a pro-American Burmese and as a liaison between the two countries. The appointment of Nyunt's wife Baw Ni Ni to Deputy Director of the USIS library in Rangoon was also evidence of the trust the U.S. placed in the Nyunts. USIS Rangoon had been engaged in information and education activities since the establishment of the Union of Burma: in 1950 the USIS sponsored an "Independence Exhibition," and by the end of 1955, the USIS's "mobile library" was circulating American books and magazines among Burmese schools and communities, and USIS films were shown to Burmese citizens (Figures 3.1, 3.2, 3.3).

Postwar U.S.–Burma relations cannot be explained without reference to the complicated decolonizing process of Burma. Burma fell under British colonialism as an "independently-governed state" toward the end of the nineteenth

Figure 3.1 Female students gathering around the USIS Rangoon "mobile library" (1960). RG306, No.61-11025, NACP.

century, and subsequent struggles for independence were primarily centered around Burmese intellectuals. Although Burma achieved nominal independence under the shadow of Japanese imperialism in August 1943, anti-Japanese resistance fermented under the oppressive rule of the pro-Japanese government. In March 1945, Major General Aung San's coup d'état toppled the pro-Japanese regime, but when Japan was defeated in the war, the British returned to rule. Aung San continued negotiating with the British and coordinating interests among ethnic minorities, but was assassinated by his political opponents. In January 1948, Burma withdrew from the British Commonwealth and officially declared independence as the Union of Burma. U Nu, the first prime minister, pursued a non-aligned path with the ultimate goal of establishing a Socialist state. His government, however, had to deal with mounting problems including armed conflicts with ethnic minorities and insurgencies of the exiled Kuomintang (KMT) faction defeated in the Chinese civil war. The Army Chief of Staff, General Ne Win, was charged with restoring public order, while U Nu implemented a democratic parliamentary system. In October 1958, U Nu entrusted the Prime Minister's position to Ne Win, but after the elections of February 1960 ended favorably, he returned to power. However, U Nu's

Figure 3.2 People reading American books and magazines at the "community room" administered by USIS Rangoon (1958). RG306, No.58-32, NACP.

"Buddhist Socialism" did not function well, and his regime was overthrown by a coup d'état led by Ne Win in March 1962. Ne Win promoted "Burmese Socialism"—a combination of extreme nationalism, Marxism, and non-aligned neutralism—and Burma gradually became isolated from the international community.[65]

The U.S. Foreign Atoms for Peace program began, therefore, exactly when Burma was seeking a third path, which was aligned neither with U.S.-led capitalism nor Soviet-led Communism, but was based on modern parliamentary democracy, under U Nu's leadership. The U Nu regime strived to nurture industry, and was attracted by the U.S. nuclear technological aid. As early as February 1955, the U.S. Embassy in Rangoon reported to the Department of State that there was "keen interest, particularly on the part of Minister of Industries U Kyaw Nyein, in the peaceful uses of atomic energy," and that "plans have been made to set up an 'Atomic Energy Center' under the State Industrial Research Institute." The Burmese government was also planning to send four physicians "for the study of atomic medicine in the United States," and one student "in the field of nuclear physics" under the educational exchange program of Smith-Mundt grants (the precursor of the Fulbright scholarship).[66]

Figure 3.3 Entrance of the Burmese Independence Exhibition sponsored by USIS Rangoon (1950). RG306, No.50-5069, NACP.

Burma showed interest in the U.S. Foreign Atoms for Peace programs so early—even before President Eisenhower's Pennsylvania State University address—likely because scientists such as U Hla Nyunt had prior knowledge in nuclear science. The fact that Nyunt had enrolled in the 1955 class of the Argonne International School of Nuclear Science and Engineering (ISNSE) also indicated how quickly Burma began to absorb U.S. knowledge of nuclear technology. The U.S. government, however, regarded the Burmese standard of scientific research as not sufficiently mature to be rendered nuclear technological aid. C. E. Barthel, an American engineer who had been appointed as director of the Burmese Applied Research Institute, replied to the U.S. Embassy's inquiry on the matter to the effect that it would be "a long time before Burma [could] make proper use of a research reactor," and therefore it was better to "head off any such request by the Burmese Government at this time." The U.S. Embassy concurred with his thought.[67]

Only two years later, however, the U.S. government suddenly became interested in providing nuclear technological aid to Burma. According to the minutes of the Inter-Agency Nuclear Coordinating Committee, a special committee established under the OCB, dated September 30, 1957, a State Department representative reported "the Burmese Government's interest in receiving U.S. assistance in connection with the construction and operation of a proposed radio-isotope laboratory." He indicated that the case of Burma had "special merit" and that the "Department would be prepared to give full support to a modest program of assistance." The meaning of this "special merit" became clear to some degree from the archival documents declassified by the author's FOIA request in 2017—this special merit was related to the discovery of Monazite ore in Burma in 1956. Monazite is a radioactive mineral containing a small amount of thorium, and occasionally uranium.[68] The declassified documents indicated that the Soviet Union was seeking "exclusive purchase rights" of "fissionable raw materials produced in Burma" in exchange for nuclear technological aid, including a nuclear laboratory, and also future installment of a reactor. "Belief was widespread in mining circles" that Burma had "deposits of uranium-bearing ores as well as monazite sand." The U.S. government wished, by any means, to prevent the Soviets from monopolizing the Burmese mineral resources.[69] The U.S. government also felt that the Soviet approach to Burma was part of the Soviet "effort to divert Burma from participation in the Asian Nuclear Center."[70] According to this U.S. interpretation, the Soviet Union was attempting to dissuade Burma from being absorbed into the S&T network constructed by the U.S., and prevent the U.S. from building a hub of nuclear R&D in Southeast Asia. In the tug-of-war between the Soviet Union and the U.S., Burma was suddenly offered nuclear technological aid from both countries. In the U.S.–Soviet competition, nuclear technological aid was used as a carrot by both parties to attract Burma into their respective camps, in order to secure mineral resources and the site for the Southeast Asian center of nuclear R&D. Offering technological aid in exchange for natural resources or political allegiance can be understood as a manifestation of hard power. However, the boundaries between the traditional hard power diplomacy and overseas information policy were actually not so clear: in fact, they were two sides of the same coin. Nuclear technological aid also involved study-abroad scholarships and training programs, through which the U.S. attempted to nurture pro-U.S. scientists. The technological aid also accompanied USIS films, lectures, exhibits, and printed materials through which positive images of American S&T were disseminated.

The U.S. government's nuclear technological aid to Burma began with the establishment of a radioisotope research laboratory. The U.S. government initially offered $8,000 worth of experimental equipment, but Freddy Ba Hli, acting director of both the Union of Burma Applied Science Research Institute and the Union of Burma Atomic Energy Center, demanded $50,000 worth. According to Ba Hli, most of the Burmese engineers had received training in the U.S. and therefore required the U.S. lab equipment that they were accustomed to. As evidence, he submitted a list of educational institutions where Burmese engineers received training, which included U.S. universities such as Leigh, Harvard, Colorado School of Mines, Montana, and so on. Ba Hli himself had received a master's degree from Lehigh University and a doctoral degree from MIT. He also explained to the U.S. Embassy that there would be "forty-three nuclear engineers for Isotope Research Institute by 1957, and more than one hundred engineers by the time Burma will introduce a power reactor."

To evaluate Ba Hli's request, the U.S. government asked Marvin C. Fox, head of the Brookhaven Asian Nuclear Center Survey Group that was visiting Rangoon on May 1–6, 1956, to examine the request in detail. Fox "carefully reviewed the list of equipment" requested by Ba Hli, and endorsed the request. Fox believed "that the list had been prepared with great care and wisdom and that the items requested would be of genuine value to the Atomic Energy Center in analyzing monazite deposits in Burma." Fox's reply indicates that the U.S. nuclear technological aid to Burma was preconditioned by the Burmese cooperation with the Asia Nuclear Center and the study of fissional ore produced in the country. The U.S. Embassy, upon receiving Fox's report, telegrammed the Department of State that "any assistance which the Department could render ... to procure the full 50,000-dollar worth of equipment would be greatly appreciated by the Union of Burma Atomic Energy Center and by the Government of Burma" and would contribute to "closer relations between the United States and Burma on atomic energy matters." It would also stimulate "Burma's interest in participating in the activities of the Asian Regional Center."[71] The telegram indicates that the U.S. nuclear technological aid was expected to elicit the appreciation of the Burmese, and thereby an inclination to build closer ties with the U.S., as a result, alienating them from the Soviet Union and drawing them into the U.S. circle of S&T. Alongside training programs and printed materials, laboratory equipment was also a tool of the governmental information program to win hearts and minds.

The Burmese Radioisotope Research Institute was built and developed under the strong influence of the U.S. government as well as private organizations. For example, Armour Research Institute was greatly committed to its operation. Armour Research Institute was a private organization engaged in a variety of research related to nuclear technology in cooperation with the U.S. government and the Armour Institute of Technology, the predecessor of Illinois Institute of Technology. Niels Beck, a researcher of the Armour Research Institute, was sent to Burma as director of the Burmese Applied Science Research Center. Beck envisaged the inclusion of Burma in the target countries of the Phoenix Project (Michigan Memorial Phoenix Project, MMPP). The MMPP was a nuclear technological aid program organized jointly by the U.S. government and the University of Michigan, and sent "nuclear consultants" to developing countries where U.S. nuclear reactors were to be installed.[72] Beck's plan indicated that Burma was one of the candidates to which a U.S. nuclear reactor should be sent in the near future. Such speculation is supported by the fact that, following Nyunt, many Burmese scientists and engineers enrolled in the ISNSE. As the next chapter will explain in detail, the ISNSE was established for the training of young scientists and engineers from the countries which were planning to import U.S. reactors.[73] However, the path Burma followed thereafter was anything but what the U.S. had expected. The dictatorship under "Burmese Socialism" completely excluded foreign capital from the domestic economy, and even the technology concerning radioisotopes introduced with the support of U.S. aid was not applied to rice production, the chief crop of Burma.[74] The U.S. financial and human investment in Burmese nuclear development was squandered.

As this chapter chiefly relied on U.S. and Japanese primary sources due to the author's language limitations, it does not capture the full scope of the South Vietnamese and Burmese local reactions to U.S. nuclear technological aid. Nevertheless, the chapter has demonstrated that U.S. nuclear technological aid was employed not only as a tool to achieve tangible goals such as new markets or natural resources, but also as a device to influence and move human thought and behavior. Different actors expected U.S. nuclear technology to promote the government's "prestige," to urge repatriation of overseas students and scientists, to nurture feelings of gratitude and strengthen the ties between the countries, and to fend off the Soviet approach. Technological aid can be a tool of so-called "hard power" diplomacy, in which aid is offered in exchange for resources or allegiance. However, it is short-sighted to think that technological aid is *only* an object to be traded in such "deals." For example, the U.S.-made lab equipment was not *just* equipment but a microcosm of America itself in which U.S.-trained scientists and engineers could continue their research in familiar settings. In retrospect, the cases in South Vietnam and Burma were diplomatic failures, but their processes demonstrate that nuclear technological aid was not just a geographical transfer of material objects, but a device of the Cultural Cold War to nurture pro-U.S. scientists, engineers, technocrats, and other specialists, to change people's perception, and to build long-lasting relationships between the countries. The next chapter will focus on one such system to produce many pro-U.S. nuclear scientists and engineers all over the world, the ISNSE, and consider the questions of how the "school" to teach U.S. nuclear technology to young foreign science elite was planned and put into practice, and what kind of results it brought about.

Notes

1 Dwight D. Eisenhower, "Address at the Centennial Commencement of Pennsylvania State University," June 11, 1955. Also see Introduction.
2 Pennsylvania State University, College of Engineering website, https://www.rsec .psu.edu/Penn_State_Breazeale_Reactor.aspx.
3 Document 238, "National Security Council Report, NSC5431/1, Statement of Policy by the National Security Council on Cooperation with Other Nations in the Peaceful Uses of Atomic Energy," August 13, 1954, *FRUS*, 1952–1954, National Security Affairs, Volume II, Part 2, https://history.state.gov/histori-caldocuments/frus1952-54v02p2/d238; Document 14, "National Security Council Report, NSC5507/2, Peaceful Uses of Atomic Energy," March 12, 1955, *FRUS*, 1955–1957, Regulation of Armament; Atomic Energy, Volume XX, https://history.state.gov/historicaldocuments/frus1955-57v20/d14.
4 According to W. Marshall, former chair of the U.K. Atomic Energy Authority, the U.S. light-water reactors became available in 1958. I would like to thank Professor Hiroshi Ichikawa of Hiroshima University for pointing this out. W. Marshall, *Nuclear Power Technology: Volume 1: Reactor Technology* (Oxford: Oxford University Press, 1984).

5 R. G. Hewlett and J. M. Holl, 1987, "A History of the United States Atomic Energy Commission, 1952–1960," Volume 3, Appendix 6, U.S. Department of Energy, Office of Scientific and Technical Information, https://www.osti.gov/servlets/purl/6150636, 1987.

6 As for the Philippines, the author has consulted Yuko Ito, "Philippine no genshiryoku hatsudensho koso to beihi kankei: white elephant no sozo" [The Philippine's Atomic Power Station Project and the U.S.-Philippine Relations: Creation of a "White Elephant"], in *Genshiryoku to reisen: nihon to ajia no genpatsu donyu [Atomic Energy and the Cold War: Introduction of Atomic Power Stations in Japan and Asia]*, eds. Tetsuro Kato and Mitsuo Ikawa (Tokyo: Kadensha, 2013), 205–234; and Shinsuke Tomotsugu, "'Ajia genshiryoku center' koso to sono zasetsu: Eisenhower seiken no tai Ajia gaiko no ichi danmen" [The 'Asia Nuclear Center' Plan and Its Collapse: One Aspect of the Eisenhower Administration's Asia Diplomacy], *Kokusai Seiji [International Relations]*, vol. 163 (January 2011): 14–27. Concerning Korea, the author has consulted Somei Kobayashi, "Nanboku chosen no genshiryoku kaihatsu: bundan to reisen no aidade" [Atomic Power Development of North and South Koreas: Between the Division and the Cold War], in *Atomic Energy and the Cold War: Introduction of Atomic Power Stations in Japan and Asia*, 167–204; Kim Sonjun, "Formation and Transition of the Korean Atomic Power System, from 1953–1980," (PhD diss., Seoul National University, 2012); John DiMoia, "Atoms for Power?: The Atomic Energy Research Institute (AERI) and South Korean Electrification, 1948–1965," *Historia Scientiarum*, vol. 19, no. 2 (2009): 170–183.

7 Concerning the developing countries' expectations for nuclear technology, see, for example, John Krige, "Techno-Utopian Dreams."

8 *Nichibei genshiryoku sangyo godo kaigi gijiroku [Minutes of Japan-U.S. Atomic Industrial Forum Joint Conference]* (1957), 210.

9 Hewlett and Holl, Appendix 6.

10 As for existing works on Atoms for Peace exhibitions, see note 34 of Introduction.

11 As for Japan's introduction of U.S. nuclear reactors, see Masakatsu Yamazaki, *The Japanese Nuclear Development, 1939–1955* (see Introduction), an oft-cited work by a science historian.

12 Department of State Instruction, From Dulles to US Embassy Tokyo, September 27, 1957, RG469, Entry 421, box 32, NACP; "AEC Tokyo jimusho o kaisetsu: shocho niwa Pennington shi" [The AEC Tokyo Office Established: Mr. Pennington Appointed Director], *Genshiryoku sangyo Shinbun [Atomic Industrial News]*, no. 54 (November 25, 1957), the National Diet Library, Tokyo, Japan.

13 Office Memorandum from R. L. Sneider to G. A. Morgan, June 12, 1956, RG59, Entry A1 3008-A, box 434, NACP.

14 From G. E. Meyer to Ambassador John M. Allison, June 11, 1956, RG59, Entry A1 3008-A, box 434, NACP.

15 From Outerbridge Horsey to Howard P. Jones, Department of State, June 15, 1956, RG59, Entry A1 3008-A, box 434, NACP.

16 Shinsuke Tomotsugu, "The Bandung Conference and the Origins of Japan's Atoms for Peace Aid Program for Asian Countries," in *The Age of Hiroshima*, eds. Michael D. Gordin and John Ikenberry (Princeton, NJ: Princeton University Press, 2020), 109–128.

17 "Nihon genshiryoku heiwariyo kikin: ichinen no gyoseki o kaerimiru" [Japanese Atoms for Peace Foundation: Looking Back the Past One Year], *Atomic Industrial News*, no. 75 (June 25, 1958), the National Diet Library.

18 "Nichibei genshiryoku sangyo tenjikai" [Japan-U.S. Atomic Industrial Exhibition], *Atomic Industrial News*, no. 35 (May 15, 1957), the National Diet Library.

19 "Bei, genshiryoku-kan o tokusetsu: Canada mo shoshaki nado shuppin" [U.S. Builds Atomic Energy Pavilion: Canada also Exhibits Irradiators], *Atomic Industrial News*, no. 105 (April 25, 1959), the National Diet Library.

20 Japan Productivity Center, *Seiansei undo 10-nen no ayumi* [*10-Year History of the Productivity Movement*] (Tokyo: Japan Productivity Center, 1965), 50–51, National Diet Library Digital Collections.

21 From USOM Japan to ICA, October 11, 1956, RG469, Entry 421, box 32, NACP.

22 Yuka Tsuchiya, "Amerika-sei keisuiro no sentaku o meguru joho kyoiku program: 1950 nendai sue no nichibei kankei" [Information and Education Programs concerning American Light Water Reactors: U.S.-Japan Relations in the End of the 1950s], *Rekishigaku Kenkyu* [*Historical Studies*], (Special Issue, October 2018): 127–136.

23 Foreign Service Despatch from Paul E. Pauly, Commercial Attache, US Embassy Tokyo to Department of State, April 5, 1957, RG469, Entry 421, box 32, NACP.

24 From Waring, ICA Tokyo to ICA, March 1, 1957; Foreign Service Despatch from Waring to Department of State, March 5, 1957, RG469, Entry 397, box 32, NACP.

25 Foreign Service Despatch from Frank A. Waring, Counselor of Embassy for Economic Affairs, Tokyo, to Department of State, March 25, 1957, RG469, Entry 421, box 32, NACP. A record of the conversation between Waring and Ishizaka is attached to the Foreign Service Despatch.

26 Tatsuo Arai, *Amerika no taigai enjo: ICA no kino to unei* [*The U.S. Foreign Aid: Function and Administration of the ICA*]. *Keidanren Pamphlet*, vol. 33 (1956): 94–95.

27 For ICA's foreign aid activities and the foundation of the Japan Productivity Center, see Koji Nakakita, *Nihon rodo seiji no kokusai kankei-shi, 1945–1964: shakai minshushugi toiu sentakushi* [*History of International Relations Concerning Japanese Labor Politics: Social Democracy as an Alternative*] (Tokyo: Iwanami Shoten, Publishers, 2008); and Go Shimada, "Sengo Amerika no seisansei kojo, tainichi enjo ni okeru nihon no hienjokoku toshiteno keikenwa nanika: minshuka, rodo-undo shien, ajia eno tenkai" [Japanese Experience as a Recipient Country in the Postwar U.S. Productivity Improvement and Foreign Aid: Democratization, Support of the Labor Movement, and Advancement to Asia], JICA Research Institute, Background Paper: A Historical Perspective, no. 2 (October 2018). As for the ICA Director's participation in the OCB meetings, see "Participation of ICA in the Work of OCB," July 8, 1955, RG469, Entry 421, box 17, NACP.

28 Foreign Service Despatch, from Frank A. Waring, "US Embassy Tokyo to Department of State," June 6, 1957, RG469, Entry 421, box 32, NACP.

29 "Obituaries," *Shikoku Shinbun* (May 4, 2001); Gerald Wendt, *You and the Atom* (New York: Whiteside/William Morrow & Company, 1956).

30 From Hoover (Acting) to American Embassy, Tokyo, November 19, 1956, RG469, Entry 421, box 32, NACP; 'The Soviets Offer Atomic Energy Agreement: Reliable Sources in Peking," *Yomiuri Shinbun*, November 1, 1956, Morning Edition.

31 Foreign Service Despatch from Paul E. Pauly to Department of State, December 17, 1956, RG469, Entry 421, box 32, NACP.

32 Memorandum of Conversation, February 27, 1957; Foreign Service Despatch from Paul E. Pauly to Department of State, March 7, 1957, RG469, Entry 421, box 32, NACP.

33 Foreign Service Despatch from Paul E. Pauly to Department of State, February 20, 1957, RG469, Entry 421, box 32, NACP.

34 In this book, the Republic of Vietnam (RVN) is referred to as South Vietnam, and the Democratic Republic of Vietnam (DRV) as North Vietnam.

35 Vietnam, placed under French colonialism from the nineteenth century, was occupied by Japan during the Second World War. After Japan's defeat in 1945, anti-colonial resistance leader Ho Chi Minh declared independence. However, as France demanded the return of its former colony, the First Indochina War broke out between France and the Viet Minh independence movement led by Ho Chi Minh. The U.S., fearing the spread of Communist influence in Asia, sided with France. In 1954, Viet Minh defeated the French troops at the Battle of Dien Bien Phu, and the peace agreement at the Geneva Conference provided that a general election should be held to create a unified Vietnam. The U.S. government, foreseeing Ho Chi Minh's victory, rejected the general election and supported Catholic nationalist Ngo Dinh Diem to establish the Republic of Vietnam (South Vietnam). President Ngo Dinh Diem's regime, however, did not win popular support, as it was smeared with nepotism and corruption, and also persecuted Buddhists. Moreover, Madame Nhu, wife of the president's brother Ngo Dinh Nhu, invited international criticism when she publicly ridiculed a Buddhist monk who burned himself to death in protest of the tyrannical rule. The U.S. government became gradually disappointed with Ngo Dinh Diem, and in 1963, he was assassinated in a coup d'état with assistance from the CIA. As to the history of U.S.–Vietnam relations, the author has chiefly consulted Hiroshi Matsuoka, *Kennedy wa Vietnam ni do mukiattaka: JFK and Ngo Dinh Diem no anto* [*How Kennedy Confronted Vietnam: Hidden Struggle between JFK and Ngo Dinh Diem*] (Kyoto: Minerva Shobo, 2015); and Hiroshi Matsuoka, Yoshikazu Hirose, and Yoshihiko Takenaka, eds., *Reisenshi: sono kigen, tenkai, shuen to nihon* [*The Cold War: Its Origin, Development, Demise and Japan*] (Tokyo: Dobunkan Shuppan, 2003).

36 Telegram from Saigon to Department of State, September 6, 1958; "Memorandum of Conversation, President Ngo Dinh Diem, Ambassador Elbridge Durbrow," August 23, 1958, RG84, Entry UD2092D, box 1, NACP.

37 Ellen J. Hammer, *A Death in November: America in Vietnam, 1963* (New York: E. P. Dutton, 1987), 48–49, 88; Jessica M. Chapman, *Cauldron of Resistance: Ngo Dinh Diem, the United States, and 1950s Southern Vietnam* (Ithaca, NY: Cornell University Press, 2013), 176–177, Kindle.

38 Telegram from Saigon to Department of State, September 6, 1958; "Memorandum of Conversation, President Ngo Dinh Diem, Ambassador Elbridge Durbrow," August 23, 1958, RG84, Entry UD2092D, box 1, NACP.

39 Concerning the Columbo Plan, the author has chiefly consulted Shoichi Watanabe, ed., *Columbo Plan: sengo ajia kokusi chitsujo no keisei* [*The Columbo Plan: Formation of the Postwar International Order in Asia*] (Tokyo: Hosei University Press, 2014). As for the Asia Nuclear Center, see Tomotsugu, "The 'Asia Nuclear Center' Plan and Its Collapse."

40 Telegram from Saigon to Department of State, September 12, 1958, RG84, Entry UD2092D, box 1, NACP.

41 Airgram from Secretary of State to US Embassy Saigon, October 5, 1958, RG84, Entry UD2092D, box 1, NACP.

42 Telegram from US Embassy Vienna to Saigon, October 2, 1958, RG84, Entry UD2092D, box 1, NACP.

43 Cach Mang Quoc Gia, "Editorial: When a Man Is Sincere," *Saigon Dailey News Round-Up*, December 3, 1958, RG469, Entry P89, box 2, NACP.

44 "Informal Paper Handed President Diem by Ambassador Durbrow," October 10, 1958; Air Pouch from US Embassy Saigon to Department of State, October 16, 1958, RG84, Entry UD2092D, box 1, NACP.

45 Memorandum of Conversation, November 26, 1958, RG469, Entry P89, box 2, NACP.

46 The Department of State Despatch, No. 200, from Arthur Z. Gardiner, Counselor of Embassy for Economic Affairs, December 5, 1958, RG469, Entry P89, box 2, NACP.

47 From Arthur Z. Gardiner to the Department of State, December 16, 1950, RG469, Entry P89, box 2, NACP.

48 From Durbrow to Secretary of State, December 16, 1958, RG469, Entry P89, box 2, NACP.

49 From Durbrow to Secretary of State, January 7, 1959, RG469, Entry P89, box 2, NACP.

50 Wolfgang Saxon, "Elbridge Durbrow, U.S. Diplomat, Dies at 93," *The New York Times*, May 23, 1997.

51 From Secretary Dulles to American Embassy Saigon, January 30, 1959, RG469, Entry P89, box 2, NACP.

52 From Durbrow to Secretary of State, February 4, 1960, RG469, Entry P89, box 2, NACP.

53 Press and Information Office, "Atomic Energy Agreement Signed Between U.S. and Viet-Nam," May 15, 1959, RG469, Entry P89, box 2; "Memorandum for Board Assistants and OCB Working Group on Southeast Asia (NSC5809)," January 13, 1960, RG469, Entry P89, box 8, NACP. As for Tran Van Chuong, see Hammer, *A Death in November*, 150–151.

54 Chalmers B. Wood, Officer in Charge, Viet-Nam Affairs, "Memorandum of Conversation," October 7, 1959, RG469, Entry P89, box 2, NACP.

55 From US Embassy Saigon to US Embassy Vienna, January 31, 1961, RG84, Entry UD2092D, box 1, NACP.

56 Foreign Service Despatch from US Embassy Saigon to Department of State, March 2 and April 30, 1961, RG84, Entry UD2092D, box 1, NACP.

57 Foreign Service Despatch from US Embassy Saigon to Department of State, April 30, August 19, November 4, 1961, RG84, Entry UD2092D, box 1, NACP.

58 From MacArthur to U.S. Embassy Vienna, July 24, 1959, RG84, Entry U2092D, box 1, NACP.

59 From Herbert Pennington to A. A. Wells, November 7, 1960; From Pennington to Department of State, October 1960, RG84, Entry UD2092D, box 1, NACP.

60 "OCB Special Report: Possible Actions to Improve the Situation in Viet-Nam," June 15, 1960, RG469, Entry P89, box 8, NACP.

61 For example, in Japanese language, Hiroshi Matsuoka, *Kennedy to Vietnam senso: hanran chinatsu senryaku no zasetsu* [*Kennedy and Vietnam War: Failure of the Counter-Insurgency Strategy*] (Tokyo: Kinseisha, 2013), 391–393.

62 *NHK BS1 Special*, "Gokuhi shirei, uran nenryo o kaishu seyo: senka no genshiro 40-nenme no shinjitsu" [Secret Mission, Recover the Uranium Fuel: Nuclear Reactor in Fire, Truth Uncovered after 40 Years], aired June 20, 2015, on NHK BS1.

63 Foreign Service Despatch from US Embassy Rangoon to Department of State, May 20, 1955, RG469, Entry 397, box 2, NACP.

64 "Distinguished Alumni: U Hla Nyunt," *Argonne National Laboratory News-Bulletin International*, vol. 1, no. 2 (April 1959): 5.

65 Hirayoshi Sakuma, *Burma (Myanmar) gendai seijishi (zohoban)* [*Burma (Myanmar) Modern Political History (Updated)*] (Tokyo: Keiso Shobo, 1993), 12–14, 60–72; Ryuji Okudaira, "Rekishiteki haikei" [Historical Background], in *Motto shiritai Myanmar* [*Learning More about Myanmar*], 2nd ed., eds. Tsuneo Ayabe and Yoneo Ishii (Tokyo: Kobundo, 1994), 1–42.

66 Foreign Service Despatch from US Embassy Rangoon to Department of State, February 1, 1955, RG469, Entry 397, box 2, NACP.

67 Foreign Service Despatch from US Embassy Rangoon to Department of State, May 20, 1955, RG469, Entry 397, box 2, NACP. There is little information about Barthel's background, but his name is included in the proceedings of the Federal Council for Science and Technology under the Kennedy administration. Federal Council for Science and Technology, *Proceedings, First Symposium, Current Problems in the Management of Scientific Personnel, October 17–18, 1963* (Federal Council for Science and Technology, 1964), vii.

68 From Raymond T. Mayer to Dr. B. A. FitsGerals, October 28, 1958; From Richard E. Usher, US Embassy Rangoon to Department of State, July 13, 1956, RG469, Entry 397, box 22, NACP.

69 From U.S. Embassy in Rangoon to Secretary of State, December 28, 1957, RG469, Entry 397, FOIA case number NW54152, NACP.

70 From U.S. Embassy in Rangoon to the Department of State, August 21, 1957, RG469, Entry 397, FOIA case number NW54152, NACP.

71 From US Embassy Rangoon to Department of State, May 10, 1956; Foreign Service Despatch from US Embassy Rangoon to Department of State, March 12, 1956, RG469, Entry 397, box 22, NACP.

72 From McCaffery, ICA Burma to ICA, September 3, 1957; From Raymond T. Mayer to B. A. FitsGerals, October 28, 1957, RG469, Entry 397, box 22, NACP. Concerning the Phoenix Project, see the University of Michigan website, http://energy.umich.edu/about-us/phoenix-project.

73 For example, From US Embassy Rangoon to Department of State, May 6, 1957; From MacCaffery to ICA, October 18, 1957, RG469, Entry 397, box 22, NACP.

74 Mitsugi Kamiya, "Biruma-shiki shakaishugi to nogyo no hatten" [The Burmese Socialism and Agricultural Development], *Quarterly Journal of Agricultural Economy*, vol. 26, no. 4 (October 1972): 175–198; Center for Research and Development Strategy, Japan Science and Technology Agency, *ASEAN shokoku no kagakugijutsu josei* [*The Current Conditions of S&T in ASEAN Countries*] (Takamatsu: Bikosha, 2015), 258–275; MYANMAR JAPON Online (November 2015), https://myanmarjapon.com/1511interview.html.

4 International Students of Atomic Energy: the Argonne International School of Nuclear Science and Engineering

The Argonne National Laboratory was established on July 1, 1946. It took over the responsibilities of the wartime Metallurgical Laboratory at the University of Chicago, a major Manhattan Project laboratory (Figure 4.1). As Chapter 1 has shown, the Metallurgical Laboratory at the University of Chicago was the center of the so-called "scientists' movement," mostly comprised of former Manhattan Project scientists who demanded nuclear disarmament and the international control of nuclear energy. Some of these scientists were in direct confrontation with the U.S. government, which was trying to develop ever-more powerful nuclear weapons. At least in part to separate nuclear research from politically opinionated scientists, the U.S. government decided to move the center from the University of Chicago to Argonne and start anew as a state-sponsored enterprise. On March 14, 1955, the International School of Nuclear Science and Engineering (ISNSE) was established within the Argonne National Laboratory. Through the "Foreign Atoms for Peace" program, discussed in Chapter 3, the U.S. government had exported research reactors and nuclear fuels to more than 30 countries by the end of the 1950s and offered them technical training as well. Many of the recipients were developing countries that did not have trained engineers and administrators to operate and manage nuclear facilities. As the Argonne National Laboratory was especially strong in reactor science, the AEC requested the laboratory accept young foreign engineers and scientists, and give them training.

Subsequently, for the five years until the ISNSE was reorganized as the International Institute for Nuclear Science and Engineering (IINSE) in 1960, 420 scientists and engineers from 41 countries received training in the ISNSE. Essentially, it was a strategic study-abroad program to nurture a cohort of pro-American "science elite" in each country, who were familiar with U.S. nuclear technology. Although most of the trainees were not strictly "students," but rather starting-level professionals who had already received higher education in their home countries, they were often referred to as "students" in U.S. government documents. After spending about seven months in the ISNSE, some trainees continued further research, or enrolled in U.S. graduate schools to obtain advanced knowledge in nuclear science. The graduates later played essential roles in developing the nuclear power policies of their home countries.

DOI: 10.4324/9781003243649-5

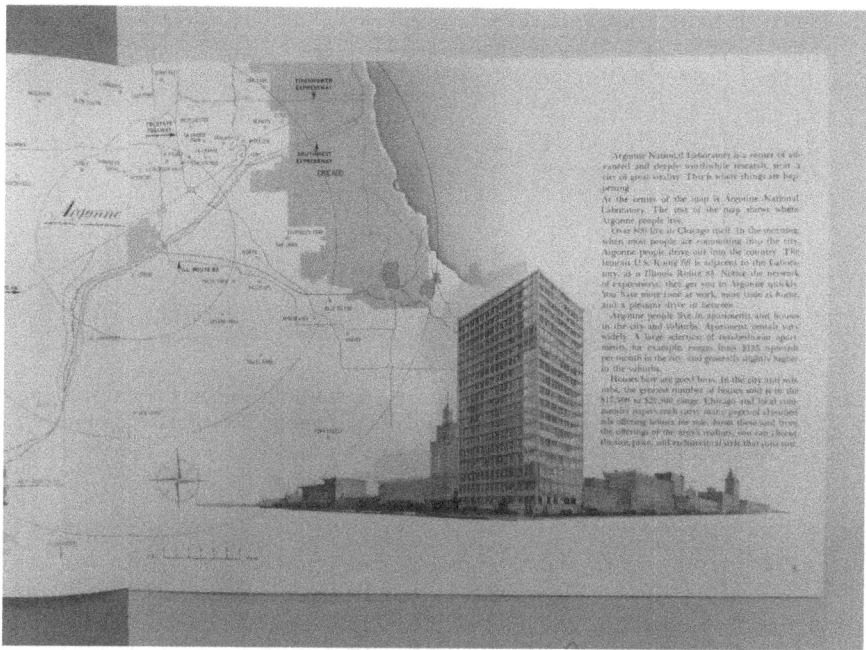

Figure 4.1 Brochure of the Argonne National Laboratory. RG326, National Archives at Chicago.

This chapter offers a glimpse of the personal experiences of young scientists and engineers who "studied abroad" in the U.S. in the late 1950s. They were important targets of the U.S. overseas information program. The government had divided the targets of the program on nuclear energy into three groups: science elite, other elite, and ordinary citizens (further information on this grouping is available in Chapter 6). This chapter focuses on the science elite, or "would be" science elite at the threshold of their careers. They shouldered heavy responsibilities and expectations from their own governments, and therefore the identities of the trainees were complicated. They had their own agency as independent scientists and engineers; however, they were also the target of the U.S. information program, and representatives of their home countries.

The U.S. government portrayed Argonne as the center from which trainees radiated U.S. technology to various parts of the world, as the laboratory's pamphlet eloquently told (Figures 4.2, 4.3). The U.S. government also viewed the ISNSE as the epitome of "scientific internationalism"—benevolently sharing scientific knowledge and skills with other countries. "Openness," in the U.S. government's view, was a unique characteristic of American science, in contrast to Communist science. However, "international students" at the ISNSE did not merely passively receive the intellectual offerings of the U.S. They had their own

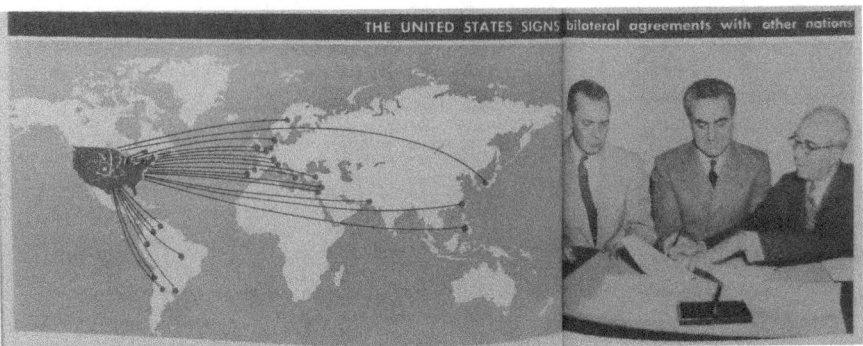

Figure 4.2 An image of American knowledge and technology radiates from Argonne to all corners of the world (Brochure of the Argonne National Laboratory). RG306, Entry A1 53, box 14, NACP.

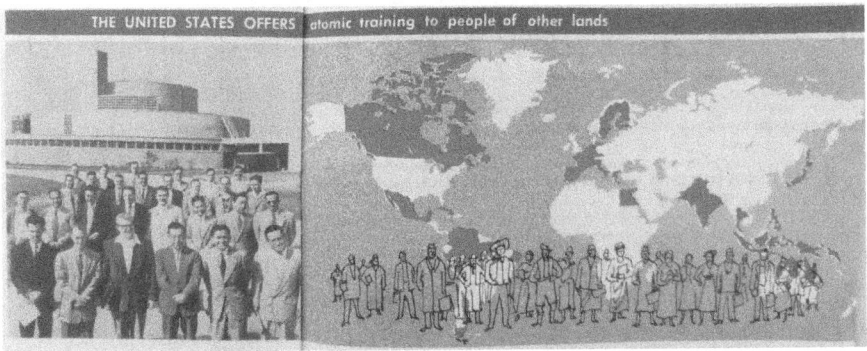

Figure 4.3 An image of American knowledge and technology radiates from Argonne to all corners of the world (Brochure of the Argonne National Laboratory). RG306, Entry A1 53, box 14, NACP.

agency as scientists and engineers, and had to meet the demands of their own governments. In fact, as Paul Kramer has argued, international students are often uncontrollable and disobedient precisely because they exercise their own agencies.[1] Although the U.S. government hoped to spread its technological hegemony among friendly nations, it was not a simple task.

During the same period, the U.K. and the Soviet Union also accepted foreign engineers to train them in their own nuclear technologies. The Atomic Energy Research Establishment (AERE) at Harwell, U.K., attracted young engineers, mostly from the Commonwealth countries, and the Soviet Union mostly accepted trainees from the People's Republic of China (hereafter, China) and East European countries. However, as the previous chapter has shown, the Soviet

Union also contacted Japanese socialist politicians concerning the possibilities of a bilateral agreement of cooperation. Therefore, the ISNSE was not the only option for foreign engineers, but one of several. Nevertheless, the ISNSE was a very attractive option, because it was a nuclear training school in the wealthiest country in the world, thus offering both high-quality living and research environments. Those who learned at the ISNSE became contact points for the U.S. government long after they returned home. Thus, in this respect, the U.S. government succeeded in cultivating a cohort of pro-American scientists in each country.

This chapter reveals the tension between scientific internationalism versus the Cultural Cold War by examining the multi-dimensional roles of the ISNSE, and the complex experiences of young scientists and engineers who studied there.

4.1 Establishment of the ISNSE

The first director of the Argonne National Laboratory was Walter H. Zinn, who was famous for his contribution to the designing of the CP-1 research reactor. Zinn divided the laboratory into basic research and pile development sections, and retained his position as director of the latter. Simultaneously, he appointed two associate directors, Harvard L. Hull and Norman Hilberry (Figure 4.4). Hull was an official with the Tennessee Eastman Corporation at Clinton during the war, and Hilberry had been Director Compton's "right-hand man" in the Metallurgical Laboratory.[2] When the AEC was established in January 1947, the Argonne National Laboratory was placed under its authority. The AEC designated that the Argonne National Laboratory was "to lead the nation in the research and design of nuclear reactors," and publicized this role on January 1, 1948. Although the AEC's decision troubled some scientists in relation to the future of the laboratory's basic research, Zinn regarded the news as good, as he was an expert on reactors, and especially interested in the development of breeder reactors.[3]

The growing importance of reactor work at Argonne resulted in heightened security, making free collaboration with universities and other research institutions difficult. Especially when the Canadian government revealed a Soviet spy network around British nuclear physicist Alan Nunn May, the security clearances for scientists became even more intensive. In contradiction to the increased security, the AEC instructed that Argonne's facilities were "to be made available for the training of students and staff members in the supporting institutions in the Chicago Area," and thus, Argonne also became a training center. In order to gather the various laboratories and research institutes scattered around the Chicago area at one site, the AEC tried to purchase a vast area of land in Du Page County, approximately 45 miles southeast of Chicago. However, as resistance from residents and landowners delayed construction, it was only in 1949 that the Argonne National Laboratory was finally built at the Du Page County site.[4]

The Atomic Energy Act of 1954 de-classified some of the nuclear technological information, enabling the Argonne National Laboratory to carry out

Figure 4.4 Norman Hilberry, Director of the ISNSE. RG326, National Archives at
    Chicago.

collaborative research projects with university and industry researchers.[5] The laboratory established an attached "reactor school," where engineers from private
companies such as General Electric Company and Westinghouse Corporation
received practical training on reactors. At that time, with the exception of a few
Belgians, no foreign national was admitted. This exception was made because
the U.S. relied on the uranium supply from the Belgian Congo, and therefore
concluded a bilateral agreement with Belgium in 1954. President Eisenhower's
"Foreign Atoms for Peace" policy changed this situation. The National Security
Council's policy paper, NSC-5431, titled "Cooperation with Other Countries
on the Peaceful Use of Atomic Energy" (August 1954), corresponding to
President Eisenhower's "Atoms for Peace" address at the U.N. General Assembly
(December 1953), proposed to offer "aid in construction of small-scale reactors," "training programs, provision of technical information, and consulting ser-
vices, to aid other countries in building their capability to use atomic energy."[6]

To implement this, the U.S. government needed to establish a training program for foreign engineers: the Department of State, the AEC, and the Operations Coordination Board (OCB) discussed the matter, and their answer was the ISNSE. As Argonne was already offering technical training to U.S. engineers and a small number of Belgian engineers, it seemed an ideal place to accommodate the ISNSE. In August 1954, J. J. Flaherty, head of the AEC's Chicago Operation Office, asked Zinn if Argonne National Laboratory was "interested in offering a training course for foreign reactor technologists." Zinn replied that he and his staff had "come to the conclusion that it [was] impossible and unthinkable" that the laboratory would *not* be interested in such an undertaking. The interaction between Flaherty and Zinn reveals that the Argonne National Laboratory had become an important research and training center of nuclear reactors in the U.S. Foreign Atoms for Peace project.[7]

The Foreign Operation Administration (FOA), the predecessor of the International Cooperation Agency (ICA), drafted the blueprint of a nuclear technological training course for foreign engineers. The FOA enjoyed a substantial degree of freedom and independence in foreign aid before it was disbanded in May 1955, and its successor, the ICA, assumed many of the FOA's functions albeit under the supervision of the Department of State. According to a memorandum prepared by Harold Stassen, director of the FOA, his organization prepared the first draft of the training program for foreign engineers, and requested comments from the Department of State, the OCB, and the AEC. The FOA's draft proposed the following:

> Through our missions in the underdeveloped countries, we will invite outstanding students with the proper scientific aptitudes to come to the United States for two years of study at one of the universities now cooperating in the Atomic Energy Commission's program through its laboratories, and upon the successful completion of two years, then to take the Atomic Energy Commission's one-year course in the peaceful use of atomic energy.[8]

In response to the FOA draft, an OCB member commented, for example, "FOA thinking embraced too many countries," and "that a regional approach would be better." A Department of State representative disagreed with the OCB's "regional" approach and argued that "a program such as this could only be carried out on a country-by-country basis." Although he understood that "nationals from some countries might need more training in basic subjects than those from the more developed countries," he believed that "the cost of the basic training should be borne by the various foreign countries." "The US should not undertake to educate the world at the taxpayer's expense," he noted. After discussing and synthesizing various comments, the blueprint was finalized and the AEC officially established the seven-month training course at the ISNSE.[9]

Although Argonne had already established a "reactor school" for U.S. engineers, accepting foreign engineers was quite a different endeavor. Because of security concerns, both new housing for these foreigners and "zoning" of lab facilities were necessary. Zinn appointed Hilberry as director of the ISNSE and charged him with transforming the "reactor school" into an international school.[10] Hilberry organized a "committee to evaluate the problem of a curriculum for the ISNSE." The committee prepared recommendations "upon the laboratory's experience in handling the training program for the Belgians," and "its experience in the training of some 150 DuPont employees," destined for service at a nuclear weapons factory at Savannah River, South Carolina. The committee also gleaned information from more than 100 foreign visitors to Argonne in order to understand the need for nuclear technologies in their countries. The committee knew from the experience with the Belgians that "all materials made available by the declassification program [were] promptly acquired and mastered by those aliens interested in this field." This was particularly true in the field of reactor physics. It would provide "a gratuitous insult," therefore, to "present an unclassified course in reactor physics to alien physicists already conversant in the field," the committee suggested. The committee recommended that the training course should also include the "gray areas"—the border zone between classified and unclassified information. Especially as Harwell Laboratory in the U.K. had just announced a "gray zone" training course for foreigners, Argonne could not afford to make their training course less attractive than that of its rival.[11]

The stated purpose of ISNSE was "to share America's scientific and technical knowledge of the peaceful applications of nuclear energy with friendly nations," and to let "engineers and scientists representing the various countries of the Free World" "join representatives of United States industry" in studying and discussing the latest developments in nuclear technology. The concrete themes covered by the training course would be the:

1. Production of reactor materials
2. Manufacture of satisfactory reactor components
3. Design, construction, and operation of nuclear power plants
4. Processing of all types of irradiated materials from nuclear reactors
5. Non-power applications of nuclear reactors
6. Utilization of radioactive products[12]

The trainees were mostly foreign nationals from countries with which the U.S. had bilateral agreements, and all needed to "pass screening by the State Department and a satisfactory file check by the FBI."[13] The annual operating cost was estimated to be 246,000 US dollars, of which 200,000 dollars (2,500 × 80 students) was covered by tuition, with the remaining 46,000 dollars paid for by the AEC. Logistical details for accepting the trainees were the responsibility of the ICA. According to an ICA document (CA-4095, December 23, 1954) signed by the Secretary of State, John Foster Dulles, the purpose of the ISNSE was "to train a cohort of qualified foreign nuclear engineers within a certain period of time so that the countries willing to receive U.S. assistance with building research reactors could secure skilled staff for operating them." In addition, the document continued, as foreign and American engineers received training together, the ISNSE would provide an ideal opportunity for "young engineers from the Free World countries and engineers in the American industry to learn from each other in their ways of thinking and doing things."[14] This document makes it clear that the training at the ISNSE was closely related to the exportation of nuclear reactors to the trainees' home countries. In short, the ISNSE was a scheme to cultivate foreign engineers who could operate U.S. nuclear reactors, and to immerse those engineers in U.S. industrial culture: the ISNSE trained young foreign engineers to absorb not only technological know-how but also the culture surrounding nuclear research and business.

The ISNSE officially opened its doors on March 14, 1955, and in 1960, it was reorganized as the IINSE. The 420 engineers and scientists from 41 countries who received training during that time eventually played important roles in the nuclear policies of each country. Therefore, from the U.S. government's perspective, the ISNSE was an overseas information program that would pour U.S. technology and philosophy directly into the minds of the foreign science elite. Simultaneously, as explained in detail later in this chapter, the process of training itself was widely propagated through TV, radio, films, and so on. In this sense, the ISNSE was a double-layered information program, a psychological campaign to win the hearts and minds of both foreign engineers as well as those who would follow their activities through media.

The first class of ISNSE (commenced in March 1955) was comprised of 31 foreign engineers and scientists from 19 countries and 10 engineers from U.S. companies. The countries of origin were Argentina, Australia, Belgium, Brazil, Egypt, France, Greece, Guatemala, Indonesia, Israel, Japan, Mexico, Pakistan, Philippines, Portugal, Spain, Sweden, Switzerland, and Thailand. They received four months of classroom lectures and three months of practical training (Figure 4.5). The training focused on unclassified information on the design, construction, and operation of the CP-5 nuclear reactor. The trainees stopped at Washington DC before proceeding to Chicago, and participated in a reception sponsored by the Department of State, luncheon with the Joint Congressional Committee on Atomic Energy members, and a meeting with President Eisenhower. The president told them, "You represent a positive accomplishment in the Free World's efforts to mobilize its atomic resources for peaceful uses and the benefit of mankind," and thanked them for this "heartening sign that we are making progress toward real international cooperation."[15]

The third class (commenced in April 1956) was comprised of 66 foreign engineers and scientists from 24 countries, and 19 engineers from U.S. companies and the AEC. From the third class, trainees took basic training courses at North Carolina State University or Pennsylvania State University before moving to Argonne; both these universities had introduced research reactors. After moving to Argonne, the trainees learned not only about reactors but also about the synthesis and the re-processing of nuclear fuels. They also mastered the handling of radioactive substances safely in a facility called "caves." A highlight of the training was an independent research project planned and carried out by each trainee using the CP-5 reactor. The trainees also visited the Oakridge National

At nine o'clock on Monday, March 14, students of   classroom in Building 203 for their first orientation ses-
the School of Nuclear Science and Engineering met in a   sion, an address by Dr. Stuart McLain.

Figure 4.5 The first cohort of ISNSE trainees in a classroom. RG326, National
Archives at Chicago.

Laboratory, the University of Michigan's Nuclear Laboratory, and Shippingport
Pressurized Water Reactor. The training course ended with a tour to the National
Reactor Testing Station in Idaho and other nuclear facilities in the West and the
graduation ceremony[16] (Figure 4.6).

4.2 Experiences of International Trainees

To examine the experiences of foreign trainees, this section will first focus on
an exemplary case of Japanese trainees, largely because of the author's literacy
in the Japanese-language sources, but also because Japanese scientists and engineers were represented in every class. They came from Japanese universities, the
Japan Atomic Energy Research Institute (JAER), and private companies such as
Hitachi and Mitsubishi. The Japanese government covered their travel expenses,
and split accommodation and living expenses with the ICA, who also made all
logistical arrangements in the U.S. In 1957, the ISNSE admitted "self-paid students," thereby increasing the number of engineers sponsored by Japanese private companies.[17] Of course, this change was not only made for Japan, but in
recruitment for all developed countries where the industries were interested in
nuclear business.

Four Japanese participants in the second class (Shigefumi Tamiya, Susumu
Suguri, Kin'ichi Torikai, and Kiyoaki Taketani) arrived in New York on October
28, 1955, via Honolulu and San Francisco, and took a train to Washington DC.

Figure 4.6 A graduation ceremony of the ISNSE. RG326, National Archives at Chicago.

After an orientation session in Washington, they moved to Chicago. Taketani, a young scientist from the University of Tokyo, contributed a series of essays to *Atomic Industrial News,* an industrial newspaper issued by the Japanese Atomic Industrial Forum, titled "Letters from the U.S." According to Taketani, the Japanese trainees entered the U.S. with international exchange visas endorsed by the ICA. In the orientation session, they received lectures on American customs, geography, history, politics, education, principle of freedom, and the "Negro problem" [sic]. They also watched some movies. When he moved to Argonne, Taketani was very impressed with the "kind and thorough teaching" and the modern "study room where the temperature was comfortably set at 75 degrees." The facilities were so modern that he felt it was "far from the traditional image of schools."[18] His report indicates that the trainees were impressed with not only American technology but also the modern lifestyle. Their training was divided into "basic training" at either Pennsylvania State University or the University of North Carolina and "practical training" at Argonne, each lasting 17 weeks. In between these two courses, a four-week tour was provided to various facilities, such as Oak Ridge National Laboratory, uranium factories, power plants, and the Tennessee Valley Authority (TVA).

As mentioned, the Department of State and the FBI carried out the security clearance of foreign trainees when they applied. Before the Japanese trainees were selected by the Japanese government, the U.S. Embassy in Tokyo had told the Japanese Ministry of Foreign Affairs that, "from the security point of view, the trainees must be free from political and ideological colors."[19] It is logical to assume that a similar notification was sent to the government of every participating country. It seems that foreign trainees at ISNSE had to be "safe" individuals, uncritical of the U.S. government or its policies. However, such criteria were not reserved only for foreigners. As in the Oppenheimer case, discussed in the previous chapter, even Americans who had access to nuclear information were to be completely loyal to the government and dissent was not tolerated.

The experiences of the foreign trainees were also reported in the Argonne National Laboratory's monthly newsletter, *Argonne National Laboratory Bulletin News*. At the ISNSE, the main nuclear reactor used for training was the "Argonaut" (short for Argonne's Nuclear Assembly for University Training). After studying the structure of nuclear reactors using the Argonaut, each trainee established his own research question, analyzed data collected from the Argonaut, and tested his hypothesis. Figure 4.7 shows a trainee from Taiwan, Rudi Yang, examining a fuel rod of the Argonaut. Yang would later become a contact person for U.S. nuclear technological aid to Taiwan and played an important role in the establishment and management of the Nuclear Research Laboratory at National Tsing Hua University.[20]

In addition to the Argonaut, an Experimental Boiling Water Reactor (EBWR) was introduced at the Argonne National Laboratory on February 9, 1957. The EBWR was mainly used for experimenting with nuclear power generation. Moreover, as the first two reactors at the site, CP-2 and CP-3, were to retire in May 1954, CP-5 commenced operations in February 1954. CP-5 used enriched uranium and heavy water to produce radiation. Furthermore, in January 1962, the Jaganaut, similar to the Argonaut but more powerful and versatile, was introduced. All these reactors were used to train foreign engineers.[21]

In one respect, the ISNSE truly represented scientific internationalism, by establishing a cosmopolitan community of trainees from various quarters of the world, nurturing friendship across national and cultural borders. The *Argonne National Laboratory Bulletin News* comprised "domestic" and "international" editions—the international edition looked almost like an alumni magazine, as it reported not only on the training but also on the graduates' whereabouts, reunions, and even their marriages and the birth of babies. The pages of the newsletters contained many photos. For example, Figure 4.8 is a snapshot from a reunion in Geneva, at the Second International Conference on Peaceful Application of Atomic Energy, held from September 1 to 13, 1958. Invitation letters were sent to graduates of the ISNSE, and 97 alumni from 27 countries participated in the reunion. Many alumni participated in the conference, and they bumped into each other "on the streets, in the hotels, in the cafes and cafeterias, and at the technical sessions," renewing old friendships. The Argonaut reactor installed in the conference site was their meeting place. The "Big All-Class meeting" was held

Rudi Yang, ISNSE former student from Formosa, holds one of the seventeen-plate fuel elements for the enriched uranium-water moderated Argonaut training reactor.

Figure 4.7 Rudi Yang, an ISNSE trainee from Taiwan. RG326, National Archives at Chicago.

onboard the S.S. Simplon cruising Lake Geneva and in the dining parlor in the hotel on the northwest shore of the lake. Hilberry, former head of the ISNSE and current head of Argonne National Laboratory, Rollin G. Taecker, new head of the ISNSE, and Congressman Sterling Cole, head of the International Atomic Energy Agency (IAEA), participated in the reunion.[22]

Newsletters also featured foreign engineers who were currently enrolled at the ISNSE. For example, the April 1959 issue featured two trainees from the eighth class on the front cover (Figure 4.9). The photo shows Kuang-Hsin Teng from Taiwan and Soetarjo Soepadi from Indonesia visiting the greenhouse and listening to John Skok explain the work of the Plant Physiology Group of the Division of Biological and Medical Research.[23] Every issue featured the everyday life and families of foreign trainees and graduates. For example, the same issue carried a photo of the happy family reunion of a fourth-class participant from Turkey. He graduated from the ISNSE, stayed seven more months in Argonne, and was

Figure 4.8 The ISNSE reunion at the 2nd U.N. Conference on the Peaceful Uses of
Atomic Energy in 1958. RG326, National Archives at Chicago.

enrolled in the University of Michigan's doctoral program when he was suddenly
recalled by the Turkish government, where he had to spend eight months living separately from his wife and children, who remained in the U.S. The article
states, however, that he was later appointed as a researcher at Argonne's reactor
engineering division on a one-year contract, and was reunited with his family[24]
(Figure 4.10).

As most trainees were young scientists and engineers from their late 20s to
30s, many began families soon after returning to their home countries. As mentioned above, the newsletter often carried articles announcing their marriages and
the birth of children; Figures 4.11 and 4.12 are examples of the wedding photos
that accompanied such newsletter articles.[25]

Graduates who took on especially important positions in nuclear development
in their native countries were featured in the "Distinguished Alumni" series. One
example was U Hla Nyunt from Burma, introduced in Chapter 2, and Bernardino
Pontes, Assistant to the Chairman of the Brazilian Atomic Energy Commission
in charge of Training and Education, was another.[26] Florencio Medina, a sec-
ond-class participant from the Philippines, was introduced as chair of the Atomic
Energy Commission in his country.[27] Among Japanese alumni, Yoshinori Ihara
from the first class, Shigefumi Tamiya from the second class, and Ryohei Kiyose
from the third class were featured in the "Distinguished Alumni" series. The
article reported on Ihara's contribution to the establishment of the Japan Atomic
Energy Research Institute, and his generosity in leading the Tokyo Argonne
Club, which organized welcome parties every time Argonne staff visited Japan.
As for Tamiya, the article described his important role as a "science attaché" at
the Japanese Embassy in London, and his remark in an interview that the most
precious experience he gained in Argonne was the "comprehensive knowledge of

ARGONNE
NATIONAL LABORATORY

NEWS-BULLETIN
International

Volume 1 Number 2   APRIL 1959

Figure 4.9 A scene of training on the front cover of the *News-Bulletin.* RG326,
    National Archives at Chicago.

nuclear engineering" and his "good friendship with scientists and engineers from
all parts of the world." Kiyose was introduced as a promising young faculty member at the University of Tokyo and the lecturer of a study group of the Japanese
Atomic Industrial Forum[28] (Figure 4.13).

Furthermore, the friendship among trainees across national and cultural borders was a favorite topic of the newsletter. In the January 1962 issue, engineers

Reunited After Eight Anxious Months

Ziya Akcasu, of Turkey, pictured at left with his wife, Melahat, 4-year-old daughter, Nur, and 9-month-old son, Feza, is back in the U. S. for a year of uninterrupted residence. After graduating from the International School's fourth session, spending another seven months at Argonne, and enrolling at the University of Michigan for Ph.D. work, he was recalled by the Turkish government. Now Ziya has a one-year appointment as Resident Research Associate in the Argonne's Reactor Engineering Division. After eight months he returned to the United States and was able to get a first look at his son.

Figure 4.10 A trainee from Turkey reuniting with his family. RG326, National Archives at Chicago.

Figure 4.11 The news of alumni's weddings. RG326, National Archives at Chicago.

from Japan, Taiwan, and Thailand were shown side by side in a graduation photo[29] (Figure 4.14). The July 1962 issue also carried a photo of foreign trainees enjoying a Chicago White Sox ball game; it named Adriaan and Anetjie du Plessis from South Africa and Hiroshi Hashimoto from Japan[30] (Figure 4.15).

Figure 4.12 The news of alumni's weddings. RG326, National Archives at Chicago.

As these examples show, the ISNSE became the frontline of scientific internationalism, where young scientists and engineers learned together and developed friendship. However, it is difficult to evaluate the degree to which such scientific internationalism helped nurture pro-U.S. feelings and disseminate American technology around the world. As introduced in the beginning of this chapter, Paul Kramer argued that while the U.S. policy for international students in the 1950s and 1960s served the "geopolitical" purposes of pursuing U.S. national interest, international students were not simply "agents" of the U.S. government because they had autonomous agency.[31] In addition, although science and technology contribute to the establishment of social hierarchy, they can also become tools to disrupt such hierarchy[32]—both of these aspects seem to apply to the ISNSE.

The foreign trainees participating in ISNSE had their own goals, and their governments also had their own objectives. Therefore, the U.S. government could not dictate who should participate nor what they should learn from their experiences. For example, while the U.S. government hoped to accept young engineers aged 22 to 25, the average age of the trainees was about 35. The U.S. government wanted to instill unforgettable experiences in young and flexible minds, while other countries sent scientists and engineers who already had some professional experience and who were expected to apply their newly acquired knowledge as soon as they returned home. Those countries did not only passively respond to the U.S. invitation to send trainees. Especially in countries where scientists already had basic knowledge of nuclear science, the governments decided where to send young scientists. For example, in 1955, the Japanese

Figure 4.13 Ryohei Kiyose introduced in a "Distinguished Alumni" article. RG326, National Archives at Chicago.

government collected information on the possibilities of nuclear technological training not only in the U.S. but also in the U.K., France, Sweden, Norway, Canada, Switzerland, and West Germany. In 1956, the Japanese Diet approved budgets for the travel and living expenses (120,000 U.S. dollars) and tuition (45,000 U.S. dollars) of 22 trainees who were sent to the U.S., 5 to the U.K., 2 each to Norway and Sweden respectively, and 1 each to Canada and France. As for the 22 trainees to be sent to the U.S., the Japanese government made specific requests about the enrolling institutions (National Laboratories such as Argonne, Oak Ridge, and universities such as Michigan and MIT), and the research fields they wanted the trainees to specialize in. Such planning was possible because Japanese scientists already had substantial knowledge about nuclear science, and had maintained overseas contacts from the prewar years.[33]

In addition, although the U.S. government presupposed concluding bilateral agreements and exporting research reactors to the native countries of the ISNSE trainees, the trainees themselves did not necessarily view this partnership with the U.S. as the only option. One Japanese trainee recollected a conversation among the trainees during a coffee break: a foreign engineer told him, "Japan already has high scientific and industrial standards, so you do not have to conclude a bilateral agreement with the U.S. Your country can develop nuclear energy on its own,

Figure 4.14 An article introducing cross-national friendship of Asian trainees. RG326, National Archives at Chicago.

Figure 4.15 International trainees enjoying a White Sox game. RG326, National Archives at Chicago.

and that would bring better results." Another foreign trainee joined in the conversation and said, "But atomic energy development needs huge capital. There is no choice other than to cooperate with the U.S." The foreign trainees had a heated debate over whether it was a good idea to rely on U.S. technology.[34] In fact, Japanese trainees were not interested only in U.S. technology; many wanted to travel to Europe and visit nuclear facilities there after finishing the ISNSE course. As the purpose of ISNSE was the dissemination of American technology, the U.S. government was reluctant to benefit its European rivals. However, as rejecting the requests of Japanese engineers might go against the spirit of scientific internationalism the U.S. was advocating, the U.S. government ultimately allowed them to go to Europe on the condition that the Japanese government funded all travel expenses and logistics.[35]

Another issue was the complaints and inconveniences resulting from the wide disparity in educational standards among the participants. For scientists and engineers from advanced countries, the basic training at Pennsylvania State University and North Carolina University was tedious, because they could receive (or had already received) similar education in their home countries. Furthermore, many of the American university faculties had not yet handled nuclear reactors, and therefore, their lectures were more theoretical than practical. Such lectures were disappointing to foreign scientists and engineers who had expected more practical training. In the words of an ICA evaluator, the trainees were divided into the "big shots" and "underdogs." Trainees from advanced countries could digest the classes' content completely, while those from developing countries often had difficulties resulting from insufficient training in their home countries. Trainees from Germany, Iraq, Norway, Belgium, Israel, and parts of France, Japan, Italy, and Turkey were categorized as "big shots," while those from Burma, Pakistan, Thailand, Egypt, Spain, and the Philippines were "underdogs," according to the ICA evaluator. U Hla Nyunt, the Burmese physicist introduced in Chapter 3, said in the group interview held on the final day of training that Burma was "setting up from nothing a little atomic energy center" and needed to learn "pertinent basic facts" and "fundamental things." To this, a nuclear physicist from Iraq made a condescending comment that he did not understand what Nyunt was talking about. A scientist from East Pakistan (present-day Bangladesh) sympathized with Nyunt and other developing countries like Burma, while a German scientist made a cynical comment, saying, "You can buy a whole research reactor in one package complete from General Electric with engineers to set it up for you" even in a country without any skilled engineers. However, regardless of whether they were "big shots" or "underdogs," the trainees all agreed that there was "no laboratory and no school in the world like Argonne, nor was there any university in the world with the facilities and staff to equal the practical research at Argonne." They "deeply appreciated" the U.S. gesture and thought that "U.S. aid in the field in the design and fueling of research reactors was imperative" in developing countries. As this comment shows, to the scientists and engineers from countries aiming to embark on nuclear power development, the experience at ISNSE was extremely important.[36]

Interestingly, while many trainees held good impressions of the American people they encountered, they felt that the U.S. government was not very successful in propagating the greatness of American science. "Ordinary citizens strained themselves to make the visitors feel at home, as did American students in the ISNSE course, and the staff of the universities and Argonne," they said. However, "from the point of view of scientific prestige Americans did not make the most of the opportunities," they felt. This impression partly came from their lack of opportunities to meet with "the great men in nuclear science" they had known.[37] Perhaps as their image of American science had been spangled with famous nuclear scientists such as Albert Einstein, Enrico Fermi, or Leo Szilard, they had expected to see those heroes in person when they visited the U.S. If so, the practical training they received at the ISNSE did not match their image of American science. Still, their favorable impressions of American people were evidence of the success of the overseas information program as the U.S. government had intended.

According to follow-up surveys by the ICA, almost all graduates of the ISNSE were involved in the nuclear policies of their native countries, indicating that the U.S. government had indeed selected the "right" people to instill with favorable impressions about the U.S. In Japan's case, Ihara, from the first class, contributed to the drafting of the Japanese first "Long-Term Development Policy for Nuclear Power" and assumed important government positions such as Deputy Director of the Atomic Energy Division, Science and Technology Agency, Director of the Atomic Energy Safety Agency, and Acting Chair of the Japan Atomic Energy Commission, and ultimately, became Assistant Minister of Science and Technology in 1979. Tamiya, from the second class, served the Japanese Embassy in London as science attaché, and was appointed Director of the Atomic Energy Division, Science and Technology Agency, in 1973. Suguri, also in the second class, became Deputy Director of JAERI's Tokai Laboratory, and Councilor at the Atomic Energy Safety Analysis Office. Taketani became a chief scientist at JAERI, and served on the Nuclear Reactor Safety Commission. Inoue, from the third class, among other roles served as the Head of Nuclear Power Division, Ministry of International Trade and Industry (MITI), Head of Power Reactor Development Division, Science and Technology Agency, Deputy Director of the Resource and Energy Agency of MITI, and Director of J-Power Company. While the knowledge and experiences gained at ISNSE were applied to the nuclear policies of their countries in differing degrees, as these examples show, the ISNSE became the career starting point for the science elite.

However, the sharing of nuclear technology with many countries later led to concerns regarding nuclear proliferation. Among those countries that sent engineers to the ISNSE, France already possessed nuclear weapons, and Pakistan, Israel, Taiwan, and Korea were pursuing nuclear armament. Among them, only Pakistan and Israel actually developed nuclear weapons (China and India succeeded in nuclear explosions in 1964 and 1974, respectively, but they did not conclude bilateral agreements with the U.S.; China did not send trainees to the ISNSE, and India did so only sporadically). It is believed today that the nuclear technological aid during the 1950s helped little in developing nuclear weapons because processing uranium into weapons-grade purity required exceedingly complex technology. However, the concern surrounding future nuclear proliferation came to annoy the U.S., especially after China's successful development of nuclear weapons, and it pushed the U.S. government to conclude the Treaty on the Non-Proliferation of Nuclear Weapons (or the Non-Proliferation Treaty, NPT) in 1968. If U.S. nuclear technological assistance had stirred the ambitions (if not the abilities) of foreign leaders for nuclear armament, the ISNSE had thus worked against the U.S. national interest.

On February 3, 1960, the ISNSE was reorganized into the IINSE, partly because there were demands for more advanced and highly specialized training, and partly because many countries had already developed domestic training programs. However, the IINSE continued training programs for foreign engineers in new forms. Those who held master's degrees and had some knowledge on nuclear science paid tuition ($1,000) and received six weeks to two semesters of training as "Participants." Those who held doctoral degrees or the equivalent did not have to pay any tuition but were expected to make scientific contributions to the Argonne National Laboratory as "Affiliates."[38] However, around 1963, as the number of foreign trainees began to decrease, the IISNE began to put more emphasis on collaboration with university laboratories within the U.S. According to the IISNE, the enrollment of foreign nationals declined because, first, industrialized countries had already developed "well organized and well supported training centers of their own," although the ISNSE had played a significant part in the "training of key personnel in many of these centers." Second, the IAEA was "taking increased advantage of the training centers abroad." Third, some foreign governments had shifted their emphasis away from nuclear energy to other fields of technology. Finally, U.S. universities had developed capacities of offering instruction in various fields of nuclear science, so that national laboratories were no longer unique in offering training courses.[39]

4.3 Overseas Information Programs Using International Trainees

The ISNSE and foreign trainees were used as the subject matter of the U.S. overseas information program. In other words, the ISNSE functioned both as an overseas technological aid program and an overseas information program. Government-sponsored media such as the USIS films and VOA radio portrayed them as the symbol of U.S. scientific internationalism, and even during the planning stage of the ISNSE, the AEC had "anticipate[d] requests for fullest exploitation by USIA," and predicted that especially the "first group of international students" would face "a heavy load of filming, recording, picture taking and interview[s]."[40] This "heavy load" media attention did in fact take place. For example, Figure 4.16 shows filming for a famous TV program *See It Now*, anchored by Edward Murrow, a prominent journalist who was later appointed as director of the USIA by President Kennedy. It is striking that he was already contributing to the government information program by reporting on ISNSE when he was not yet affiliated with the USIA. The *See It Now* program featuring the ISNSE was broadcast on April 5, 1955.[41]

While *See It Now* was a domestic TV program, the ISNSE was also introduced overseas through USIA-sponsored media such as USIS films and VOA radio. *Training Men for the Atomic Age*, a USIS film released in February 1958, featured a cohort of foreign trainees participating in the fall 1956 class of the ISNSE. It was a 15-minute black and white documentary, prepared both as a 35 mm theater film and 16 mm mobile projector film, and was also broadcast on TV. The U.S. National Archives and Records Administration (NARA) in College Park, Maryland, stores an English-language "test print" and the film's script.

The film opened with a close-up of a nuclear reactor, and accompanying narration about how it was a helpful device for the welfare of human beings.

> This is a nuclear reactor, a machine of the Atomic Age. Inside of it, a controlled chain reaction of splitting atoms can take place. It is a strange machine, for it makes no noise and you can see no moving parts. All around the heavily shielded pile are scientific instruments, for this reactor is primarily for research. But one thing is true of all reactors. They can liberate atomic energy for man's good.[42]

Figure 4.16 Filming for the TV program *See It Now*. RG326, National Archives at Chicago.

After explaining the applications of nuclear energy in medicine, industry, and power generation, the film pointed out "an acute shortage in the world ... of specially trained men and women to work with the atom and develop its peaceful uses." "Training men for the atomic age," according to the film, was "an international challenge," and thus should be solved "through international cooperation." The film then conveyed the fact that the ISNSE was established for this purpose. Students of the ISNSE were characterized as "mature and intelligent" engineers who already had substantial scientific knowledge before joining the ISNSE, but they were expected to study "in many fields in addition to the one that was their specialty." The camera zoomed in on individual trainees.

> Let's meet a few of them. Bergua of Spain is a chemical engineer. Srinivasan is a chemical engineer from India. Azad of Iran is a skilled physicist. Chemistry is the specialty of Ishihara of Japan. Saffioti earned his doctorate of chemistry in Brazil. Cheng of China is an electrical engineer. So is Akcasu of Turkey. Diederichs of Germany is a mechanical engineer. Lascaris of Greece is a specialist in physics and electronics.[43]

Study hours were long, and even during coffee breaks, the students helped each other to better understand the classes' content. In doing so, they nurtured genuine friendships. The film showed the trainees receiving basic training at the university, visiting Oak Ridge National Laboratory and Shippingport Nuclear Power Plant (still under construction), and finally embarking on the practical training in Argonne. On one day, students were grouped into several teams to manufacture uranium fuel plates. On another occasion, the students experimented with the methods of reprocessing spent fuel to lower the cost of power generation. They also learned how to use gamma rays to destroy bacteria, and how best to transfer heat from a reactor to an electronic generator. The film further focused on the Argonaut, a pioneer reactor for educational purposes which was "an extremely valuable instrument" for the students, who would go on to teach others when they returned to their countries. Argonne's largest research reactor, CP-5, was also introduced.

The film emphasized that the students would "spread their knowledge to others in ever widening circles." This was exactly the image portrayed in the Argonne National Laboratory's pamphlet shown in Figures 4.2 and 4.3. The films cited concrete examples of such "widening circles" of knowledge as follows:

> In Tokyo, Japan, Mr. Susumu Suguri is at work in the Electro-Technical laboratory teaching other students and researchers. At Santiago, Dario Moreno is teaching at the University of Chile's School of Engineering. His subject: the practical application of atomic studies in the field of nuclear power. At a laboratory near Oslo, where Norwegian and Dutch scientists are working together, Koren Lund, another former student of the International School, explains a pilot plant for reprocessing uranium. In Milan, Italy, is nuclear engineer, Lorenzo Roseo. He too returned recently from the Argonne

Laboratory and is leading a new project at the Italian center for nuclear research. In Buenos Aires, engineer and physicist Ernesto Schonfeld is teaching men and women of Argentina ... In Cairo, Egypt, Mikhail Saad and Effat Kamal are two more graduates of the International School. They, too, carry on the same tradition of disseminating knowledge in the field of atomic energy.

These ISNSE graduates were "on their way to becoming leaders in nuclear science," and they were "bound together in a kind of United Nations of science," the film emphasized. In the USIA's view, scientific internationalism and the spreading of U.S. technologies did not contradict, but supported each other. In this way, the ISNSE served as a dual information program, targeting the science elite on the one hand, and the foreign audiences of the USIS films (and VOA radio) on the other. The ISNSE was an extraordinary "school" that was placed on the frontline of the Cultural Cold War. It was a device to propagate scientific internationalism as a unique feature of the United States. However, as mentioned throughout this chapter, it is also important to understand the agencies of the trainees and their home countries' governments. The individual scientists and engineers took advantage of the opportunity to study abroad, cultivate their careers, and build international networks, and the governments of their home counties used the trainees' knowledge for modernization and nuclearization. This agency notwithstanding, the knowledge and skills offered by the U.S. government filtrated into key policies of each country as the former ISNSE trainees became involved in policy-making in education, science, and technology. The U.S. succeeded in drawing these countries closer to the U.S.-centered circle of science and technology. Although the life span of the ISNSE was short, it played a role that cannot be ignored in the scientific theater of the Cultural Cold War.

Notes

1 Paul A. Kramer, "Is the World Our Campus? International Students and U.S. Global Power in the Long Twentieth Century," *Diplomatic History*, vol. 33, no. 5 (November 2009): 775–806.
2 Holl, *Argonne*, 51.
3 Holl, 60–63.
4 Holl, 62–66.
5 Holl, 129.
6 Document 238, "National Security Council Report, NSC5431/1, Statement of Policy by the National Security Council on Cooperation with Other Nations in the Peaceful Uses of Atomic Energy."
7 "Appendix A, Background and Discussion," from Hilberry to Flaherty, September 30, 1954, RG326, ANL Miscellaneous Correspondence & Reports, box 17, National Archives at Chicago.
8 From Harold E. Stassen to Lewis L. Strauss, September 13, 1954, RG59, Entry A1 3008-A, box 401, NACP.

9 From Philip J. Farley to George, October 26, 1954; From Special Assistant to the Secretary to GCS, PJF, October 30, 1954, RG59, Entry A1 3008-A, box 401, NACP.

10 Holl, *Argonne*, 134.

11 From Hilberry to Flaherty, September 30, 1954, RG326, ANL Miscellaneous Correspondence & Reports, box 17, National Archives at Chicago.

12 "International School of Nuclear Science and Engineering," September 22, 1954, RG326, ANL Miscellaneous Correspondence & Reports, box 17, National Archives at Chicago.

13 "Draft: AEC, Training in Reactor Technology, Report to the General Manager by the Director of Reactor Development," RG326, ANL Miscellaneous Correspondence & Reports, box 17, National Archives at Chicago.

14 CA-4095, December 23, 1954, RG59, Entry A1 3008-A, box 400, NACP.

15 Holl, *Argonne*, 136; Argonne National Laboratory, Nuclear Engineering website, http://www.ne.anl.gov/About/hn/news961012.shtml.

16 Holl, 137.

17 From ICA Tokyo to ICA, January 22, 1957; From USOM Japan to ICA, December 14, 1956, RG469, Entry 421, box 3, NACP.

18 "Taketani Kiyoaki's Report from the U.S." [Taketani Kiyoaki-shi no beikoku dayori] No. 1, *Genshiryoku sangyo Shinbun* [*Atomic Industrial News*], no. 3 (November 25, 1955); no. 5 (January 25, 1956).

19 "Beikoku ni okeru gaikokujin genshiryoku kagaku gijutsusha kunren keikaku" [The U.S. Plan to Train Foreign Nuclear Scientists and Engineers], January 1955, Record No. C'.4.1.1.1-4, Diplomatic Archives of the Ministry of Foreign Affairs of Japan.

20 For details, see Yuka Tsuchiya, "The Michigan Memorial Phoenix Project and Taiwan: Scientific Internationalism and Cold War" in *Knowledge as Diplomacy: U.S. and Asia*, eds. Yuka Tsuchiya et al. (Kyoto University Press, forthcoming in 2022). "The Argonaut: Argonne's Nuclear Assembly for University Training," *Selected Topics from the Argonne News*, March 1957, RG326, ANL Publications, box 5, National Archives at Chicago.

21 "Two Decades of Growth, Progress," *The Argonne News; A Retrospective Issue*, July 1966, RG326, ANL Publications, box 5, National Archives at Chicago.

22 "Reunion at Geneva," *Argonne National Laboratory News-Bulletin International*, vol. 1, no. 1 (January 1959): 4–5, RG326, ANL Publications, box 5, National Archives at Chicago.

23 *Argonne National Laboratory News-Bulletin International*, vol. 1, no. 2 (April 1959), RG326, ANL Publications, box 5, National Archives at Chicago.

24 "Recent Events at Argonne," *Argonne National Laboratory News-Bulletin International*, vol. 1, no. 2 (April 1959): 12, RG326, ANL Publications, box 5, National Archives at Chicago.

25 "News of the Sessions," *Argonne National Laboratory News-Bulletin International*, vol. 3, no. 3 (July 1961): 20, RG326, ANL Publications, box 6, National Archives at Chicago.

26 "Distinguished Alumni: Bernardino Pontes," *Argonne National Laboratory News-Bulletin International*, vol. 1, no. 4 (October 1959): 11, RG326, ANL Publications, box 5, National Archives at Chicago.

27 "Distinguished Alumni: Florencio Medina," *Argonne National Laboratory News-Bulletin International*, vol. 2, no. 1 (January 1960): 11, RG326, ANL Publications, box 5, National Archives at Chicago.

28 "Distinguished Alumni: Ryohei Kiyose," *Argonne National Laboratory News-Bulletin International*, vol. 1, no. 3 (July 1959): 11; "ISNSE Alumni: Yoshinori Ihara," vol. 4, no. 2 (April 1962): 16; "ISNSE Alumni: Shigefumi Tamiya,"

vol. 4, no. 3 (July 1962): 13, RG326, ANL Publications, box 5 & 6, National Archives at Chicago.

29 "News of the Sessions," *Argonne National Laboratory News-Bulletin International*, vol. 4, no. 1 (January 1962): 19, RG326, ANL Publications, box 6, National Archives at Chicago.

30 "News of the Sessions," *Argonne National Laboratory News-Bulletin International*, vol. 4, no. 3 (July 1962): 20, RG326, ANL Publications, box 6, National Archives at Chicago.

31 Paul A. Kramer, "Is the World Our Campus," 782–783, 799.

32 Eiko Tsuchida, "Tekunoroji ga tsukuru kokumin, ethnicity: bunka-teki icon to shiteno kagaku gijutsu to shudan identity" [Technology Creates Nation and Ethnicity: Science and Technology as Cultural Icons and Group Identity], in *"Hate" no jidai no Amerika-shi: jinshu, minzoku, kokuseki o kangaeru [American History in the Age of "Hate": Race, Ethnicity, and Nationality]*, eds. Ayumu Kaneko and Yoshiyuki Kido (Tokyo: Sairyusha, 2017), 112.

33 "Ryugakusei haken keikaku" [Plan to Send Students]; "Beppyo 2. Ryugakusei ukeire yoteikoku oyobi sono gaikyo" [Attached Table 2. Countries Receiving Foreign Trainees and Their Overview]; From Yoshio Fujioka to W. F. Libby, June 6, 1956, Record No. C'.4.1.1.1-4, Diplomatic Archives of the Ministry of Foreign Affairs of Japan.

34 Oyama Akira, "Genshiro gakko seikatsuki" [The Reactor School Diary], *Yomiuri Shinbun*, August 11, 1955.

35 From Waring to ICA, April 4, 1957, RG469, Entry 421, box 32, NACP.

36 "Evaluation Meeting with Graduates of Second School of Nuclear Science and Engineering at the Argonne National Laboratory, Lemont Ill., June 1, 1956"; "Evaluation Meeting at the Argonne National Laboratory, Lemont, Ill., January 9, 1957, with Participants of the Third Session, ISNSE, April 1956 to January 1957," RG59, Entry A1 3008-A, box 287, NACP.

37 "Evaluation Meeting with Participants of the Fifth Session (January 28, 1957–November 6, 1957) of the University-ISNSE at the Argonne National Laboratory, Lemont, Ill., October 25, 1957," RG59, Entry A1 3008-A, box 288, NACP.

38 Division of International Affairs, United States Atomic Energy Commission, "International School of Nuclear Science and Engineering to Become International Institute of Nuclear Science and Engineering," September 9, 1959 (date on the cover letter), Michigan Memorial Phoenix Project, (hereafter MMPP), box 15, Bentley Historical Library, University of Michigan.

39 "The Institute of Nuclear Science and Engineering," November 7, 1963, MMPP, box 20.

40 From Shelby Thompson to John A. Hall, August 24, 1954, RG326, ANL Miscellaneous Correspondence & Reports, box 17, National Archives at Chicago.

41 "Two Decades of Growth, Progress," *The Argonne News; A Retrospective Issue* (July 1966), RG326, ANL Publications, box 5, National Archives at Chicago.

42 "Training Men for the Atomic Age," movie script, RG306, Entry A1 1098, box 46.

43 "Training Men for the Atomic Age."

5 Contradictions in the Overseas Information Program: Nuclear Tests in the Pacific and Compensation Negotiation with Japan

In the previous chapters, we have discussed how nuclear technology was used not only as a form of technological aid policy, but also as a means of moving people's hearts and minds, and as a tool for shaping the image of the state and its leaders. However, at the same time, the nuclear tests being conducted in the Pacific Ocean by the U.S. government were provoking international criticism and causing damage to the nation's image. This final chapter of Part I focuses on the circumstances surrounding the nuclear tests, which, in contrast to *Atoms for Peace*, compelled the U.S. government to restrict and/or manipulate information disseminated abroad.

When the Japanese tuna fishing boat, *Lucky Dragon 5* (*Daigo Fukuryū-maru*), was exposed to radiation at Bikini Atoll in the Pacific Ocean in March 1954, it sent shockwaves through Japanese society. When Aikichi Kuboyama, the ship's radio operator, died six months later, a large-scale anti-nuclear movement, involving ordinary citizens, trade unions, and religious organizations, broke out. The U.S. government, wishing to resolve the situation, paid $2 million as solatium to the Japanese government, which accepted the payment and tentatively settled the matter. However, during Operation Redwing (5 May–24 July 1956, with the danger area in effect from 20 April to 11 August) and Operation Hardtack (danger area in effect from 5 April to 8 September 1958 for Eniwetok Atoll, and 25 July to 25 August 1958 for Johnston Atoll), also carried out in the Pacific Ocean, a number of Japanese fishing boats, particularly long-liners for pelagic tuna fishery, sailed in the surrounding waters. If a "second Bikini incident" were to occur, it would intensify the anti-nuclear and anti-American movements, and the Soviet Union and other Communist countries would exploit the occasion as propaganda, inevitably undermining the U.S. national image in the international community. While the U.S. government attempted to take all possible measures to prevent exposure, it was still almost impossible to ensure the complete safety of all ships sailing in the vicinity of the test sites. The U.S. government therefore attempted to dispel the negative publicity of the H-bomb tests by controlling and manipulating information. The Japanese government was equally keen to avoid a "second Bikini incident," and a common interest in regard to information policy was thus naturally established between the U.S. and Japanese governments.

DOI: 10.4324/9781003243649-6

There is a large body of prior literature on the Bikini Incident (*Lucky Dragon 5* Incident). On the subject of Japanese tuna fishing boats, there are, for instance, the first-hand testimony of former crew member of the *Lucky Dragon 5*, Matashichi Oishi, the writings of Masatoshi Yamashita, an activist who has followed the issue of radiation exposure on tuna fishing boats for many years, and a collection of documents collated by Yasuo Miyake, among others. In addition, the exposure of Marshall Islanders to radiation is discussed in the works of Seiichiro Takemine and Hideki Sasaki. Also important is the work of Hiroko Takahashi, who discusses how the U.S. government investigated the long-term effects of radioactive fallout.[1] In the field of diplomatic history, Akira Kurosaki discusses the Japan–U.S. relationship after the Bikini Incident primarily from the perspective of Japanese domestic politics; he points out that Japanese conservative forces, caught in a struggle between U.S. nuclear strategy and domestic public opinion, attempted to win over domestic public opinion by advocating opposition to nuclear testing from the viewpoint of "humanitarianism," without denying the need for possession of nuclear weapons.[2] Toshihiro Higuchi of Georgetown University also argues that the Japanese government's appropriation of the anti-nuclear initiative from the left-wing forces caused the leftist Japan Council against Atomic and Hydrogen Bombs (Gensuikyō) to shift its focus from the issue of radioactive contamination to anti-U.S.-Japan Security Treaty, and as a result, the Japanese left failed to establish solidarity with ordinary citizens whose primary concern was safety from radioactivity.[3] In his latest research, Higuchi points out that the Bikini Incident was the first time that not only direct exposure to radioactive fallout, but also long-term exposure to radiation on a global scale was brought to the attention of both experts and the public, sparking a debate about acceptable levels of exposure.[4] Despite this wealth of prior research, no empirical studies have discussed the issue of compensation between the U.S. and Japan for the nuclear tests that continued after the $2 million solatium provided in relation to the Bikini Incident. With the exception of Higuchi's work on the "clean bomb,"[5] there has also been little research into what information programs were put in place by the U.S. government during this period to dispel the increasingly negative image resulting from growing international public criticism of the nuclear tests.

In the first section of this chapter, we focus on the records of behind-the-scenes negotiations surrounding the Japanese government's request for compensation for Operation Redwing in 1956, after the $2 million Bikini Incident settlement, to examine the difficulties the U.S. government faced in response. The U.S. government was concerned not only that refusing compensation would damage U.S.-Japan relations, but also that it would call America's moral responsibility into question in the international community, and cause damage to the national image. However, before the negotiations for compensation related to Operation Redwing could be completed, plans for the subsequent Operation Hardtack began to materialize. The next section deals with the process by which the U.S. government planned an image campaign during Operation Hardtack to use a "clean bomb" with low radioactive fallout, and to allow media and scientists from various countries to observe the tests. However, when the project was abandoned, the U.S. government directed international public attention to the U.S. displays at the 2nd U.N. International Conference on the Peaceful Uses of Atomic Energy held in Geneva at the same time. However, when Operation Hardtack was actually launched, a much-dreaded incident of radiation exposure occurred. In addition, the ships that were exposed to a large amount radioactive rain and subsequently examined by the U.S. Navy medical team were Japanese Coast Guard vessels conducting oceanographic surveys as part of the International Geophysical Year (IGY), a year of international cooperation through science. Through examination of official U.S. and Japanese documents, Section 5.3 considers the response of the U.S. government, who feared the incident would escalate into a "second Bikini," and Section 5.4, the response of the Japanese government, who shared the same fears as the U.S. The final section reveals how the U.S. government, in response to the anti-nuclear movement, which by the time of Operation Hardtack had spread internationally, attempted to suppress the spread of information.

While the previous chapters have focused on overseas information programs that actively publicized science and technology, the theme of this chapter is what could be called a "negative overseas information program" that aimed to improve the national image abroad through the withholding or manipulation of information. The U.S. and Japanese governments responded with grave concern to the human and economic damage caused to hundreds of tuna long-liners and other vessels in the Pacific Ocean during Operation Redwing and Operation Hardtack. However, the behind-the-scenes negotiations between the two countries and the records of damages used in them were never made public, and in the end no compensation was paid to either the seafarers themselves, or to the shipowners. Thus, by examining the process by which information about nuclear weapons was controlled and manipulated rather than publicly released, this chapter highlights the inextricable link between programs to release and promote information and programs to control and manipulate information.

5.1 Commencement of Operation Redwing and the Question of Compensation

A review of U.S. and Japanese diplomatic documents shows that, prior to Operation Redwing in 1956, the Japanese government sought compensation for the cost of fuel needed for ships to bypass the danger area for the hydrogen bomb tests, and that there was a view in the U.S. government that compensation should be provided. The U.S. Embassy in Tokyo, in particular, insisted to the Department of State that some form of compensation was necessary for the sake of U.S.–Japan relations and America's national image in the world at large. Documents also show that the Japanese requests for compensation were continued by the Kishi government, which was formed in 1957.

In January 1956, when plans for new nuclear tests in the Pacific were announced, the Japanese government, in a written statement dated 25 January, requested that "if the tests are carried out and the Japanese nation and people incur damages, the U.S. government will provide full compensation." In a cable to Secretary of State Dulles, dated February 9, Ambassador John M. Allison noted that both the ruling and opposition parties in Japan were united in their protest against the U.S. hydrogen bomb tests, and that even if no radioactive contamination were to occur, fishermen would likely "demand compensation for time and expenses incurred by detours around the test area, and possibly also for exclusion from fishing within the test area." The ambassador also suggested cooperation with the Japanese in terms of, for instance, providing prior notification, periodic consultations on safety precautions, and joint scientific surveys after tests.[6] On February 22, two days before the "danger area" was to be announced, J. Graham Parsons, Counselor[7] at the U.S. Embassy, wrote a cable to Secretary of State Dulles warning of "snowballing anti-nuclear test publicity" and pressing the U.S. government to urgently consider compensation for the fishing industry. He said that compensation "should be dealt with on political level and divorced from question of legal liability" to avoid "revival of Japanese bitterness and hysteria characteristic of [the] Bikini incident and serious difficulties in our relations with [the] Japanese government." Parsons divided the compensation into two categories: (1) losses resulting from the inability to fish in restricted areas and from the need to navigate in order to reach other areas; and (2) losses resulting from the contamination of fishing grounds and catch. He advised, given the former damages could be estimated before a nuclear test, that it would be wise to show understanding of such damages and to pay an "ex-gratia basis lump sum" in advance, rather than calculating the compensation after receiving claims. Such a move would show U.S. "recognition of inconvenience caused to fishermen" as well as "demonstrate to [the] public US concern and [their] humanitarian attitude in contrast to Soviet callousness." In the event that (2) "contamination of fishing grounds and catch" were to be discovered, the Japanese reaction would likely be "immediate and explosive" and any response by the U.S. government would "very likely be almost completely unsuccessful." So, Parsons suggested that a "scientific survey group composed of both U.S. and Japanese scientists" be formed in advance to prepare for this eventuality. If the investigation were to indeed reveal radiation damage, Parsons said, the U.S. government should "compensate quickly—again on ex gratia basis—using findings of survey to determine amount of lump sum payment." He further proposed that the danger area, which was to be announced in two days' time, on February 24, be accompanied by the following statement "from viewpoint of meeting [the] political and propaganda problem":

(1) 1956 tests will be on [a] substantially smaller scale than those of 1954 and U.S. government and scientific authorities [are] taking maximum precautions [to] avoid injury ...

(2) [The] U.S recognizes certain Japanese fishing boats will inevitably be forced to detour from test area and therefore U.S. is providing sum to Japanese government ...

(3) [The] U.S. invites Japanese to participate in [a] scientific survey to check contamination of fish or fishing grounds and will be guided by finding[s] of [the] survey in considering further ex gratia compensation.[8]

In this way, Parsons advised the Secretary of State to control anti-nuclear public opinion in Japan and the world through timely dissemination of information. However, the reply from Secretary of State Dulles was that while the problem of compensation was "under active examination," it was not possible to "make [a] simultaneous statement [regarding compensation and danger areas as] suggested by [the] Embassy." Dulles further suggested that the Embassy and the USIS should make the utmost effort to "explain to influential Japanese and [the] public" the fact that "these tests [we]re in [the] interest [of the] free world."[9] On February 28, Ambassador Allison responded to Dulles, "Although Embassy/ USIS [were] continuing [to] do [their] utmost to explain to influential Japanese and [the] public that tests are in [the] interest [of the] free world," this fact did not appear to be "acceptable to [the] Japanese as sufficient justification in light [of their] past experience." The ambassador cited the results of a private firm poll commissioned by the USIS in January to illustrate how deep-rooted Japanese opposition to nuclear testing was. The majority of respondents held negative feelings toward nuclear weapons and atomic energy, with 30% feeling that the U.S. was promoting military, rather than peaceful, uses. Only 9% believed that the elimination of nuclear weapons would benefit the Communist countries, and 61% thought that nuclear weapons should be banned, even if doing so would benefit Communist countries. Moreover, support for the peaceful use of nuclear energy plummeted from 87% before the announcement of Operation Redwing, to 42% afterwards. After explaining this situation in Japan, Ambassador Allison outlined that the while the Japanese were "sufficiently realistic to know [that] there [was] no hope of calling off tests," they nevertheless believe they had the "right to expect compensation for damage to their alleged legitimate interests." The Japanese fishing industry, which was directly affected by the tests, was "particularly difficult [to] deal with." Allison advised that the "propaganda value in Japan [would be] immeasurable" and the "impact [on the] free world [would be] favorable also," if the U.S. government were to publicly announce that it would provide compensation, with "technicalities [regarding] amount, source funds, measure of damages, etc. [to] be worked out later."[10]

The U.S. government announced the danger area on February 24 as planned, but the Japanese government did not announce it to the public until 2 March. In the meantime, on February 29, Minister Shigeru Shima of the Embassy of Japan in Washington, the Science Attaché, Takashi Mukaibo, and several others discussed with six officials from the Department of State and the AEC "the implications of the tests to Japanese fishing" and "the contamination of the fish." The Americans "spread out a chart showing the danger area," and the Japanese, "a chart showing catches by area and the danger area." The Americans said they "wanted to know what kind of losses would be involved and how they would be evaluated," and asked three specific questions: how many pounds of fish were caught in this area each year; how many ships caught these fish; and how many extra miles would a ship have to travel to get from Japan to the southeast corner of the danger area by rerouting around it. The Japanese side replied that they would contact the Federation of Japan Tuna Fisheries Co-operative Associations to obtain an estimate.[11] (Figure 5.1 shows an example of the map of the danger area though this particular map might not be the one used in the meeting above.)

The announcement of the danger area in Japan on March 2, as expected, provoked "considerable furor and criticism," and was reported on the front pages of all major newspapers. The criticism was directed at the "absence [of] reference to compensation and failure [of the] U.S. to take cognizance of Diet resolution on urging international agreement [to] ban A and H-bomb tests." After outcry from members of the Japan Socialist Party and public backlash, the Japanese government was forced to promise that, although it could not force the U.S. government to suspend its nuclear tests, they would at least seek compensation.[12] Ambassador Allison told Secretary of State Dulles that the U.S. Embassy had also been inundated with letters of protest from various quarters, including a letter of petition from the Seamen's Union, which was a "friendly, strongly anti-communist group directly affected by tests." The letter "express[ed] appreciation for safeguards [to] prevent repetition of Fukuryu-maru [*Lucky Dragon*] incident" and acknowledged that the development of weapons to counter Communism was a "necessity," but expressed concern that nuclear testing would damage

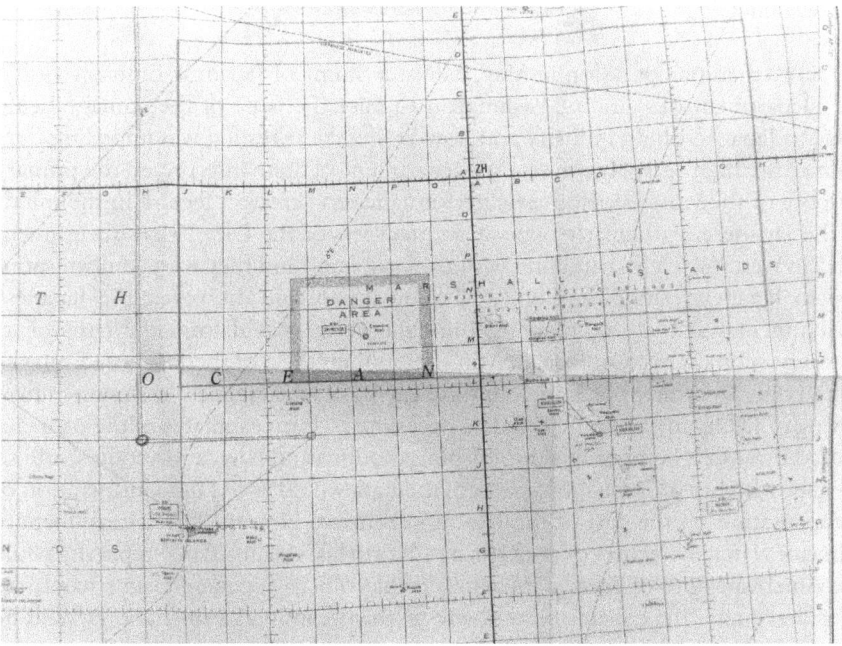

Figure 5.1 A map of the danger area stored at the U.S. National Archives. RG59, Entry 3008-A, box 428, NACP.

America's reputation and contribute to enemy propaganda. Ambassador Allison urged Secretary of State Dulles to visit Japan and explain clearly to the Japanese "what to expect on [the] compensation issue," for if the U.S. government recognized that "our friends, such as Seamen's Union" were in a difficult position because of nuclear testing and responded to their "moderate demands," the friendly groups would have "ammunition to counter the less restrained outbursts of more extreme elements which communists [we]re exploiting."[13] On March 14, Ambassador Allison sent Dulles two telegrams in one day (at 1:26 and 6:33), advising him that the U.S. government should formally reply to the Japanese parliamentary resolution (calling for a ban on nuclear testing).[14] Following repeated telegrams from Ambassador Allison informing Dulles of the tense situation in Japan over the nuclear tests and requesting a response, the U.S. government finally decided to issue an official position statement on the parliamentary resolution and the compensation issue. In a written statement to the Japanese government issued on the occasion of Secretary of State's visit to Japan and his meeting with Prime Minister Hatoyama on March 19, Acting Secretary of State Herbert Hoover Jr. said that the U.S. government was prepared,

> if after the test series has ended, any evidence is officially presented that substantial economic losses for Japan or Japanese nationals have been incurred as a result of establishment of the danger area and the tests, to give further consideration to the question of compensation in the light of any such evidence.[15]

On more than one occasion in May, Minister Shima of Japanese Embassy visited the Department of State in Washington to raise the issue of the damage being done to Japan's fishing industry and how U.S. nuclear testing was impeding freedom of the high seas. However, the Department of State interpreted the primary purpose of the visits as a domestic performance to "create a feeling in the minds of the Japanese, particularly opposition members of the Diet," that government officials were "actively struggling with this problem" and they were "fully responsive to Japanese public opinion."[16] During this period, however, the Japanese government was busy aggregating the data which would form the basis of its compensation claims. On June 8, Minister Shintaro Seki of the Japanese Embassy visited the Department of State to inform them that the Japanese government was having difficulty in collecting the data necessary to form estimates of the damages caused by vessels being shut out of fishing grounds and navigation routes, and to enquire whether prompt submission of the data would speed up consideration of compensation on the part of the U.S. government. Howard L. Parsons, Deputy Director (later, Director) of the Office of Northeast Asian Affairs, replied: "while the American note of March 19 had said that official Japanese claims would be considered after the conclusion of the tests, the preliminary explanation would be helpful." Seki also asked during this meeting if there was any truth to the rumors that the U.S. government was measuring the radiation levels of fish landed on the U.S. West Coast. James V. Martin, Jr., Japan Desk Officer, replied that routine spot checks by the Food and Drug Administration (FDA) were being carried out, but that they were not limited to Japanese tuna. He also stated that the same question had been asked by Mukaibo, Science Attaché at the Japanese Embassy. Seki explained that he feared that "the leftists in the Ministry of Welfare in Japan" would try to exploit such information to damage U.S.–Japan relations.[17] The issue was later discussed between the Ministry of Foreign Affairs (MOFA) and Ambassador Allison, and MOFA told Allison that, although the Japanese government understood that the FDA's "radioactivity test on tuna [was] routine procedure and [could not] not be stopped," it hoped that the "tests [could] be kept confidential."[18] If the fact was made public that the U.S. was concerned about radioactive contamination in fish consumed domestically even though it was reluctant to officially admit any hazards from nuclear testing, the Japanese public, and some "leftist" government officials, MOFA feared, would fiercely criticize the U.S.

Having calculated the damages to fisheries as requested by the Americans, the Japanese Government, through Consul Tanaka at the Japanese Embassy in Washington, informed the Department of State privately on November 13, 1956, after the end of Operation Redwing, that a formal claim for compensation would soon be submitted. The total figure for damages was 97.34 million yen, including 70 million yen for fuel expenses incurred by 51 merchant ships and about 50 fishing boats that had to navigate around the danger area, and 27.34 million yen for the dispatch of the research vessel *Shunkotsu-maru* (on this document there was a handwritten conversion of yen into dollars by a Department of State official: @ 360 = $270,388). Tanaka also reported that the claim could increase, as some vessels that had departed to fish in Samoa had not yet returned. The Department of State was surprised at the claim that the Japanese had made for the research ship *Shunkotsu-maru*.[19] In December, the Japanese government presented the U.S. Embassy with a further compensation claim that had ballooned to 146 million yen (about $400,000). In response, the U.S. Embassy's Acting Ambassador, Outerbridge Horsey, stated that "in general [the] Japanese government ... made effort [to] present [a] reasonable and well documented claim for compensation." He went on to explain that although the Japanese government had not yet publicly announced that it would make a demand for compensation, once the Diet resumed in January, "fishing and merchant marine interests w[ould] start building up pressure on [the] Diet for actual compensation and [the] issue [would] become public," and that the "Japanese government [would] have little choice but take lead in urging US payment." Horsey advised that the U.S. government should promptly consider "compensation [of] all or [a] substantial portion [of] Japanese claims, on ex gratia basis if possible." Any delay would result in an "almost certain increase" in the cost of settlement, and failure to consider compensation would likely "cause serious disturbance [in] U.S.-Japan relations." If the situation were to drag on, and develop into a major controversy, the Japanese government, "however reluctantly, [would] appeal to UNGA [United Nations General Assembly]." Giving the above reasons, Acting Ambassador Horsey urged the U.S. government to pay compensation.[20]

Meanwhile, as Ambassador Allison and others had feared, post-Operation Redwing, Japanese public opinion had become even more critical of the thermonuclear tests. A survey of 1,275 Japanese citizens over the age of 20 conducted shortly after the tests by the USIA under contract with the Tokyo-based "Central Research Agency," "with no indication to the respondents of American interest in the survey," showed that an overwhelming majority (94 percent of the university-educated, and 86 percent of the general public) said that they "did not approve" of the tests, a figure that was significantly higher than the 48 percent of Europeans who reported they did not approve in a similar survey. On the subject of the peaceful use of nuclear energy, there was also a marked deterioration when compared with a survey conducted in January of the same year, with the "NET favorable" (the difference of those who thought atomic energy was "more of a boon to mankind" minus those who thought "more of a curse" in percentage) dropping by more than 10 points to negative 58 points (i.e., in the latter survey, only 13% answered favorably whereas 71% responded it was "more of a curse") among the general public.[21] Entry/exit surveys conducted at the Peaceful Uses of Atomic Energy Exhibition held in Osaka in May 1956, during Operation Redwing, also showed that the visitors in Osaka were distinctly "tougher" than at the Tokyo exhibition held six months earlier, and that their negative feelings were "far less susceptible to change" by the exhibition than they had been in Tokyo.[22]

In 1957, the U.S. Embassy in Tokyo reported to the Secretary of State that a "reply to Japanese approach on compensation for nuclear tests in 1956 [was] becoming matter of increasing urgency," as the Japanese Diet was now in session. In light of a British nuclear test scheduled for that spring, the issue of compensation was particularly likely to garner increased attention. If pressed in parliament, the Japanese government would likely be compelled to "admit [that the] initial request for compensation [had] already [been] made to us" including the "amount of [the] request," in order to demonstrate the government's commitment. The U.S. embassy feared that this would "freeze" the Japanese government's claim for compensation and "stimulate [a] press campaign to force our agreement." In the margin of the received telegram, written by hand is the following, "Reluctant to make another 'exgratia' settlement—since 2 exgratia settlements equal an obligation," which was likely to have been penned by either the Secretary of State or one of his entourage.[23] The U.S. government did not reject the possibility of compensation from the outset, and although it was willing to let Japan calculate damages and consider their claims, it feared that a second payment of compensation would routinize the U.S. practice of providing compensation.

The Department of State considered what approach it should take toward the compensation issue. Parsons, Director of the Department of State's Office of Northeast Asian Affairs, argued strongly that the U.S. government should make a prompt lump-sum payment, saying that "[t]hough much of the [Japanese] claim is not properly substantiated, as a whole it deserves serious consideration." This was because the Japanese people were very sensitive regarding nuclear issues, and it would be better to settle the matter "quietly" before it became a big issue, and in the case that they refused compensation, the Japanese government could bring the matter to the International Court of Justice.[24] There was support for this view within the Department of State: a Departmental memorandum sent to Parsons and Robertson, Director of the Bureau of Far Eastern Affairs Office, by an unknown author, stated that "Despite the weakness of the Japanese claims," it might be "politically advisable" to pay the amount equivalent to the cost of the "fishing boat detour." The memorandum further pointed out that the U.S. Embassy in Tokyo had been insisting, even before Japan's demand for compensation was made, that "any moderate Japanese claim [should] be paid promptly," and that unofficial information had been received from the Japanese Embassy in Washington that "if the Department should pay somewhat less than the Japanese Government has asked it would probably be satisfactory," and that "[r]efusal to pay at all might well result in a Japanese decision to take the case to the International Court of Justice," in which case "it would be legally as well as politically impossible for the U.S. to refuse to appear and respond." The memorandum concluded that "avoidance of payment should be accomplished if possible while retaining the goodwill of the Japanese," but if this were not possible, "payment should be made promptly and as a matter of obligation rather than of grace."[25]

As there were strong grounds for both affirmative and negative arguments regarding compensation, the Department of State felt that "the decision [wa]s not an easy one." Since the U.S. might endanger "its relations with Japan by refusal to pay," the stance that "a prompt payment [was] desirable" was understandable. However, as the U.S. government had already paid a lump sum for the 1954 tests, the concern that "the second ex gratia payment would tend to confirm our liability" and "form a precedent for the future" was also persuasive. A working group was established with representatives from the Department of State, the Department of Justice, the Navy, the U.S. Atomic Energy Commission (AEC), and others to consider the Department's position on the compensation issue.[26]

In parallel with the discussions on compensation, a project was underway in which Japanese and U.S. scientists would collect samples of fallout and seawater from Operation Redwing and share the results of their analyses. As mentioned earlier, the Americans believed that involving Japanese scientists would help to defuse the Japanese opposition to nuclear testing. From the point of view of Japanese scientists, the sharing of samples had the advantage of furthering basic research on radioisotopes and of allowing them to approximate the composition of the nuclear fuel used by the U.S. In January 1957, the Japanese Embassy in Washington proposed to the Department of State that data on fallout and marine contamination from Operation Redwing be exchanged and that Japanese and American scientists hold a conference. However, the Department of State responded that they could not provide the Japanese with samples collected near Eniwetok and Bikini Atolls, presumably because they might include sensitive information about fissionable materials.[27]

In the meantime, the Department of State was becoming more inclined to reject the request for compensation. Ambassador Allison, a proponent of compensation, was also replaced by a new Ambassador, Douglas MacArthur II, nephew of the former Supreme Commander of the Allied Powers, Douglas MacArthur. On March 13, 1957, the Department of State handed the Japanese a reply stating that "the data submitted failed to show that the tests and danger area had caused economic losses" and that "the Japanese had no case" concerning the cost of the *Shunkotsu-maru* survey "because their participation in the survey was voluntary."[28] The following day, March 14, Richard Sneider of the U.S. Embassy visited MOFA to provide a more detailed account of the American response. Kazuyoshi Inagaki, Deputy Director of the European and American Affairs Bureau,[29] "expressed considerable disappointment" that none of the Japanese claims for compensation had been accepted. MOFA officials also explained that it was "very difficult ... to document the specific fishing losses suffered by each fishing vessel" as was required by the U.S. The following day, Sneider had lunch with Inagaki to informally gather information about the Japanese government's internal situation. Inagaki told Sneider that the Japanese government intended to proceed with gathering additional evidence, but that there was a widespread feeling in the Ministry of Foreign Affairs that "further arguments on legal grounds [we]re pointless." Inagaki said that a "political solution" needed to be found, as some were of the opinion that the matter "could only be resolved in the International Court."[30] MOFA also explained the difficulties in collecting the data to Ambassador MacArthur; for instance, the difficulty in accurately calculating the catch lost by each fishing vessel, and in the case of merchant vessels, the inevitability that the ships would divert their courses to an unspecified degree to take a wide berth from the danger area. The "deputy chief American affairs bureau" (presumably referring to Inagaki, Deputy Director of the European and American Affairs Bureau) once again emphasized the necessity of "political solution," and suggested, as his personal opinion, that this would mean "half [the] amount mentioned in papers submitted" as compensation. Ambassador MacArthur telegraphed Secretary of State Dulles that their "negative reaction may discourage" the MOFA "from resubmitting note in present form," but that the Japanese government would certainly be forced to seek compensation if pressed in the Diet.[31]

The Japanese side filed a formal claim for compensation, including additional data, on May 22. The amount was approximately 48 million yen in damages resulting from the detours of merchant vessels, 27 million yen for fishing vessel detours, 27 million yen in survey costs for the *Shunkotsu-maru* and other vessels, and 1 million yen in disaster prevention costs for the Federation of Japan Tuna Fisheries Co-operative Associations, coming to a total of 103 million yen. This was the first day that Japan formally filed a claim for compensation although, as we have already seen, behind-the-scenes negotiations had already been ongoing. The Department of State's response to this was curt. A telegram sent by Secretary of State Dulles to the U.S. Embassy in Tokyo began with the criticism that the newly submitted tables and maps did "not differ materially from those submitted last December" and that "none of material submitted meets [the] objections" that the U.S. had raised after the first claim. The fisheries claim was carefully reviewed by the Department of State as the most important point; however the newly submitted documents presented neither "normal catches" nor "normal routes," and therefore failed to establish what was actually lost. The Department of State met with representatives of the Japanese Embassy and the Fisheries Agency on February 29, 1956, and informally advised that "these very points should be covered in submitting any claims for losses." At that time, the representatives stated they had statistics on the "catches in previous years." Nonetheless, the "fact that no statistics have been furnished strongly suggest[ed]" that the Japanese were unable to show that the "fishing industry suffered any actual losses because Japanese fishermen did not fish in danger area."[32] The official response from the U.S. government which turned down the Japanese claim was delivered on September 13, the content of which was as follows: the detours of the merchant ships, the *Shunkotsu-maru*'s survey, and the preventive measures were all "conducted voluntarily by the Japanese government or individuals concerned, and cannot be determined to be directly caused by the setting of the danger area or the tests," and also, the detour of the fishing vessels was not compensable on the grounds that "the submitted data did not establish evidence of actual loss."[33]

However, even as the negative response was being drafted, there was much debate within the Department of State. On August 5, Parsons, the head of the Office of Northeast Asian Affairs, decided it would be best to "say 'no' to the Japanese as diplomatically as possible" and avoid discussing each claim in detail in order to "discourage any further attempts to pursue the matter." It was based on this idea that the response to the Japanese government was drafted.[34] The U.S. Embassy in Tokyo's Horsey, however, took direct issue with this. While he did not dispute the conclusion that the "Japanese claim [wa]s not justified by the facts submitted," he ventured to

> reiterate ... that the expenditures for the Shunkotsu Maru and protective equipment and the wide berth given the danger zone by Japanese merchant vessels and fishing boats helped significantly to quiet Japanese fears about radioactive fish, etc. and thereby kept the 1956 tests from becoming a major international issue.

While these measures were indeed taken at the initiative of the Japanese for their own benefit, they were also taken for the benefit of the U.S. "The Japanese have gone to considerable efforts to document their case," and failure on the U.S. side to explain its reasons for rejection with the same level of care would be interpreted as "a lack of adequate consideration of the Japanese case."[35] Having said this, Horsey argued that the U.S. side should also explain, based on objective evidence, why the Japanese claim for compensation should not be granted. The Department of State's Bureau of Far Eastern Affairs echoed Horsey's sentiments, arguing that "the Embassy's point [wa]s that our principal problem [wa]s to minimize adverse public reaction" in Japan, and dismissing Japan's request without any explanatory materials would "leave us in a bad position from the point of view of public relations."[36] There was also the opinion that it was necessary to "elicit detailed information from the Japanese" to carefully examine whether the detours of fishing boats were in fact necessary before simply dismissing their claim.[37] Gerard C. Smith, Special Assistant to the Secretary of State, pointed out that the Department of State had no data to support its argument against the Japanese, and in order to find its own data on the Japanese fishing industry, he requested William Neville of the Tuna Research Foundation, which had been set up in Tokyo in relation to "the purchase of Japanese-caught tuna for export to the United States for canning" to collect necessary information.[38] In sum, while the U.S. responded that it could not compensate for the damages, they were continuing their efforts to gather facts, including data from Japan, about whether or not damages were indeed incurred, especially with regard to fisheries.

The U.S. Department of State had requested that the Japanese government not make public the series of U.S.–Japan negotiations over compensation.[39] On the Japanese side as well, the negotiation was kept secret. Even in a later MOFA document, dated January 9, 1960, that summarized the chronological outline of the Japan–U.S. negotiations on nuclear testing, the section on "claims for damages" was marked as "this section, top secret."[40] The Japanese government preferred to keep the negotiation process confidential likely because of the fear of public criticism if the negotiations broke down, and also because revealing the details of the compensation negotiations would stimulate the fishing industry and others and make their claims uncontrollable.

On January 21, 1958, the Japanese government handed a letter to Ambassador MacArthur, urging the U.S. government to reconsider its claim for compensation for damages resulting from Operation Redwing. At that time, the Japanese government said that the previous U.S. reply indicating they would not offer compensation had "not become public knowledge" in Japan, and therefore, any debate on the matter was "avoided during November Diet session." If any question should be raised in the forthcoming Diet session, the Japanese government "would say [the] matter [was] still under negotiation" with the U.S. government. Ambassador MacArthur interpreted that "the main purpose of Japanese note [was] probably to give plausible grounds for avoiding public conflict with U.S. on [the] issue which [could] be very sensitive politically." Since it was also in the U.S. interest "to avoid such conflict," MacArthur continued, "unless this issue becomes inflated and there is need to explain our position publicly," U.S. interest would be "best served by withholding our reply" and thus it would be "delayed for a couple of months."[41] Although MacArthur recommended to delay the U.S. response for "avoiding public conflict," his cable delivered to the Department of State had the following typewritten memo attached:

[S]ince we are on the threshold of a new series—issuance of info[rmation] on the danger area, don't you think that we should ask MacArthur to reconsider[?]. It's a rather messy situation to go into another series without first clearing up the debris arising from the one two years ago.[42]

It is not clear by whom it was written, but the memo indicates that, even at this point, there was still no unanimity of opinion within the Department of State, and at least some officials were trying to work out a settlement concerning the Japanese claim. However, in the end, with no answer from the U.S. government, the danger area for Operation Hardtack was announced in February.[43]

Around the same time, in January 1958, the Japanese government had filed a claim for compensation with the United Kingdom for "damages incurred by Japanese merchant ships and fishing vessels as a result of detours or changes of fishing grounds" in response to the nuclear test conducted by the United Kingdom on Christmas Island in May 1957.[44] Although the U.K. government's response was simply that the Japanese claim was "under consideration," the U.S. Embassy in Tokyo received the information that the U.K. had a concrete plan of compensation "for the diversion of merchant and fishing vessels around the Christmas Island danger area," but that "no compensation would be paid for loss of fishing catch" and that the compensation would be "a single sum roughly equivalent to the size of the claims on an ex gratia basis, without admitting any legal responsibility." Upon receiving this information, the U.S. Department of State reaffirmed its policy as follows: (1) "We would not pay compensation for detour costs for merchant vessels, since it appears that the alleged detours were optional as normal merchant routes did not pass through our danger area," and (2) "We would not pay compensation for detour costs for fishing vessels, since it has not [been] shown that any economic losses were actually suffered." However, they recognized that if the U.K. were to pay compensation and the U.S. continue to refuse, that would put the U.S. in a "somewhat embarrassing position in Japan."[45]

In the Japanese Diet, the government was repeatedly pressed by Socialist Party politicians on the progress of the compensation negotiations, and finally responded on July 9, 1958, that (1) the Japanese government had not requested amounts for specific, detailed damages, (2) the request submitted in the spring of 1957 was rejected by the U.S. in September for insufficient evidence, and (3) it was resubmitted in January 1958 and was still under consideration by the U.S. government. MOFA informed Ambassador MacArthur that they had previously rejected the Socialist Party's demands for information, but were forced to release this information due to the persistence of the demands. Ambassador MacArthur told Secretary of State Dulles that "as yet there ha[d] been no consequent reference to test compensation issue in press," but recommended that some kind of response be sent to the Japanese government on compensation before the Diet would adjourn.[46] Dulles' response to this was that the Japanese government presented "no new information [to] substantiate [their] claim." Accordingly, the U.S. reply to the Japanese government should state that "after careful examination" the U.S. maintained their "view expressed [in the] Embassy note [on] September 13, 1957." Dulles also believed that it was "unnecessary" to provide any further explanation of the U.S. position.[47]

A close examination of the U.S. and Japanese diplomatic documents has demonstrated that the U.S.–Japan negotiation on the compensation for damages from nuclear tests was a much longer, and a much more complicated process than had previously been known. The Japanese government first raised the issue of compensation during the Hatoyama administration, which advocated "independent diplomacy" and restored diplomatic relations with the Soviet Union, but negotiations continued behind the scenes even after the more anti-Communist and pro-U.S.–Japan-alliance Kishi administration took office. Although the U.S. side always maintained its official stance that it could not pay compensation due to insufficient evidence, there were diverse opinions within the U.S. government, including those who advocated a prompt payment of a lump sum. At the very least, the U.S. side collected and examined objective data on the existence of damages to the fisheries. However, the U.S. government, fearing that compensation would become the norm, did not publicize these diverse internal opinions. In addition, the U.S. government likely had to take into consideration the fisheries industry in the U.S., as, at the time, there was trade friction between the U.S. and Japan over Japan's frozen tuna exports. U.S. fisheries had been dealt a blow by the large amount of cheap frozen tuna from Japan, resulting in frequent strikes by fishermen and dockworkers in Southern California, where major canned tuna factories were located.[48] Under such circumstances, the U.S. government would have wanted to avoid publicizing negotiations on compensation for Japan's fishing industry. The Japanese government also kept the negotiations with the U.S. secret for fear that the demands from the fishing industry would spiral out of control. In the end, the Japan–U.S. negotiations over fishery compensation continued to stagnate without being discussed publicly. However, as will be discussed later, a situation that would further set the negotiations back would arise during Operation Hardtack.

5.2 Overseas Information Programs on the "Clean Bomb" and Their Setbacks

In March 1958, around the commencement of Operation Hardtack, President Eisenhower held a press conference and announced that a "clean bomb," which would emit much less radioactive fallout than in the past, would be used for the tests. Since the Bravo nuclear test in 1954, which caused the *Lucky Dragon 5* incident, had made the dangers of radioactive fallout widely known, the AEC had continued to research and develop nuclear explosions that emitted less radiation, and Edward Teller and Ernest O. Lawrence, who were proponents of nuclear testing, strongly supported this research. Also, for the United States, which was seeking a way out of its "mass retaliation strategy," the "clean bomb" was a promising military technology that opened up the possibility of small-scale tactical nuclear weapons.[49] Hiroko Takahashi demonstrates that "Operation Chariot," a "clean bomb" experiment in Alaska, was pushed from 1957 to the early 1960s, but abandoned due to opposition from local residents.[50] However, at the nuclear test sites in the Pacific Ocean, the tests were carried out. During Operation Redwing in 1956, the AEC had also held a press conference about the "clean bomb," although at that time, there was controversy within the AEC about announcing the "clean bomb," primarily from the perspective of maintaining secrecy. In the *Bulletin of the Atomic Scientists*, physicist Ralph Lapp, who was also a member of the Chicago Scientists' Movement (see Chapter 2), criticized the idea of "clean" nuclear weapons, stating, "Part of the madness of our time is that adult men can use a word like humanitarian to describe an H-bomb."[51] However, the U.S., which was on the defensive in the propaganda war due to the Soviet Union's unilateral suspension of nuclear testing before Operation Hardtack in 1958, once again resorted to "clean bombs" as countermeasures against the Soviet peace offensive.

The President's "clean bomb" press conference was a carefully prepared scenario by the Operations Coordination Board (OCB). The OCB, which reported to the National Security Council (NSC) and oversaw interdepartmental psychological policies, had played an important role in the 1954 *Lucky Dragon* incident, including the payment of the "solatium,"[52] and also explored ways to avoid damaging America's national image during Operation Redwing. On March 19, 1958, in anticipation of Operation Hardtack, the OCB decided to draft a presidential statement in cooperation with the Department of State, the Pentagon, the AEC, the USIA, and the Special Assistant to the President for National Security Affairs, on the grounds that it would present an opportunity for the United States to gain a "psychological advantage" if President Eisenhower held a press conference to announce a nuclear explosion with low radiation emissions.[53] Early drafts described a "large nuclear detonation, from which the radioactive fallout will be only 5% of that resulting from such a detonation prior to this time," and that "such advances by American scientists can have tremendous significance in the peaceful uses of atomic energy." Later, however, objections arose from within the OCB and USIA as to the overemphasis on "cleanliness," and the manuscript was revised to draw attention to the active cooperation with the international community and disclosure of information, rather than "cleanness" of the bomb.[54]

At the president's press conference on March 26, it was announced that, in order to showcase the "clean bomb" internationally, scientists and media representatives from 14 countries of the U.N. Scientific Committee on the Effects of Atomic Radiation (Argentina, Australia, Belgium, Brazil, Canada, Czechoslovakia, Egypt, France, India, Japan, Mexico, Sweden, the Soviet Union, and the United Kingdom) would be invited to observe the nuclear tests. The plan was for the guests to gather at the Radiation Laboratory of the University of California in Berkeley for a briefing on nuclear testing, and then travel to Honolulu before proceeding to the Eniwetok nuclear test site. After observing the test, the scientists would return to Berkeley to analyze the fallout and other samples from the test.[55] The Department of State was to fund the stay in Honolulu, and the AEC was to cover all other travel and accommodation expenses.[56] Later, at the invitation of the U.S. Navy, four Southeast Asia Treaty Organization (SEATO) countries (the Philippines, Thailand, Pakistan, and New Zealand) and military personnel from South Korea, Taiwan, Iran, Iraq, and Spain were added to the invitation list. At the OCB meeting, Department of State representative Gerard C. Smith expressed the Department of State's view that such an invitation was desirable for "political and psychological reasons," to which AEC Chairman Strauss agreed.[57]

However, in an OCB meeting of June 4, "uneasiness was expressed over the proliferation of testing sites" in terms of "public relations aspects,"[58] and the public release of the "clean bomb" was also to be reviewed. Furthermore, at the June 11 OCB meeting, the AEC, who had initially been proactive regarding the publicizing of the "clean bomb," remarked that the invitation program was a "State Department 'show'" and that the AEC would neither take responsibility for, nor fund it.[59]

The Soviet Union and India had already declined the invitation, and Japan followed suit. Soviet scientists had demonstrated that the "clean bombs" were not really that "clean." The foreign information program surrounding the "clean bomb," which was designed to showcase to the world America's openness and scientific advancement, had shown, on the contrary, that the world was not convinced by the rhetoric of "clean" nuclear explosion. For example, Japanese newspapers published critical articles with headlines such as "The Lie of Clean Hydrogen Bombs," and the Ministry of Foreign Affairs gave a similarly terse analysis, remarking that the "clean bomb" project was "presumably aimed to primarily garner political and psychological effects at future disarmament conferences and with world public opinion in general," but that "originally, nuclear weapons were effective as 'deterrents' due to their great destructive power, and the idea of 'clean' weapons of mass destruction is self-contradictory," and that "current world opinion is past the stage of consenting to nuclear tests free of international control, simply based on claims of low radioactivity tests and the potential for peaceful uses."[60] Sensing this cold reaction from the international community, on July 30, the U.S. government finally decided to cancel the public test of the "clean bomb" (originally scheduled for August 25), ostensibly because the test coincided with the 2nd U.N. Conference on the Peaceful Uses of Atomic Energy, and therefore invited guests would not be able to attend the U.N. conference. Instead, they announced that scientists from various countries would be invited to the Berkeley Lab at a later date, where they could see the data obtained from the test. The official telegram from U.N. Ambassador Matsudaira to Foreign Minister Fujiyama, which reported the above, includes the handwritten message, "If this is the case, our participation will not become a problem," followed by the minister's response, "I agree."[61] The U.S. government's display at the 2nd U.N. Conference on the Peaceful Uses of Atomic Energy in Geneva was so lavishly staged, with displays of the latest nuclear fusion technology and models of giant power reactors, that the conference was virtually a "spectacular American show."[62] This display was in part an attempt to recover from the "Sputnik Shock," but it was also meant to cover the failure of the "clean bomb" overseas information program outlined here.

The public testing of the "clean bomb" and the invitation of scientists to analyze its data was an attempt by the U.S. government to dispel the negative image of nuclear testing and turn it into a positive foreign information program, but the campaign failed miserably. Governments and the public around the world reacted critically, viewing both the notion of a "clean nuclear explosion" as a logical contradiction, and the U.S. attempts to use nuclear tests in overseas information programs as cynical. Thus, the U.S. government abandoned the positive information dissemination on nuclear tests, although it needed to continue to control and downplay the information to make the tests less conspicuous. In particular, the issue of radioactive contamination of ships was a matter that required the utmost attention from the U.S. government, in light of the *Lucky Dragon 5* incident.

5.3 Operation Hardtack and the *Takuyo* and *Satsuma* Exposure Incidents

In order to prevent another *Lucky Dragon 5* incident from occurring, the U.S. government dropped cylinders from the sky with warning messages to prevent Japanese pelagic fishing boats and other vessels from approaching the danger area, urging them to leave. A warning message preserved in the U.S. National Archives states in rudimentary and roughly written Japanese, "You are now entering a dangerous zone. Please immediately adjust your course to _____. Do not go near the Bikini and Eniwetok Atolls." The same message is written in several languages (Figure 5.2).[63] Many Japanese fishermen received such warnings, as the danger area overlapped or was adjacent to good tuna fishing grounds. For example, the *Hoko-maru* 18, a long-line tuna fishing boat, left Misaki Port in Kanagawa Prefecture—the largest port for tuna fishing boats in Japan—on February 5, 1956, and was sailing to American Samoa on April 2 through the waters off Fiji when a U.S. patrol plane circled overhead and dropped a "communication cylinder" on the boat warning it to leave the area. The *Hoko-maru* left the area at full speed and returned to Yokosuka Port on June 1, after which it reported its route in detail to the government.

The Japanese Embassy in Washington reported this to the U.S. Department of State, also remarking that, in light of international law, the "Japanese government [could] not agree with" the U.S. stance that "vessels entering the danger zone [were] incurring risks for which the U.S. [could] not assume responsibility."[64] Also in May, the U.S. Navy, which was patrolling near the danger area, sent information to the Department of State that a Japanese fishing boat had entered the area. According to the report, about an hour before the hydrogen bomb test, "the vessel, No. KNI-79, in position, latitude 17-34 N., longitude 170-40 E., was located by patrol aircraft and directed to clear the area." The vessel was safely out of the danger zone by the time of the detonation, but the Navy asked the Department of State to report the incident to the Japanese government and issue a stern warning. KNI-79 indicates a Kanagawa-flagged first-class vessel (a power fishing vessel of 100 tons or more), presumably another tuna fishing boat that sailed from Misaki Port in Kanagawa Prefecture.[65] The U.S. government was concerned that such incidents might one day lead to a claim for compensation from Japan. George C. Spiegel, Special Assistant to the Secretary of State, suggested that "it would be well to have data on the fishing vessels [found in the danger area] here in Washington and readily available."[66]

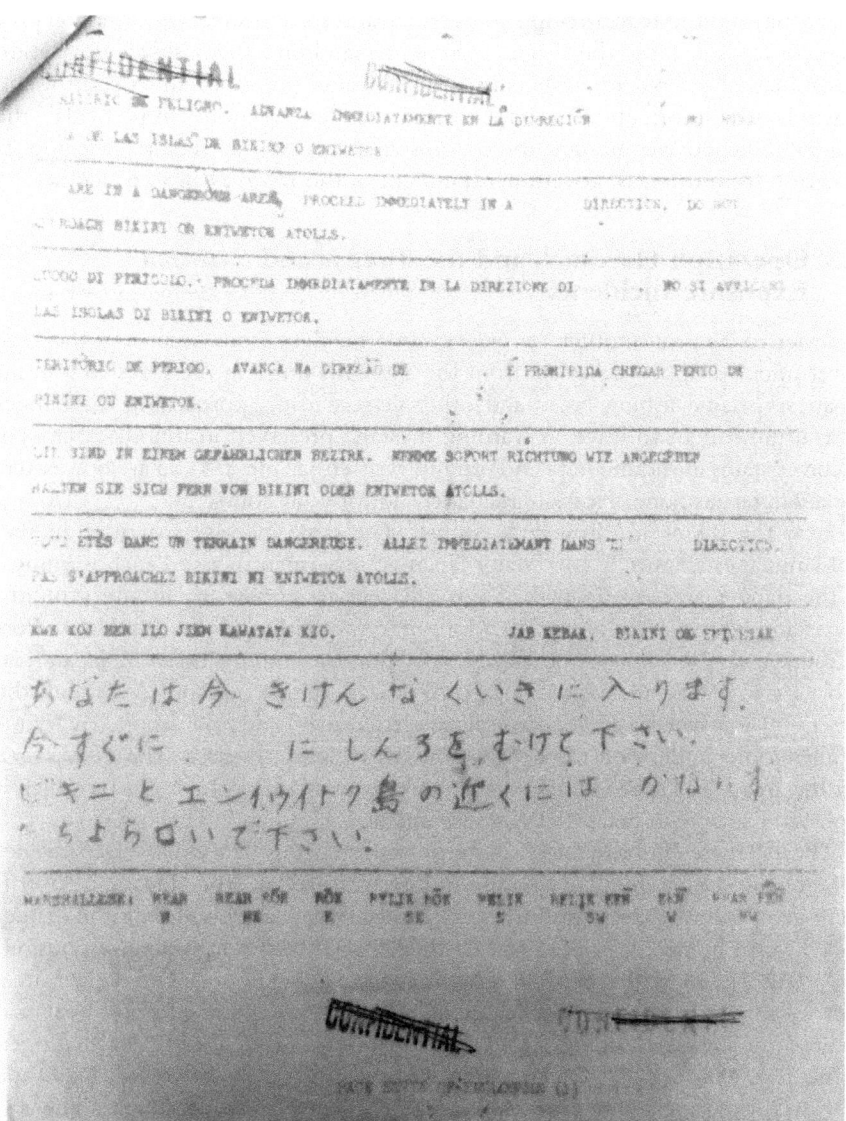

Figure 5.2 A warning message written in several languages instructing vessels to move out of the danger area. RG59, Entry 3008-A, box 427, NACP.

The Japanese, too, feared that another *Lucky Dragon 5* incident might occur. The Japanese government voluntarily submitted to the U.S. Embassy a list of Japanese merchant ships operating in the vicinity of the danger area, and requested that "the local commander directly issue a notification to ships operating in the vicinity immediately before and after the tests."[67] However, Secretary of State Dulles replied that, because the schedule changed according to climatic conditions, it was "operationally impossible" for the U.S. to notify the Japanese vessels of the times of "individual detonations," and "in any event such information would be of no value [to] Japanese vessels if they remain[ed] outside [the] danger areas" during the testing.[68]

In practice, however, there was no shortage of cases in which ships outside the danger area were exposed (or were suspected of being exposed) to radioactive contamination. For example, on June 5, 1956, the Sumitomo Metal Mining Co. cargo ship *Mizuho-maru*, which had returned to Japan after sailing in waters close to the danger area, was found to be radioactive at 1,150 counts per minute, and the crew's white blood cell count was low. Japanese newspapers reported that the exposure was due to "a third surprise thermonuclear experiment reportedly carried out by U.S. in the Bikini atoll area [on the] morning [of] May 28." Dulles immediately sent a telegram to the U.S. Embassy in Tokyo to confirm this, and the Japanese government submitted the crew's white blood cell count data to the embassy.[69] Ambassador Allison then informally obtained the Ministry of Foreign Affairs' final report on the *Mizuho-maru*, which he forwarded to the Department of State. The report stated that the "average radioactive contamination of [the] ship [was] 140 counts per minute" and therefore "not deemed directly hazardous to [the] human body," and that the drop in the crew's white blood cell count was "not considered serious or attributable to radiation." On basis of this report, the Japanese government considered the "matter closed" and did not plan to issue any "public statement."[70] There was also news that Kawasaki Kisen Co.'s *Seizan-maru*, en route to Australia to load iron ore, was hit by a suspected radioactive squall.[71] In addition, all major Japanese newspapers reported that the Soviet research vessel *VITYAZ*, which was taking part in the International Geophysical Year (IGY), had evacuated on June 7 after detecting strong radiation 2,000 miles west of the danger area. Ambassador MacArthur consulted with the Department of State, the AEC, and the USIA on how to deal with the story, which could continue to attract attention "if [the] ship visits [a] Japanese port," even though, according to the information available to him, "none of the ship's crew appeared to have been affected physically."[72] The Soviet ship actually docked at Dejima in Nagasaki on June 9 and stayed until June 14, after which it left for Vladivostok, during which time the crew gave interviews to the media. According to the report, the "vessel was 475 miles west of Eniwetok on May 29 when it recorded 'radio activity count of 70,000 in rain'" but the press had "not given [the] news much prominence" to the relief of the U.S. Embassy.[73]

These reported cases demonstrate that it was very difficult to prevent exposure incidents completely, no matter how much effort the U.S. Navy put in to keeping ships out of the danger area. Also, tuna long-liners were even more likely to be exposed to fallout than merchant ships, as they had no fixed route and would readily change course to follow schools of tuna. However, the exposure of the fishing boats was rarely made public. The fact that a citizens' group in Kochi Prefecture conducted interviews to discover more about the radiation exposure of tuna fishermen, and that former seafarers and their families subsequently filed a lawsuit against the state for damages (the case was dismissed at the second hearing in 2019; another case seeking recognition of work-related injuries was subsequently brought and was still ongoing as of 2021), illustrates how many seafarers were exposed to danger. Their health problems were never addressed in the U.S.–Japan compensation negotiations.[74]

In the midst of this tense situation, a major incident occurred. At 10:24 p.m. on July 16, 1958, Ambassador MacArthur sent an urgent telegram to Secretary of State Dulles. According to information from the Commander in Chief of the U.S. Pacific Fleet (CINCPACFLT), two Japan Coast Guard (JCG) ships "engaged in [P]acific survey projects in connection [with the] IGY have reported high levels of radioactivity in vicinity of Truk." Onboard readings showed "19,000 counts per minute on [the] scintillation counter," rainwater at 100,000 counts per minute per liter, and seawater at 247 counts per minute per liter. The JCG official "told [the] naval attaché that crews on both ships [were] very worried about radioactivity" and instructed the ships to divert to Rabaul in Papua New Guinea for fresh-water decontamination (Figure 5.3).[75] Ambassador MacArthur would continue to keep Dulles informed moment-by-moment of the Japanese situation regarding this case.

The two vessels concerned were the JCG observation ship *Takuyo* and the JCG patrol vessel *Satsuma*, with the *Takuyo* recording particularly high levels of radiation. As will be discussed in detail below, the fact that the Japanese government ultimately made no claim for compensation in this case had the effect of making it more difficult to claim compensation for the fishing boats.

On July 14, 1958, *Takuyo* was conducting an IGY survey in the South Pacific, when it encountered a squall while sailing 160 nautical miles west of the nuclear test zone at Eniwetok Atoll, and detected 100,000 counts of radiation per minute in the rainwater, at which point it ceased operations and evacuated south. The *Satsuma*, which stood by at 300 nautical miles west of the danger area, evacuated south with the *Takuyo*. On receiving their report, the JCG ordered the two ships to enter Rabaul for decontamination and medical examinations. A blood test conducted by the ship's doctor on July 18 on a voluntary sample of 15 crew members showed a maximum white blood cell count of 4,900, a minimum of 2,000, and an average of 3,300 (it was noted that "6,000 to 8,000 is normal"), suggesting "an abnormally lowered white blood cell count."[76] A medical examination by Australian Health Department staff in Rabaul found that 10 *Takuyo* seafarers had suffered a 30% to 40% drop in their white blood cell count to between 2,000 and 4,000 in the 3 days until March 18. Seven other crew members had readings between 4,000 and 4,900.[77] Blood tests carried out on the entire crew at Rabaul also showed that the *Takuyo* had values outside the normal range, from a low of 3,000 (4 crew members) to a high of over 10,000 (4 crew members).[78] However, the reports to the U.S. Department of State about the exposure of the crew were conflicting. A July 22 report stated: "radio-activity and effects on personnel are not as serious as reported … No one on ships is showing palpable signs of illness, or radiation burns, or giddiness, or losing hair." "[The] Medical Officer has examined seven from each ship of whom only five altogether

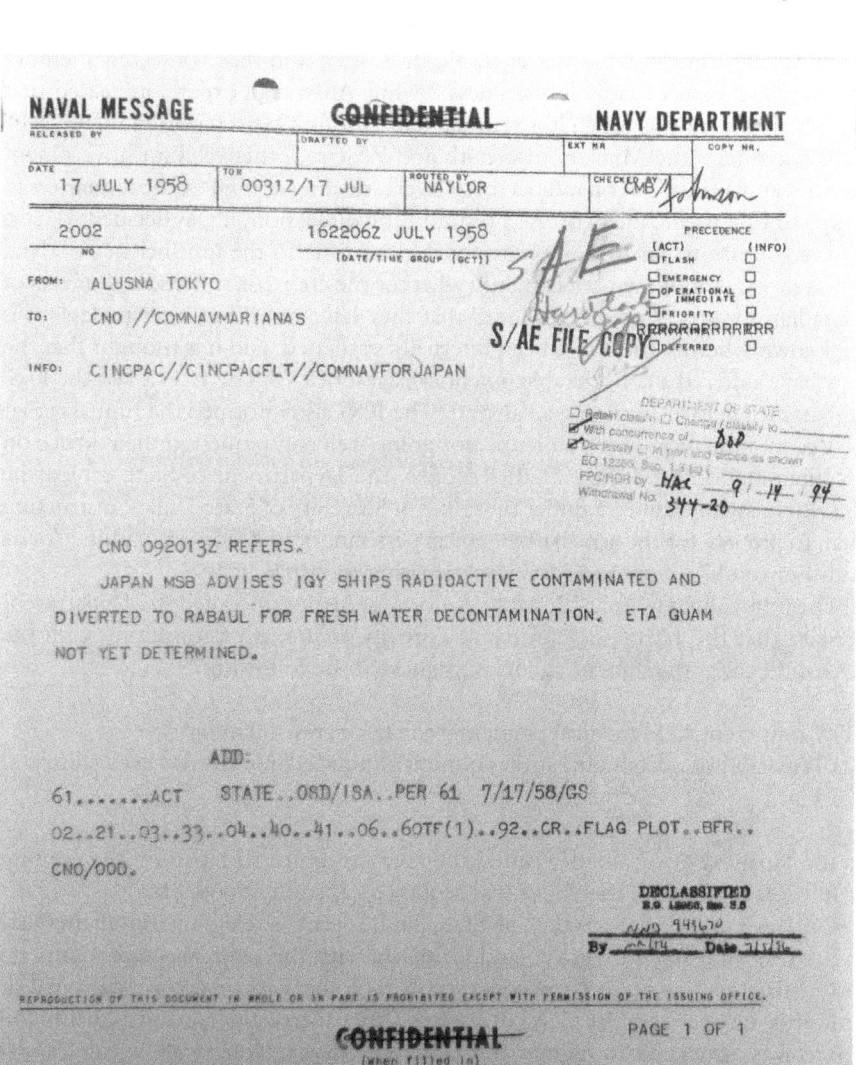

Figure 5.3 A State Department copy of the Navy telegram report on the accident. RG59, Entry 3008-A, box 427, NACP.

have [a] white blood count of less than 5,000. In the view of the Medical Officer [in] Rabaul, none of these five need[ed] treatment." However, a telegram dated July 23 stated that "14 seamen aboard had been exposed to radiation" and "their white corpuscle count was 3500, whereas a healthy person has a count of 7,000 to 8,000." Furthermore, according to a report submitted to the Secretary of State on the same day, 23 July, the JCG stated that the "white blood count of some of the crew members [was] as low as 2,000 or 3,000" and that "one crew member was showing signs of radiation sickness," while Australian experts indicated that the "the problem was much less serious than" the JCG had reported."[79]

At the time, "the Ministry of Health and Welfare, other relevant government agencies and authorities on radioactivity" were meeting at the JCG headquarters in Tokyo to discuss the incident. As a result of the discussion, it was decided to have the crew return to Japan as soon as possible, as "due to the insufficiency of data, although we cannot say with certainty whether the crew has suffered any effects of the radiation, it is at present believed that they have not," but nevertheless "it is not known whether they have been internally irradiated, and it is thought that the crew have suffered a considerable psychological shock."[80] The JCG asked the U.S. Embassy for help in arranging an aircraft. The JCG also informed the Embassy that the decontamination of the ship was not going well and requested their advice on decontamination procedures.[81] In response, the Department of State cabled the U.S. Embassy in Canberra under the name of Secretary of State Dulles, instructing them to prepare for the arrival of "medical personnel and decontamination experts from Eniwetok" whom the AEC was preparing to dispatch.[82]

The following day, July 22, Ambassador MacArthur informed the Department of State that the Japanese Ministry of Foreign Affairs, after consulting with the JCG and other government agencies, requested the following:

(1) Competent U.S. medical personnel to check crew members;
(2) Provision of decontamination equipment and technicians to check ship; and
(3) Eight pocket dosemeters.

As the Japanese government wanted the crew to return to Japan on the evening of July 24, dispatch of the above was requested as soon as possible.[83]

On the same day, Secretary of State Dulles sent a telegram simultaneously to the U.S. embassies in Tokyo and Canberra with the same message. Canberra was instructed to obtain permission for U.S. Joint Task Force 7 (stationed at Eniwetok nuclear test site) aircraft carrying the doctors to land in Rabaul, and Tokyo was instructed to request the Japanese Government to allow the *Takuyo* to remain in Rabaul until the U.S. team arrived.

> In view [of the] complication that arouse [sic] in 1954 in connection with [the] Fukuryu Maru, we feel we should take every possible precaution [to] check this reported incident immediately. If reminded of [the] 1954 incident, [the Department] believe[s the] Australian Government will understand [the] potential explosive character [of] this type [of] situation.[84]

Dulles thus emphasized the seriousness of the situation. Ambassador MacArthur replied on the following day, the 23rd, that the Japanese Government had agreed to keep the *Takuyo* in Rabaul, but that they wanted a medical team to arrive as soon as possible, and that the JCG had instructed the two ships to fully cooperate with the Americans. Furthermore, according to MacArthur, late on the evening of the 22nd, JCG in Tokyo received a report from Rabaul that one of *Takuyo*'s crew members was showing symptoms of radiation sickness. The MOFA, fearing that the incident might attract media attention, repeatedly reminded the U.S. Embassy to "avoid any publicity on U.S. actions."[85] In response, Dulles instructed MacArthur as follows:

> We too [are] most anxious [to] avoid any press references [on] this matter. If there is [a] news leak or press query ... we would merely confirm [that the] U.S. Government has offered [a] service team [of] U.S. experts [to] investigate [the] reported occurrence of increased radioactivity. [The] Japanese Government [has] accepted this offer.[86]

This shows that both the Japanese and U.S. governments attempted to manipulate the information to avoid the incident becoming public.

However, a telegram to the Department of State from the U.S. Embassy in Canberra stated: "Although we will make every effort [to] avoid publicity [of] U.S. action" and had requested the Australian Ministry of External Affairs to "handle [the] matter on [a] confidential basis," it would become difficult to "suppress speculative press reports from Rabaul with [the] arrival [of] the U.S. experts there." Therefore, the Department suggested that the "Japanese Embassy [in] Canberra be instructed [to] avoid publicity and be prepared [to] take [the] position outlined" in the previous telegram, i.e., Japanese acceptance of the U.S. "service team."[87] MOFA followed suit, saying that

> at the request of the Americans, no special newspaper announcements will be made, and the policy will be to give as little publicity as possible. In the event of an enquiry from a journalist or other person, we will respond that "we have accepted an offer from the United States to help with the situation."[88]

However, because two newspaper reporters, Keisuke Konno of the *Minami Nihon Shinbun* (Kagoshima Prefecture) and Eiji Katayanagi of the *Yomiuri Shinbun*, were on board the *Satsuma*,[89] Japanese and American newspapers began to cover the incident. The *Yomiuri* was particularly bold in continuing to report on the case. An article that appeared on the front page of the *Yomiuri Shinbun* morning edition on July 25 heightened Ambassador MacArthur's sense of urgency. In a telegram to Secretary of State Dulles, the USIA, the AEC, and the Pentagon, the ambassador said that "Until today, [the] press has made only brief, back page references to [the] possibility [that] some crew members may have radiation illness," but the *Yomiuri* had published a lengthy article revealing that a team of American doctors and specialists had been sent to Rabaul. The *Yomiuri*'s feature suggested that: "crew members may be seriously sick"; "Possible Government remonstrations to U.S. due 'heavy radioactivity' outside danger zone"; and "Socialist plans to interpellate government on incident when lower House Foreign Affairs Committee meets July 28."[90] MOFA asserted that the "government remonstrations" in the *Yomiuri* article were entirely fictitious, but Ambassador MacArthur worried that "pressures must be beginning [to] develop as result this story." In response to the article, both MOFA and the U.S. Embassy received enquiries from the press, and responded in the manner previously agreed upon. MacArthur advised that the results of testing by the team of American experts be sent to the Japanese Government as soon as possible, as

> top Japanese radiation experts have refrained from alarmist statements and have taken [a] responsible position so far, but we believe they will be less cooperative if reports of our experts come to them first through press rather than official channels.[91]

In addition to the *Yomiuri Shinbun*, other articles in Japanese papers included "Fearful Escape from Ashes of Death: 100,000-Count Squall" (*Mainichi Shinbun*, evening edition, July 15), "All Crew Members Have Lowered White Blood Cell: *Takuyo*'s Report to JCG" (*Nihon Keizai Shinbun*, evening edition, July 21), and in the "Current Events" column of the *Yomiuri Shinbun* on July 27, it was reported that meteorologist Yasuo Miyake, who worked on the research ship *Shunkotsu-maru* after the *Lucky Dragon 5* incident, argued that it was "serious enough to be called a second Lucky Dragon incident," and that it was necessary to "seek full compensation from the U.S. for damages incurred." In the United States, Oregon's *Capital Journal*, in a piece entitled "Doubly Unlucky Dragon," recalling the *Lucky Dragon* incident, criticized U.S. government experts for "repeating the same mistake twice."[92]

The American medical team was originally scheduled to arrive in Rabaul on the 23rd, and the Japanese government had planned to return the crew to Japan on the 24th, but the medical team did not actually arrive until the 25th.[93] Upon arrival in Rabaul, the AEC and naval medical teams gave a definitive diagnosis of the extent of the crew's exposure to radiation: "it appears quite possible that the rainfall that occurred on the Takuyo Maru did contain radioactive fallout from the tests," although "[the] amount of this fallout was not sufficient to have caused radiation sickness in the crew members." "The medical examinations showed no evidence of radiation injury. The lowest white blood count of 3,350 is below average but is not considered serious and counts in this range are occasionally seen under normal circumstances."[94] The Japanese government asked the Americans to send their report in confidence and as soon as possible, preferably before *Takuyo* and *Satsuma* returned to Japan. In order to avoid controversy, the Americans intended to wait for the AEC to analyze its findings before releasing them to the Japanese. After repeated requests from the U.S. Embassy to send the report to Japan as soon as possible, it was sent to Japan on August 4.[95]

Ambassador MacArthur felt that this incident had rekindled public opposition to nuclear testing in Japan, where there had previously been a "growing and somewhat fatalistic acceptance of testing of atomic weapons by the United States" among the Japanese people. In particular, the fact that the vessels were exposed to fallout even as they were sailing well outside the danger area entirely discredited the American claim to have developed a "clean bomb." On the whole, however, the tone of the Japanese press was "less violent thus far than we would have anticipated," and it was thought that if the incident had occurred a year earlier, the reaction would have been much more intense. Nevertheless, the ambassador's conclusion was that there was no doubt that American diplomacy with Japan had been damaged by the incident.[96]

Following the arrival of the report, the Japanese began to examine it. To obtain information on the Japanese government's move, the U.S. Embassy invited Dr. Tsuzuki, expert on "A-bomb sickness" and Japanese representative to the U.N. Scientific Committee, to the embassy. According to Tsuzuki, a special government committee met on August 5 and decided to carefully inspect the crew and the ship. Over the course of a week or more, the entire crew was to be inspected, as well as the ship's hull and the water and food carried on board. The inspection was to be conducted under the supervision of Dr. Masanori Nakaizumi of the Atomic Bomb Casualty Commission (ABCC).[97]

When the *Takuyo* and *Satsuma* arrived in port on August 7, a large number of reporters and radiation experts were waiting for them, but the tone of the coverage was much more restrained than at the beginning of the incident. Tests revealed that the white blood cell counts of the two crew members were still abnormally low, and the JCG informed the U.S. Embassy of this, but "this information [was] not made public." As a result, the newspapers reported on the return of the two ships to port, but generally in small articles, describing the extent of the radiation exposure as "not serious." Rather, the articles relayed that the crew expressed gratitude for the "treatment given them by U.S. doctors at Rabaul" and the ship's doctor stated that "U.S. counts [were] most probably more accurate since U.S. doctors had better equipment available." The press also carried "reassuring statements issued by Japanese radiation experts" saying that crew was "not harmed as much as we anticipated" and were "free from any serious radiation effects."[98] The Japanese government, who had carried out a detailed examination of the crew, published their results, which were consistent with the American findings: "There [are] no current indications of radiation damage" and "there [is] no need for further detailed examination of the crew."[99] Following the government's announcement, the major newspapers reported favorably that the crew were "safe" and "unaffected by the exposure." Katayanagi, the *Yomiuri* reporter aboard the *Satsuma*, produced a newsreel entitled "Sea of Death," which the U.S. Embassy found to be generally "balanced" and acceptable.

In addition to the *Takuyo* and *Satsuma*, there were other incidents during this period, such as the Tokyo University of Fisheries training ship that turned back after measuring 2,000 counts of radiation in the rain while sailing off the coast of Taiwan during an IGY survey, and reports that two crew members on a Japanese cargo ship became ill with abnormally low white blood cell counts. However, as news coverage of *Takuyo* and *Satsuma* had died down, the U.S. and Japanese governments concluded that so would the coverage of other cases. Nevertheless, a number of newspapers continued to comment critically in their editorials. The *Mainichi Shinbun*, for instance, deplored that *Takuyo* and *Satsuma* were forced to abandon their IGY research, noting that their "exposure to radiation was outside [the] designated danger zone," and concluding that test suspension was "essential if ships could be exposed so far away from [the] already large danger area," and that the Japanese government should "seek compensation from U.S." Even the "generally pro-American" *Sankei Shinbun* argued the incident proved that there is "no safety zone on earth any longer" and "called for renewed efforts for test ban[s]."[100]

On August 13, Ambassador MacArthur sent a telegram to the Department of State summarizing the *Takuyo* and *Satsuma* incidents as almost "concluded" as follows:

> Noteworthy has been very responsible manner with which GOJ [government of Japan] has handled [the] question of possible radiation sickness among crew members, in [the] face [of] tremendous pressures from press and [the] opposition to build up this incident into another FUKURYU MARU case and create major controversy with us ... As result, efforts by [the] press and leftists to build up to atmosphere of hysteria and thereby dramatize Japanese demands for nuclear test suspension have fallen flat ... However, as responsible Japanese officials have told us, their efforts would not have succeeded were it not for full and expeditious cooperation of US Govt.[101]

Although the U.S. side recognized that the incident had "concluded," the Japanese government had not, at this point, given up on the possibility of seeking compensation for the incident. On July 25, the MOFA sent a memorandum to the U.S. government calling attention to "the fact that radiation exposure forced the cancellation of the observation of ocean currents conducted as part the International Geophysical Year project" and to "the fact that the radioactive contamination occurred outside the danger zone," given that the Japanese government had long pointed out that "there is no scientific proof that radiation damage can be adequately prevented either inside or outside the danger zone" and in view of the possibility of "compensation claims that may arise in the future."[102] Although MOFA told the U.S. Embassy that they were not "prejudging results of [the] current investigation by US experts" and that the "memorandum [was] not [a] 'protest note' or request for compensation," Ambassador MacArthur suspected it might "lay [an] initial basis" for a later request for compensation, but at the same time, he speculated that the MOFA "considered some written communication to US" was essential "to provide [a] basis for reply to anticipated Diet enquiries." The MOFA did not make public the contents of this memorandum.[103]

In the meantime, the Japanese government began to consider the basis on which actual damages should be claimed. Damages were divided into "direct damages" and "indirect damages." Regarding "direct damages," even if direct harm to the crew could not be proved, medical treatment or changes to the ships' navigation "on the basis of sufficient reason to believe that harm had occurred in view of the effects of radioactive substances on human health" could form the basis for a claim for compensation. Furthermore "the detection of 100,000 counts of radioactivity in the squall" was held to constitute "sufficient reason to believe that damages had been incurred." The primary "indirect damage" was "the impossibility of continuing the [IGY] survey." If it could be proved that the ships were "compelled to make port at Rabaul," and that even if "the U.S. team of doctors judged that the ship had … no concern regarding radioactivity," it was necessary to send the ship back to Japan for re-examination, this state of affairs could also be the basis for a claim for compensation. Among other considerations was the fact that, out of concern for their families, the crew "had to be ordered to return to port," although this was judged to be "too subjective to be a sufficient basis for a claim for damages."[104]

Telegrams between the U.S. Embassy and the Department of State show that the MOFA was attempting to reconcile differences of opinion within the Japanese government. Just as the Japanese government was debating the merits of their claim, the U.S. Embassy called an official from the North American Section of the Ministry of Foreign Affairs (established in May 1958 under the American Bureau) and asked when the U.S. government should make public its response to the pending claim for compensation for Operation Redwing in 1956. As noted in Section 5.1, the U.S. government had replied on September 13 to the Japanese Government's resubmitted request for compensation on January 21, 1957, stating that it could not provide compensation because the documents submitted by the Japanese government failed to establish what was actually lost. Later, in January 1958, the Japanese government had submitted another request for compensation, which had met with the same result. However, before that response could be made public, the *Takuyo* and *Satsuma* incidents occurred, and the MOFA asked the Americans to "withhold delivery [of the] formal note." Now that the *Takuyo* and *Satsuma* incidents had been "closed" in the eyes of the U.S. government, on September 8, Ambassador MacArthur again asked MOFA when they should publicize the response. "Deputy Director of American Bureau [of MOFA] again suggested" that the U.S. should "delay delivering [the] note," since the American Bureau was "*attempting to secure agreement*" *within the government so that* "*no compensation claim be made*" for the *Takuyo* and *Satsuma* incidents (italics by the author).[105]

The internal government coordination by the American Bureau of MOFA, and the eventual decision by the JCG not to seek compensation, eliminated the possibility of any compensation claims for *Takuyo* and *Satsuma*, and the precedent it set essentially put a stop to other compensation claims. In addition, other ministries and agencies, in light of the fact that even the compensation claims for Operation Redwing remained unresolved, expected that negotiations for further claims would tread a similar course, and the mood prevailed that it would be futile to make any further requests for compensation. This is illustrated by the record of the "Inter-Ministerial Liaison Meeting on the 1958 U.S. Nuclear Tests in the Pacific Region," held on September 17, 1958, with representatives from MOFA, the JCG, the Civil Aviation Bureau of the Ministry of Transport, the Shipping Bureau of the Ministry of Transport, the Fisheries Agency, and the Ministry of Health and Welfare. At the meeting, which was held because "the Diet will resume in October and we need to consider our stance towards compensation," the Chief of the JCG's Management Division stated that there was "no intention to claim compensation" for the damage to the *Takuyo* and *Satsuma*, based on "the conclusion that there was no direct damage observed either to the human crew or the ships themselves." In response, an assistant director of the Fisheries Agency's Marine Affairs Division 2 said: "We believe that the establishment of an expansive danger area in the Pacific Ocean will certainly impact our fisheries" and that fishing vessels "are instructed to report any substantive damage," indicating a proactive attitude towards (or at least acknowledging the need for) claims for compensation in relation to fishing vessels. At the same time, however, the Fisheries Agency admitted that it had found it difficult to "prepare data to calculate damages in a way that would satisfy the U.S." The Ministry of Transport's Civil Aviation Bureau stated that four Japan Airlines flights had been delayed because of the loss of communication caused by the nuclear tests, estimating damages to "the company caused by delays at one million yen." Although the *Japan Times* reported that the Ministry of Transport "was to make a claim for compensation," this report was later rescinded and in fact the Ministry had "no intention of lodging a claim." The Shipping Bureau of the Ministry of Transport also stated that there were

> indirect losses, including the cost of diverting ships, but as claims for compensation for 1956 had not been settled and that, without a reasonable possibility of compensation, shipping companies had no incentive to produce the time-consuming documents necessary for proof of damages.

The Ministry of Health and Welfare doubted whether "contamination of *Takuyo* and *Satsuma*" would be "regarded damages" by the U.S. since even "the concept of danger differed greatly between our side and the United States." The Ministry feared that appeals for compensation might end in vain. MOFA's Legal Affairs Division suggested to "keep the calculation of indirect damages pending" until the final decision of the Japanese government was made, for "if no compensation is claimed for the JCG's official ships while claims for fishing vessels and merchant ships are filed, the inconsistency will weaken our position." Finally, the Director of the North American Affairs Division of MOFA summarized the Japanese government's stance as, "no direct damages have been established, and the existence and extent of indirect damages is still under consideration." At the same time, however, the ministries were asked to continue to consider the "strength of the demand for compensation from fishing vessels, shipping companies and other parties involved," as well as public opinion.[106] In August, the Japanese government had thought it had grounds to demand compensation for both direct and indirect damages, but we can see here that its attitude towards the issue has regressed significantly. This was due to the precedent set by the JCG of not making compensation claims, and a shared sense of resignation among the various agencies that the U.S. would not respond to their demands for compensation after all.

On August 3, 1959, about a year after the radioactive contamination of the *Takuyo* and *Satsuma*, the *Takuyo*'s chief engineer, Hirokichi Nagano (aged 33 at the time of the incident) died of acute myeloid leukemia. However, as a result of the autopsy, the Ministry of Health and Welfare concluded that the "radiation dose he had received was so small that, in the view of current medical science, it was difficult to establish a direct link between the exposure and his leukemia."[107] The results of previous medical examinations of the rest of the *Takuyo* crew were also collected and reviewed, but it was similarly concluded that there were "no findings suggestive of effects of radiation." All major newspapers carried articles alleging a connection with radiation exposure, such as "Chief Engineer Nagano Dies of Leukemia—One Year after Eniwetok Radiation Panic" (*Yomiuri Shinbun*, August 5), "Crew Member Dies of Leukemia—Takuyo Exposed to 'Ashes of Death' Last Summer" (*Asahi Shinbun*, August 5), and "Chief Engineer Nagano Dies of Leukemia – One Year after Takuyo Exposed to Ashes of Death" (*Mainichi Shinbun*, August 4). The *Yomiuri Shinbun* (August 6) also reported, based on interviews with former crew members and others involved, that the findings of the tests carried out after the crew's return to Japan as "so far there is no obvious connection" had been publicized with the phrase "so far" deleted, and that many Japanese journalists had viewed the government's attitudes to keep the incident "secret and private" as problematic. The article also revealed that "despite the government's assertion that the crew did not require further examinations, in fact there were 16 members on the ship who needed special medical attention."[108] However, this media criticism did not shake the policy of the Japanese government, which had already decided not to pursue compensation claims.

In the end, the Japanese government made no claim for compensation for the *Takuyo* and *Satsuma* incidents, nor for the entirety of Operation Hardtack. The official reason for this was that "in spite of consultations with the relevant ministries, no reports of damages to merchant ships or fishing vessels have been received, and the fact that the previous compensation issue remains pending."[109] The radioactive contamination of *Takuyo* and *Satsuma* consequently raised the hurdle for Japan in regard to compensation claims generally. The need for uniformity between ministries also made it difficult to lodge claims for compensation for pelagic fisheries that had lost fishing grounds or been forced to make detours.

5.4 The Relationship between Demands for Nuclear Test Bans and Claims for Compensation

It has been pointed out in previous studies that the Japanese government's repeated requests to the U.S. government to suspend nuclear testing were intended to capture public opinion, to divide the anti-nuclear movement led by left-wing forces, and to avoid being pursued by the opposition in the Diet. However, the Japanese Government itself was considering whether the demands for a halt to nuclear testing should be merely a gesture to sway public opinion, or whether there should be other "real reasons" behind them, and what balance should be struck between the two. In April 1957, soon after the establishment of the Kishi Government, the issue of opposition to nuclear testing was discussed with the aim of "contributing to the establishment of a clear, basic stance in the Ministry of Foreign Affairs." First of all, discussions iterated that "because of our nation's position as part of the free nations," the possession of nuclear weapons could not be disavowed; therefore the need to "separate testing from nuclear possession, and argue that testing is either unnecessary or should be limited" was confirmed. The following discussion then took place on the above-mentioned point regarding the nature of opposition to nuclear testing:

> It is true that the government's advocacy of a ban on testing began with a superficial "gesture" (author's note: appealing to the domestic public opinion), but it has now advanced to the stage where it is no longer merely a matter of simply "gesturing," and has come to have significance as part of the "autonomous diplomacy" of Japan, and thus we share the opinion that we should give much less weight to mere gestures and concentrate our efforts on more substantial diplomatic strategies, not just appealing to the emotion of the general public.[110]

The arguments raised for the anti-nuclear testing stance being not merely "gestural" included the following aspects: (1) emotional (as a victim of atomic bombs), (2) legal (e.g., freedom of the high seas), (3) direct damages (e.g., to fisheries, and forced navigational changes), and (4) long-term harm to humanity. However, it was acknowledged that (1) to (3) were "generally weak" and that (4) "should form the strongest moral argument in diplomacy." The reason given for the weakness of (3) was that "if they say they will compensate us in full, the matter is settled," although in practice, the possibility of such compensation was almost nil.[111] Damages to fishing boats and merchant ships, the above arguments show, were acknowledged as one of the "real reasons" for the Japanese government's opposition to nuclear testing, but that was judged to be a weak diplomatic argument.

However, the texts of numerous requests to the U.S. government for a halt to nuclear testing clearly stated that the Japanese government would reserve the right to make a claim for compensation in the event of economic loss. For example, an appeal made through the Japanese ambassador to Washington, Koichiro Asakai, when the danger area for Operation Hardtack was announced, stated that the Japanese government was gravely concerned because the zone was "in close proximity to the navigational and operational waters of Japanese merchant ships and fishing vessels." The appeal stated:

> The Japanese government makes clear that the U.S. government is responsible for the compensation for losses and damages inflicted on Japan and Japanese citizens by nuclear tests, including economic losses incurred by the

establishment of the danger area, and it reserves every right to demand a complete compensation for said losses and damages.[112]

Taken together, the Japanese government's appeals for a halt to nuclear testing were partly a "gesture," and partly genuine, and its response of protesting and seeking compensation for the damage caused to its own fisheries and shipping industry, while "weak" as an argument for diplomatic negotiations, was well recognized as necessary by the Japanese government.

The U.S. government showed an understanding and discerning attitude towards the "gesture" element. For instance, the U.S. government, noting that the commencement of Operation Hardtack coincided with the Japanese general election, showed care not to weaken the position of the LDP government. On May 3, 1958, Walter S. Robertson, Assistant Under-Secretary of State, advised in a memorandum to Philip J. Farley, Special Assistant to Secretary of State in charge of AEC, that a nuclear test should not be conducted just before the Japanese general election scheduled for May 22. The reason given was that, if a test were to be carried out after the announcement of the election, "Socialists will seize upon" the opportunity "to attack Mr. Kishi and the Liberal Democratic Party." Robertson argued that electoral considerations should be taken into account because it was "definitely in the interests of the U.S." to "help create public opinion favorable to the LDP which has proclaimed cooperation with the U.S. as a basic tenet." In response, the Department of State informed the Pentagon and the AEC that if it would be difficult to postpone, they should at least limit testing to a small scale.[113] The AEC accepted and gave instructions that no nuclear tests of more than 200 kilotons be conducted until after 11:00 p.m. (local time on Eniwetok Island) on May 22. The AEC had established a rule that "only detonations of more than 200 kilotons in yield are announced to the public," so tests of less than 200 kiloton yield were not made public, and as a result, no information about nuclear tests would be released during the Japanese election.[114] After Operation Hardtack, the Department of State expressed gratitude to the AEC for its "decision to defer certain 'Hardtack' shots until after the Japanese elections," a result of which was that nuclear testing "did not become an election issue despite the efforts of the Socialists to make it so."[115] The U.S. government's control of information about nuclear testing extended even to the domestic politics of other countries.

However, the U.S. and Japan did not share the same cohesiveness when it came to the substantive, non-gesture reasons of the Japanese for opposing nuclear testing. The Nikkatsu-ren (the Federation of Japan Tuna Fisheries Co-operative Associations), the Japan Shipowners' Association, the Yaizu City Fish Processing Association, and other pelagic fishing-related organizations had sent petitions to both the U.S. and Japanese governments calling for a ban on nuclear testing. Among these, the Shipowners' Association petitioned the Japanese government to pressure the U.S. government to "pay compensation."[116] The Japanese government was thus well aware of the pushback from the fishing industry, but as the overall momentum of compensation claims waned, it was unable to channel the voices of the fishing industry into diplomatic negotiations. After the end of Operation Hardtack, voices within the U.S. government advocating that damages to the fisheries should be objectively investigated subsided, too.

5.5 Information Control on Protest Ships

"Protest ships" that dared to venture into the danger areas of the hydrogen bomb tests were also subject to information control by the U.S. government, as media coverage of the work of anti-nuclear activists could re-emphasize the continuing H-bomb testing and potentially stir up anti-nuclear and anti-American sentiment around the world. The U.S. government's actions surrounding the protest ships, described below, can be viewed as a "negative overseas information program" designed to prevent media coverage of the nuclear tests and protests, and to keep them as low profile as possible.

In January 1958, Albert Bigelow, a former U.S. Navy officer who had retired to protest nuclear testing and who had formed the Committee for Non-Violent Action with other Quaker anti-nuclear activists, notified the U.S. government that he and his cohort planned to enter the danger area aboard the protest ship *Golden Rule*. On February 11, Ambassador MacArthur telegraphed the Secretary of State that Japanese newspapers were "featuring reports and pictures of departure from U.S." of the *Golden Rule* and that a Japanese anti-nuclear group (the Japan Council against Atomic and Hydrogen Bombs, hereafter Gensuikyō) was considering sending a protest ship to the waters. MacArthur pointed out that in the year before, a Japanese anti-nuclear group had planned to send "protest ships" to the Christmas Island area, the site of the aforementioned British nuclear test, but had eventually abandoned the plan due to opposition from the Japanese government, and that a "lack of any official U.S. comment on [the] Golden Rule" may "encourage the Council to carry through with the protest fleet plans this year."[117]

The OCB discussed countermeasures on a number of occasions, fearing that the sailing of the Golden Rule would provoke protests in Japan, including those by Gensuikyō. At a meeting on February 19, 1958, for example, consideration was given to issuing a statement which would "work as a deterrent" to other protests, and it was agreed that a statement should be issued to the effect that "the U.S. government would stop Golden Rule [from] enter[ing] the danger zone by all means." At the same time, the OCB, through the USIA, sent a notice to USIS branches in each country stating that it "would oppose any event or public relations activities concerning Golden Rule."[118] In other words, they attempted to suppress information so that the Golden Rule would not become a talking point in countries around the world.

However, there was no singular approach within the U.S. government in regard to controlling information about these protests. The Department of State was hesitant to make such a strong statement of opposition, given that the U.N. was in the midst of deliberation on the principle of "freedom of the high seas." This is because the danger area in which the nuclear tests would take place was within the "high seas" under the trusteeship of the United Nations, and international public opinion had already turned against the U.S. for establishing a danger area within international waters in which to conduct nuclear tests. To issue a statement condemning the *Golden Rule* at such a time would have been counterproductive in terms of the overseas information program. The Navy, however, was critical of the Department of State's stance. In particular, CINCPAC argued that a "decisive action" should be taken "here in Hawaii" before the *Golden Rule* sailed into the danger area because "the more obliquely we meet the problem the weaker our legal and moral position will appear and the greater will be the likelihood of arousing misguided sympathy for the ketch and her crew."[119] The AEC also sent a letter to the Secretary of State urging him to "use any authority at your disposal to prevent these individuals (author's note: crew of the *Golden Rule*) from interfering with" Operation Hardtack.[120]

On April 19, the *Golden Rule* arrived at Yacht Harbor in Honolulu, from where she made a statement that she was heading for Eniwetok Atoll, where Operation Hardtack would take place. After consultation between legal advisers from the AEC and the Department of Justice, it was decided that it would be best to prohibit the *Golden Rule* from entering the danger area in accordance with the Atomic Energy Act, "both from a legal and a public relations viewpoint," and the crew were temporarily detained by the Honolulu District Court.[121] Bigelow and his colleagues appeared in court, where they reiterated their intention to sail into the danger area, contrary to AEC rules. There they were arrested, tried, and convicted, given a 60-day suspended sentence and placed on probation for one year.[122] Bigelow would have preferred 60 days' imprisonment over probation, but this did not eventuate. The decision against Bigelow and crew was based on the AEC's declaration in its statement of April 11, 1958, when it established the danger area, "prohibiting unauthorized entry into a prescribed Danger Area of the Eniwetok Proving Ground by American citizens and other persons subject to the jurisdiction of the United States." Conversely, there was no legal basis for the U.S. government to crack down on foreigners who entered the danger area on protest boats if they were not subject to U.S. legal jurisdiction. This may have been one of the reasons why the U.S. government feared that foreign anti-nuclear activists, such as those in Japan, would enter the danger area.

The news that Captain Bigelow and his crew had been arrested traveled around the world. In London, supporters held a silent vigil, protesting the arrest of the *Golden Rule* crew in spite of them having broken no laws, and calling on the U.S., the U.S.S.R., and the U.K. to cease nuclear testing and production of nuclear weapons. The U.S. Embassy in London reported in detail to the Department of State the names and biographies of the groups and leaders who took part in the vigil.[123] In the United States, H. Alexander Smith, a member of the Senate Foreign Relations Committee, received a series of letters questioning the legitimacy of the arrests.[124] Also, A. L. Wirin, the lawyer representing the *Golden Rule* crew, flew to Japan immediately after representing the crew in court in Honolulu to meet with Japanese anti-nuclear activists. Wirin was a human rights lawyer who also represented Japanese-Americans who had been incarcerated during the

Second World War. When the American Nobel Prize-winning chemist Linus Pauling filed a lawsuit in Washington DC on April 4 of the same year, claiming that the danger of radioactive fallout from nuclear testing violated his fundamental constitutional rights, Wirin worked for his legal team, including the Japanese social reform activist Toyohiko Kagawa and three pelagic tuna fishermen from Muroto City, Kochi Prefecture. Wirin flew to Japan primarily to discuss the case, but through him the news of the *Golden Rule* reached the anti-nuclear activists in Japan.[125] The Department of State was also apprehensive about Wirin's movements. Ambassador MacArthur telegraphed Secretary of State Dulles twice, once when Wirin arrived in Japan (May 9) and again when he left (May 15), to inform him of his movements in Japan. In particular, Wirin's claim that Japanese fishermen were being harmed by nuclear testing was marked in pencil, presumably by Secretary of State Dulles himself.[126]

Inspired by the *Golden Rule*, an anti-nuclear activist yacht called the *Phoenix* sailed into the danger area in protest. On board were Earle Reynolds and his wife, their son Ted (15 at the time of sailing), daughter Jessica (10), and 3 Japanese, including Hiroshima University graduate and sailor Niichi Mikami. Reynolds was a doctor working for the Atomic Bomb Casualty Commission (ABCC) in Hiroshima, and in October 1954, after completing his term of office, he set off from Hiroshima with the aforementioned members on a round-the-world voyage. After three and a half years of touring India, Africa, and Central and South America, the *Phoenix* made port in Hawaii, where the trial of the *Golden Rule* was the talk of the town. Reynolds researched the AEC and nuclear testing in the Honolulu library and became convinced that nuclear testing was inhumane. He then sailed for the danger area, accompanied by his family and Mikami.[127] The yacht was captured by the coastguard at Kwajalein Atoll and the captain, Earl Reynolds, was arrested and transported by plane to Honolulu with his family. Reynolds was released on bail in Honolulu, but Mikami and the Reynolds' son were interned along with their yacht, and Mikami was later taken to a U.S. naval base while on his way back to Japan. A letter from Mrs. Reynolds explaining how this happened was published in the *Mainichi Shinbun*, and Earl Reynolds' memoirs were translated by Mikami and others and serialized in the *Chugoku Shinbun*. Ambassador MacArthur was also quick to inform the Department of State of this sequence of events.[128]

Information about nuclear testing and protests in the Pacific was a subject that the U.S. government wished to keep as low-profile as possible, in stark contrast to overseas information programs such as Atoms for Peace. This was based on a strong sense of caution, fearing a repeat of the *Lucky Dragon 5* incident, which had become a major blot on the U.S. overseas information program. The same sentiment ruled the negotiations between the U.S. and Japanese governments on compensation for Operations Redwing and Hardtack. Although there was conflict within the U.S. government between those who argued that a lump sum should be paid in order to bring about a "quiet" resolution, and those who argued that no compensation should be paid, the diverse opinions on nuclear tests were kept confidential. The Japanese government has also kept the negotiations secret for fear of public criticism and of galvanizing leftist forces in the event compensation claims were unsuccessful. Furthermore, the incidents of radiation exposure on the *Takuyo* and *Satsuma*, which occurred in the middle of the negotiations, made the Japanese and U.S. governments even more secretive. As a result of this control of information by both Japan and the U.S., the issue of compensation for the nuclear tests faded away without much attention or debate in Japan.

In an attempt to improve its image, impaired by nuclear tests, the U.S. government organized a foreign information program around their "clean bomb," but this was not well received internationally. As protests against the nuclear tests intensified both at home and abroad, the U.S. government was compelled to control, rather than disseminate, information. This history illustrates how nuclear testing weakened the U.S. position in the Cultural Cold War, and how difficult it was for overseas information programs to outweigh the political damage. On August 23, just ten days after Ambassador MacArthur reported to the Department of State that the *Takuyo* and *Satsuma* incidents had been "brought under control," President Eisenhower announced a unilateral suspension of nuclear testing. This change in U.S. policy on nuclear testing in 1958 is analyzed in detail in Itsuki Kurashina's article, which focuses on the initiative of Secretary of State Dulles. The paper reveals that Dulles, believing that "some significant gesture" was needed to change "the image of the United States in international public opinion," proposed the suspension of nuclear testing as a psychological measure to prevent deterioration in relations with allies.[129] As Dulles' idea had already been made when information was received in March 1958 that the Soviet Union was to declare a unilateral suspension of nuclear testing, the *Takuyo* and *Satsuma* incident did not directly influence the decision to suspend nuclear testing. However, the "potentially explosive issue" of the radiation exposure of Japan Coast Guard vessels, precisely at a time when the government was discussing a pause in nuclear testing, led by Dulles, likely made the option of declaring a halt to nuclear testing even more strategically inevitable. In October 1958, after the end of Operation Hardtack, the United States, the United Kingdom, and the Soviet Union held a conference in Geneva to suspend nuclear testing, and negotiations began which eventually resulted in the Partial Test Ban Treaty (PTBT), banning nuclear testing in the atmosphere. Higuchi's recent research has also pointed out that in the background to the PTBT was increasing global concern surrounding radioactive contamination. Perhaps it was the painful reminder that no amount of disinformation would prevent the damage to America's national image caused by atmospheric core testing that brought Eisenhower to the negotiating table over Teller and other AEC stalwarts who insisted that testing continue. As we shall see in the following chapters, the U.S. overseas information program concerning S&T subsequently shifted away from nuclear energy, and began to focus more on medicine and space exploration.

Notes

1 Matashichi Oishi, *Bikini jiken no shinjitsu: inochi no kiro de* [*The Truth about the Bikini Incident: At the Crossroads of Life*] (Tokyo: Misuzu Shobo, 2003); Masatoshi Yamashita, *Kaku no umi no shogen: Bikini jiken wa owaranai*

[*Testimony of the Nuclear Sea: The Bikini Incident Has Not Concluded*] (Tokyo: Shin Nihon Shuppan-sha, 2012); Daigo Fukuryumaru Heiwa Kyokai, ed., *Shinsoban: Bikini suibaku hisai shiryoshu* [*New Edition: Collection of Materials on the Bikini Thermonuclear Disasters*], with the supervision by Miyake Yasuo, et al. (Tokyo: University of Tokyo Press, 2014); Seiichiro Takemine, *Marshall shoto: owarinaki kakuhigai o ikiru* [*The Marshall Islands: Living the Endless Nuclear Victimization*] (Tokyo: Shinsensha, 2015); Hideki Sasaki, *Kaku no nanmin: Bikini suibaku jikken 'josen' gono genjitsu* [*Nuclear Refugees: Realities after the 'Decontamination' of the Bikini Thermonuclear Tests*] (Tokyo: NHK Publishing, 2013); Hiroko Takahashi, (*Shintei zoho-ban*) *Fuin sareta Hiroshima, Nagasaki: bei kakujikken to minkan boei keikaku* [(*Updated and Enlarged Edition*) *Sealed up Hiroshima and Nagasaki: U.S. Nuclear Tests and Civil Defense Program*] (Tokyo: Gaifusha, 2012), 182–188.

2 Akira Kurosaki, "Amerika no kakusenryaku to nihon no kokunai seiji no kosaku: 1954–60 nen" [The Intersection of U.S. Nuclear Strategy and Japanese Domestic Politics: 1954–60], *Chose-hanto to nihon no dojidaishi* [*The Contemporary History of Korea and Japan*], ed. Dojidaishi Gakkai (Tokyo: Nihon Keizai Hyoronsha), 189–233.

3 Toshihiro Higuchi, "An Environmental Origin of Antinuclear Activism in Japan, 1954–1963: The Government, the Grassroots Movement, and the Politics of Risk," *Peace & Change*, vol. 33, no. 3 (July 2008): 333–367.

4 Toshihiro Higuchi, *Political Fallout: Nuclear Weapons Testing and the Making of a Global Environmental Crisis* (Stanford, CA: Stanford University Press, 2020).

5 Toshihiro Higuchi, "'Clean' Bombs: Nuclear Technology and Nuclear Strategy in the 1950s," *The Journal of Strategic Studies*, vol. 29, no. 1 (February 2006): 83–116.

6 Telegram from Allison to Secretary of State, February 9, 1956, RG59, 711. 5611, box 2876, NACP.

7 J. Graham Parsons is different from Howard L. Parsons, Director of the Office of Northeast Asian Affairs, Department of State.

8 Telegram from Parsons to Secretary of State, February 22, 1956, RG59, 711. 5611, box 2876, NACP.

9 Telegram from Dulles to Embassy in Tokyo, February 24, 1956, RG59, 711. 5611, box 2876, NACP.

10 From Allison to Secretary of State, February 28, 1956, RG59, 711. 5611, box 2876, NACP.

11 Memorandum of Conversation, February 29, 1956, RG59, Department of State, Miscellaneous Lot Files, box 10, NACP.

12 From Allison to Secretary of State, March 6, 1956, RG59, 711. 5611, box 2876, NACP.

13 Telegram from Allison to Secretary of State, March 12, 1956, RG59, 711. 5611, box 2876, NACP. It is speculated that the "Seaman's Union" refers to Zen-nihon Kaiin Kumiai [All-Japan Seaman's Union], a rightist labor union established in 1945.

14 Telegram from Allison to Secretary of State, March 14, 1956, RG59, 711. 5611, box 2876, NACP.

15 "For the Press, No. 158," March 23, 1956, RG59, 711.5611, box 2876, NACP.

16 Memorandum of Conversation, May 3, 1956: Office Memorandum, May 4, 1956, RG59, 711.5611, box 2876, NACP.

17 Memorandum of Conversation, June 8, 1956, RG59, 711. 5611, box 2876, NACP.

18 Telegram from Allison to Secretary of State, June 21, 1956, RG59, 711. 5611, box 2876, NACP.

19 Memorandum of Conversation, November 13, 1956, RG59, Entry A1 3008-A, box 428, NACP. This survey by *Shunkotsu-maru* was not the 1954 one carried out soon after the *Lucky Dragon* incident, but the second survey held in 1956. Concerning the first survey, the NHK ETV Special, *Umi no hoshano ni tachimu-katta nihon-jin: Bikini jiken to Shunkotsu-maru* [*The Japanese Who Confronted the Radiation at Sea: The Bikini Incident and Shunkotsu-maru*] (aird, September 28, 2013) and Satoru Okuaki, *Umi no hoshano ni tachimukatta nihon-jin: Bikini kara Fukushima eno dengon* [*The Japanese Who Confronted the Radiation at Sea: A Message from Bikini to Fukushima*] (Tokyo: Junpo-sha, 2017), have empha-sized the Japanese scientists' heroic battle against the U.S. government's atti-tudes which underestimated the radioactive contamination of sea water. By contrast, the second *Shunkotsu-maru* survey was a U.S.–Japan collaborative research project. Upon Ambassador Allison's proposal, the U.S. government tried to tame the Japanese fear of radioactive fallout by carrying out a joint sur-vey by American and Japanese scientists. Ambassador Allison's telegram to the Secretary of State on June 15, 1956, for example, urged the U.S. authorities in Guam to issue landing permits for the *Shunkotsu-maru* crew promptly because failure to do so could lead to resentment on the part of "the very Japanese sci-entists we are counting upon to moderate Japanese emotionalism and fear about radiation problem." From Allison to Secretary of State, June 15, 1956, RG59, 711. 5611, box 2876, NACP.
20 From Horsey to Secretary of State, December 22, 1956, box 2877, RG59, 711. 5611, box 2877, NACP.
21 Far Eastern Public Opinion Barometer, "Japanese Reactions to U.S. Nuclear Tests," Report #11, August 28, 1956, RG59, Entry A1 3008-A, box 428, NACP.
22 Foreign Service Despatch from USIS Tokyo to USIA, May 23, 1956, RG469, Entry 421, box 32, NACP.
23 From Horsey to Secretary of State, February 9, 1957, RG59, Entry A1 3008-A, box 430, NACP. A copy of this telegram is also filed in RG59, 711.5611 box 2877, where it is dated February 8, 1957.
24 From Parsons to Kearney, January 25, 1957, RG59, General Records of the Department of State, Miscellaneous Lot Files, box 10, NACP.
25 To Parsons and Robertson, March 1, 1957, RG59, General Records of the Department of State, Miscellaneous Lot Files, box 10, NACP.
26 From Robertson to Herter, February 13, 1957, RG59, General Records of the Department of State, Miscellaneous Lot Files, box 10, NACP.
27 From Japanese Embassy to Department of State, January 18, 1957; From Department of State to Japanese Embassy, March 1, 1957, RG59, 711.5611 box 2877, NACP.
28 Office Memorandum, August 2, 1957, RG59, Entry A1 3008-A, box 430, NACP.
29 Inagaki worked for the Domei Tsushin, Japanese government-affiliated news agency during the war, and for the MOFA in the postwar years. In February 1955, he was Acting Director of the European and American Affairs Bureau of MOFA, and in January 1957, appointed to Deputy Director. In April 1957, the Bureau was divided into the North American Bureau and the European and Asian Bureau, and Inagaki headed the First Section of the North American Bureau. "Dai 26 kokkai, shugiin bunkyo-iinkaigiroku dai 14-go" [The 26th Congress, Lower House Education Committee Meeting Minutes No. 14], March 29, 1957, Kokkai kaigiroku kensaku shisuemu [Database of the Minutes of Japanese Diet Meetings], https://kokkai.ndl.go.jp/simple/detail?minId=102605077X0 1419570329&spkNum=37#s37; Gaimusho Hyakunenshi Hensan Iinkai, ed.,

Gaimusho no hyakunen [*One Hundred Years of the Ministry of Foreign Affairs*], vol. 2 (Tokyo: Hara Shobo, 1969), 767; "Sengo gaimusho jinji ichiran, obei-kyoku (1951–1957)" [Postwar Ministry of Foreign Affairs Personnel, European and American Affairs Bureau, 1951–1957], Sengo gaikoshi kenkyu-kai, ed., Database of Japanese Diplomatic History, https://drive.google.com/file/d /0B_wk3O1slLl7amdfOUVqSVp5alU/view?resourcekey=0-MRoV9PgzStK 3Q8McA03eew.

30 Memorandum of Conversation, March 15, 1957; Memorandum for the Record, March 15, 1957, RG59, General Records of the Department of State, Miscellaneous Lot Files, box 10, NACP.

31 From MacArthur to Secretary of State, March 22, 1957, RG59, 711.5611, box 2877, NACP.

32 Telegram from Dulles to Embassy Tokyo, August 30, 1957, RG59, 711.5611, box 2877, NACP.

33 Office Memorandum, August 2, 1957, RG59, General Records of the Department of State, Miscellaneous Lot Files, box 10, NACP; Bei hoku shiryo dai 62-4-go, "Bei gensuibaku jikken ni tomonau hosho mondai ni kansuru taibei sessho keii" [American Bureau Report No. 62-4, The Chronological Outline of the Negotiation with the U.S. concerning Compensation for the U.S. Thermonuclear Tests], March 7, 1962, Record No. C'.4.2.1.1-1-1 (microfilm C'-0005), Diplomatic Archives of the Ministry of Foreign Affairs of Japan.

34 From Parsons to Horsey, August 5, 1957, RG59, General Records of the Department of State, Miscellaneous Lot Files, box 10, NACP.

35 From Horsey to Parsons, August 19, 1957, RG59, General Records of the Department of State, Miscellaneous Lot Files, box 10, NACP.

36 From Kearney to Parsons, August 28, 1957, RG59, General Records of the Department of State, Miscellaneous Lot Files, box 10, NACP.

37 Memorandum from John H. Pender to Spiegel, August 23, 1957, RG59, Entry A1 3008-A, box 430, NACP.

38 Memorandum from Gerard C. Smith to Edward Gardner, September 3, 1957; Memorandum from Edward R. Gardner to Philip J. Farley, November 5, 1957, RG59, Entry A1 3008-A, box 430, NACP.

39 Memorandum for the Record, March 15, 1957; From Robertson to Parsons, March 1, 1957; From Fender and Kearney to Bell and Parson, February 7, 1957, RG59, General Records of the Department of States, Miscellaneous Lot Files, box 10, NACP.

40 "Beikoku no taiheiyo ni okeru kakujikken ni kansuru ken" [U.S. Nuclear Tests in the Pacific], January 9, 1960, Record No. C'.4.2.1.1-1-3 (microfilm C'-0006), Diplomatic Archives of the Ministry of Foreign Affairs of Japan.

41 From MacArthur to Department of State, January 22, 1958, RG59, Entry A1 3008-A, box 429, NACP.

42 A memo attached to Ambassador MacArthur's telegram (signed "Gu" and addressed to "Phil," presumably Philip J. Farley, Special Assistant. to the Secretary, Department of State), RG59, Entry A1 3008-A, box 429, NACP.

43 "Kakujikken ni kansuru chronology (1958-nen beikoku kankei)" [Chronology concerning Nuclear Tests: 1958 U.S.], September 4, 1958, Record No. C'.4.2.1.1-1-1 (microfilm C'-0005), Diplomatic Archives of the Ministry of Foreign Affairs of Japan.

44 "Kakujikken ni kansuru hosho seikyu no keii ni kansuru ken" [Chronological Outline of the Compensation Request for the Nuclear Tests], July 5, 1958, Record No. C'.4.2.1.2 (microfilm C'-0009), Diplomatic Archives of the Ministry of Foreign Affairs of Japan.

45 From Robertson to Bane, November 6, 1958, RG59, General Records of the Department of State, Miscellaneous Lot Files, box 10, NACP.

46 From MacArthur to Secretary of State, July 9, 1958, RG59, 711.5611, box 2879, NACP.

47 From Dulles to Embassy Tokyo, July 30, 1958, RG59, 711.5611, box 2879, NACP.

48 Yuka Tsuchiya, "Maguro enyo-gyogyo to tuna-kan sangyo o meguru nichi-bei kankeishi: 1950–60 nendai no boeki-masatsu, suibaku jikken, soshite sen-zenki karano renzokusei" [U.S.-Japan Relations through Tuna Fisheries and Canneries: The Trade Conflict of the 1950s–60s, Nuclear Tests, and Continuity from the Prewar Era], *Chushikoku American Studies*, vol. 8 (2017): 111–131.

49 Toshihiro Higuchi, "'Clean' Bombs," 83–116.

50 Takahashi, Sealed Up Hiroshima and Nagasaki, 188–191.

51 Hewlett and Holl, XII-16-17.

52 Office Memorandum from Col. B. B. Hovell to Leland Randall, November 4, 1954, RG469, Entry 421, box 17, NACP.

53 Minutes of OCB Meeting, March 19, 1958, RG59, Entry 3008-A, box 427, NACP.

54 "Draft of Presidential Press Conference Statement," March 20, 1958; From Watson to Lodge, March 24, 1958; From Spiegel to Berding, March 25, 1958, RG59, Entry 3008-A, box 427, NACP.

55 Information Memorandum, April 24, 1958; From Robertson to Under Secretary, May 10, 1958; From CINCPAC to CNO, May 20, 1958, RG59, Entry 3008-A, box 427, NACP.

56 Telegram from Dulles to USUN New York, July 16, 1958, RG59, 711.5611, box 2879, NACP.

57 From Arthur L. Richards to Farley, May 21, 1958, RG59, Entry A1 3008-A, box 427, NACP.

58 From Arthur L. Richards to Farley, June 4, 1958, RG59, Entry A1 3008-A, box 427, NACP.

59 From Donelan to Farley, June 11, 1958, RG59, Entry A1 3008-A, box 427, NACP.

60 Sei dai 1434-go (shikyu joho), Asakai taishi yori Fujiyama daijin e [Sei Dai No. 1434 (urgent information), from Ambassador Asakai to Minister Fujiyama, March 28, 1958, C'.4.2.1.1-1-3 (microfilm C'-0006), Diplomatic Archives of the Ministry of Foreign Affairs of Japan; Higuchi, *Political Fallout*, 104–105, 107–108, 128–129.

61 Showa 33, 13876, hei koku ren Matsudaira taishi yori Fujiyama daijin e [1958, No. 13876, hei koku ren, From Ambassador Matsudaira to Minister Fujiyama," July 31, 1958, C'.4.2.1.1-1-3 (microfilm C'-0006), Diplomatic Archives of the Ministry of Foreign Affairs of Japan.

62 Hewlett and Holl, XVI-13.

63 "Message for Communication Cylinder," RG59, Entry A1 3008-A, box 427, NACP.

64 Telegram from Embassy/USIS Tokyo to Secretary of State, June 8, 1956; From Embassy of Japan to the Department of State, July 17, 1956, RG59, Entry A1 3008-A, box 428, NACP.

65 Memorandum from H. D. Riley to Officer in Charge, Japanese Affairs, Office of Northeast Asian Affairs, Department of State, May 24, 1956, RG59, 711.5611, box 2876, NACP.

66 Memorandum from Spiegel to Musick, August 6, 1958, RG59, Entry A1 3008-A, box 427, NACP.

67 Ho kei 2 Dai 24-go, "Taiheiyo ni okeru beikoku no kakubakuhatsu jikken ni tomonau fukin shuko senpaku no kigai boshi sochi nitsuite" [Ho kei 2,

No. 24, "Precautions against Hazards for the Vessels in the Vicinity of U.S. Nuclear Tests in the Pacific], March 26, 1958, Record No. C'.4.2.1.1-1-3 (microfilm C'-0006), Diplomatic Archives of the Ministry of Foreign Affairs of Japan.

68 From Dulles to MacArthur, April 15, 1958; From MacArthur to Dulles, April 9, 1958, RG59, Entry A1 3008-A, box 427, NACP.

69 "Tokyo Kyodo in English" June 5, 1956, RG59, Entry A1 3008-A, box 428; From Dulles to Embassy Tokyo, June 5; Telegram from Embassy/USIS Tokyo to Secretary of State, June 8, 1956, RG59,711.5611, box 2876, NACP.

70 Telegram from Allison to Secretary of State, June 21, 1956, RG59, 711.5611, box 2876, NACP.

71 "Seizan-maru, fuyoi na koko: Geiger-kan motazu kiken suiiki fukin o toru" [Seizan-maru, Careless Navigation: Passing by the Danger Area without a Geiger Counter], *Sankei Shinbun*, July 26, 1958.

72 Telegram from MacArthur to Secretary of State, June 9 & June 11, 1958, RG59, 711.5611, box 2879, NACP.

73 Telegram from MacArthur to Secretary of State, June 11, 1958, RG59, 711.5611, box 2879, NACP.

74 "Bikini kokubai sosho nishin mo genkoku no seikyu kikyaku: Takamatsu kosai hanketsu" [The Bikini State Compensation Suit, Dismissed the Second Time: Takamatsu High Court Decision], *Mainichi Shinbun*, December 12, 2019.

75 From MacArthur to Secretary of State, July 16, 1958, RG59, 711.5611, box 2879, NACP.

76 "Hoshano kankei keika gaiyo, 33.7.14-33.7.21 Asa, Takuyo" [Outline of the Incident Concerning Radioactivity, from July 14, 1958 to July 21, 1958 Morning, *Takuyo*], Record No. C'.4.2.1.1-1-3-1 (microfilm C'-0006), Diplomatic Archives of the Ministry of Foreign Affairs of Japan.

77 Telegram from Embassy Tokyo to Secretary of State, July 21, 1958, RG59, 711.5611, box 2879, NACP.

78 "Hakkekkyu sokutei-chi" [White Blood Cell Count], Record No. C'.4.2.1.1-1-3-1 (microfilm C'-0006), Diplomatic Archives of the Ministry of Foreign Affairs of Japan.

79 Naval Message from ALUSNA TOKYO to COMNSVMSRISNAS, July 17, 1958; from CNO to CINCPACFLT, July 21, 1958; From Parsons to US Embassy Tokyo, July 22, 1958; From FBIS to Agency Offices, July 23, 1958, RG59, Entry A1 3008-A, box 327, NACP.

80 Kaijo hoan-cho, "Sokuryo-sen Takuyo oyobi junshisen Satsuma no hoshano osen ni tsuite no keii" [The Japan Coast Guard, Chronological Outline of the Radioactive Contamination of the Survey Ship *Takuyo* and Patrol Ship *Satsuma*], September 10, 1958, Record No. C'.4.2.1.1-1-3 (microfilm C'-0006), Diplomatic Archives of the Ministry of Foreign Affairs of Japan.

81 Telegram from Embassy Tokyo to Secretary of State, July 21, 1958, RG59, 711.5611, box 2879, NACP.

82 From Dulles to Embassy Camberra, July 21, 1958, RG59, 711.5611, box 2879, NACP.

83 From MacArthur to Secretary of State, July 22, 1958, RG59, 711.5611, box 2879, NACP.

84 From Dulles to Embassy Canberra, July 22, 1958, RG59, 711.5611, box 2879, NACP.

85 From MacArthur to Secretary of State, July 23, 1958, RG59, 711.5611, box 2879, NACP.

86 From Dulles to Embassy Tokyo, July 23, 1958, RG59, 711.5611, box 2879, NACP.

87 From Sebald to Secretary of State, July 24, 1958, RG59, 711.5611, box 2879, NACP.

88 Denso dai 10877-go, "Takuyo, Satsuma no hoshano osen ni kansuru ken" [Telegram No. 10877, Radioactive Contamination of *Takuyo* and *Satsuma*], July 24, 1958, Record No. C'.4.2.1.1-1-3-1 (microfilm, C'-0006), Diplomatic Archives of the Ministry of Foreign Affairs of Japan.

89 "Takuyo, Satsuma, hisai ni kansuru ken" [The Accident Involving *Takuyo* and *Satsuma*], July 25, 1958, Record No. C'.4.2.1.1-1-3-1 (microfilm C'-0006), Diplomatic Archives of the Ministry of Foreign Affairs of Japan.

90 "Kansokusen Takuyo, Satsuma: seifu, hoshano jiken o jushi, chikaku taibei moshiire" [Survey Ships *Takuyo* and *Satsuma*: Government Takes It Seriously, Will Remonstrate against U.S.], *Yomiuri Shinbun*, July 25, 1958, Morning Edition.

91 From MacArthur to Secretary of State, July 25, 1958, RG59, 711.5611, box 2879, NACP.

92 Dai 240-go, "Zai Portland Nihon ryojikan Imashiro ryoji kara Kishi sori-daijin e" [No. 240, From Consul General Imashiro in Portland to Prime Minister Kishi], August 15, 1958, Record No. C'.4.2.1.1-1-3-1 (microfilm C'-0006), Diplomatic Archives of the Ministry of Foreign Affairs of Japan.

93 From Embassy Canberra to Secretary of State, July 23, 1958, RG59, 711.5611, box 2879, NACP.

94 From Starbird to Farley, August 1, 1958, RG59, Entry A1 3008-A, box 427, NACP.

95 Telegram from MacArthur to Secretary of State, July 26 & August 4, 1958, RG59, 711.5611, box 2879, NACP.

96 From MacArthur to Secretary of State, August 4, 1958, RG59, 711.5611, box 2879, NACP.

97 Telegram from MacArthur to Secretary of State, August 6, 1958, RG59, 711.5611, box 2879, NACP.

98 From MacArthur to Secretary of State, August 8, 1958, RG59, 711.5611, box 2879, NACP.

99 The Japan Coast Guard, "Chronological Outline of the Radioactive Contamination of the Survey Ship *Takuyo* and Patrol Ship *Satsuma*"; From MacArthur to Secretary of State, August 11, 1958, RG59, 711.5611, box 2879, NACP.

100 From MacArthur to Secretary of State, August 8, 1958, RG59, 711.5611, box 2879, NACP.

101 From MacArthur to Secretary of State, August 13, 1958, 711.5611, box 2879, NACP.

102 "Takuyo, Satsuma no hoshano osen ni kansuru taibei moushiire no ken" [Request to the U.S. Government Concerning Radioactive Contamination of *Takuyo* and *Satsuma*], July 24, 1958; "Takuyo Satsuma no hoshano osen ni kansuru ken" [Radioactive Contamination of *Takuyo* an *Satsuma*], July 25, 1958, Record No. C'.4.2.1.1-1-3-1 (microfilm C'-0006), Diplomatic Archives of the Ministry of Foreign Affairs of Japan.

103 Telegram from MacArthur to Secretary of State, July 26, 1958, RG59, 711.5611, box 2879, NACP.

104 "Takuyo Satsuma-maru no hi-hoshano jiken ni tsuite" [Radioactive Contamination Incident of *Takuyo* and *Satsuma*-maru], August 19, 1958, Record No. C'.4.2.1.1-1-3-1 (microfilm C'-0006), Diplomatic Archives of the Ministry of Foreign Affairs of Japan.

105 Telegram from MacArthur to Secretary of State, September 8, 1958, RG59, 711.5611, box 2879, NACP.

106 "Beikoku no 1958-nendo taiheyo chiiki kakujikken ni kansuru kankei kakusho renrakukai ni kansuru ken" [Inter-Ministerial Liaison Meeting on the 1958 U.S. Nuclear Tests in the Pacific Region], September 17, 1958, Record No. C'.4.2.1.1-1-3 (microfilm C'-0006), Diplomatic Archives of the Ministry of Foreign Affairs of Japan.

107 Dai 1132-go, "Takuyo norikumi-in no shibo ni kansuru ken" [No. 1132 "Death of a Takuyo Crew Member"], August 5, 1959; "Takuyo hisai no ken" [The Accident Involving Takuyo], August 1, 1959; Genbaku higai taisaku ni kansuru chosa kenkyu renraku-kyogikai igaku-bukai, "Ketsuron" [Medical Subcommittee, the Liaison Council for the Investigation of Measures against Atomic Disaster, "Conclusion"], March 28, 1960; "Takuyo norikumi-in, Nagano hoankan no shibo ni kansuru ken" [Death of Coast Guard Officer Nagano, a Crew Member of *Takuyo*], March 31, 1960, Record No. C'.4.2.1.1-1-3-1 (microfilm C'-0006), Diplomatic Archives of the Ministry of Foreign Affairs of Japan.

108 "Takuyo no kenko kanri: yo-chui-sha o 'sokai'" [Health of *Takuyo* Crew: Those in Watch List 'Evacuated'], *Yomiuri Shinbun*, August 6, 1959, Morning Edition.

109 American Bureau Report No. 62-4, The Chronological Outline of the Negotiation with the U.S. concerning Compensation for the U.S. Thermonuclear Tests.

110 "Dai 20-kai kanji-kai (4-gatsu 3-ka) kiroku" [No. 20 Board Meeting, April 3, Minutes], Record No. C'.4.2.1.2 (microfilm C'-0009), Diplomatic Archives of the Ministry of Foreign Affairs of Japan.

111 "No. 20 Board Meeting, April 3, Minutes."

112 "Gaimusho joho-bunka-kyoku happyo: Eniwetok-suiiki ni okeru kaku-jikken ni kansuru taibei moushiire ni tsuite" [Ministry of Foreign Affairs Information and Cultural Bureau: Request for the U.S. Concerning Nuclear Tests in the Eniwetok Area], February 20, 1958, Record No. C'.4.2.1.1-1-3 (microfilm C'-0006), Diplomatic Archives of the Ministry of Foreign Affairs of Japan.

113 Memorandum for Herbert B. Loper and Paul F. Foster, May 3, 1958; From Walter S. Robertson to Philip J. Farley, May 3, 1958, RG59, Entry A1 3008-A, box 427, NACP.

114 From Paul F. Foster to Philip Farley, May 16, 1958, RG59, Entry A1 3008-A, box 427, NACP.

115 From Robertson to Farley, June 4, 1958, RG59, Entry A1 3008-A, Box 427, NACP.

116 Incoming Telegram from Tokyo to USIA, July 13, 1956, RG59, Entry A1 3008-A, box 429, NACP; "Senshu haku dai 45-go" [Senshu haku No. 45], March 31, 1956, Record No. C'.4.2.1.2 (microfilm C'-0009), Diplomatic Archives of the Ministry of Foreign Affairs of Japan.

117 From MacArthur to Secretary of State, February 11, 1958; From AEC Chairman to Dulles, March 25, 1958; From William B. Maccomber, Jr. to Alexander Smith, May, 17, 1958, RG59, Entry A1 3008-A, box 427, NACP.

118 OCB Minutes, February 19, 1958; April 2, 1958, White House Office, NSC Staff Papers, OCB Secretariat Series, box 14, Dwight D. Eisenhower Presidential Library, Abilene, Kansas.

119 Memorandum for Strauss, March 28, 1958; Naval Message, April 12, 1958, RG59, Entry A1 3008-A, box 427, NACP.

120 From Acting Chairman to John Foster Dulles, March 25, 1958, RG59, Entry A1 3008-A, box 427, NACP.

121 From Louis Strauss to John Foster Dulles, May 2, 1958, RG59, 711.5611, box 2879, NACP.

122 From William B. Macomber, Jr., Assistant Secretary to Senator Smith, May 17, 1958, RG59, 711.5611, box 2879, NACP.

123 Foreign Service Despatch from Embassy London to Department of State, May 6, 1958, RG59, 711.5611, box 2879, NACP.

124 From Smith to William B. Macomber, Jr., May 17, 1956, RG59, 711.5611, box 2879, NACP.

125 "Kagawa-shi ramo genkoku ni: beishimin ga kakujikken hantai no sosho okosu" [Mr. Kagawa and Others Join the Plaintiff: U.S. Citizens File a Lawsuit against Nuclear Tests], *Yomiuri Shinbun*, March 26, 1958, Morning Edition; "Kakujikken kinshi ni zensekai ga kyodo hotei-toso: Wirin-shi kisha-kaiken" [The Whole World Collaborate in the Court Case to Ban Nuclear Testing: Mr. Wirin at Press Conference], *Yomiuri Shinbun*, May 10, 1958, Morning Edition; Densho Encyclopedia, http://encyclopedia.densho.org/A.L._Wirin/.

126 Telegram from MacArthur to Secretary of State, May 12 & May 16, 1958, RG59, 711.5611, box 2879, NACP.

127 "Hiroshima no fushicho mouichido umi e" [The Phoenix of Hiroshima Returns to the Sea], *Asahi Shinbun*, February 13, 2018, Morning Edition; Earl Reynolds, "Phoenix-go no boken" [The Adventure of the Phoenix], no. 5–10, trans. Niichi Mikami and Hiroshi Matsumoto, *Chugoku Shinbun*, October 14–18, 1958; "Nihon-sei yotto, sekai o meguru" [A Japan-Made Yacht Navigates across the Globe], *OFFSHORE*, vol. 101 (August 1983): 8.

128 From MacArthur to Secretary of State, July 23, 1958, RG59, 711.5611, box 2879, NACP.

129 Itsuki Kurashina, "John Foster Dalles to gunbi kanri: 1958–59 kakujikken kinshi joyaku kosho o chushin ni" [John Foster Dalles and the Arms Control: PTBT Negotiations, 1958–59], *The Hitotsubashi Journal of Law and International Studies*, vol. 2, no. 3 (November, 2003): 1167–1193.

Part II

New Development in the Overseas Information Programs

6   From Atoms for Peace to Science for Peace

The golden age of Foreign Atoms for Peace did not last long. Although there were complex and combined reasons behind the decline of Foreign Atoms for Peace, one of the primary factors was the negative impact of the U.S. nuclear tests on international public opinion. This may at first seem contradictory, given that the early Atoms for Peace campaign in Japan was at least partly motivated by the U.S. desire to curb the Japanese fear of nuclear energy cultivated through their experiences of Hiroshima, Nagasaki, and the Bikini incident: the Atoms for Peace exhibitions and films emphasized that nuclear energy was not always harmful, but could be used for "peaceful" purposes. However, in late 1957, such a naïve outlook was no longer accepted by the international society. Nobel Prize-winning chemist Linus Pauling and his wife Ava Helen Pauling were circulating a petition among scientists to stop nuclear testing, and their peace activism was attracting international attention. As many anti-nuclear citizens' groups organized throughout the Western world, the appeal of the atom began to wane.

Another reason was that some countries, including Japan, had already imported research reactors from the U.S. and were seriously considering the introduction of power reactors. Although the U.S. had not yet produced commercial power reactors in 1957, it was simply a matter of time before they succeeded, and U.S. private firms were receiving positive interest from overseas markets like Japan. As foreign leaders and markets had therefore already been convinced of the usefulness of nuclear energy, it was no longer necessary to propagate the rosy picture of nuclear energy through films and exhibitions. At the same time, while this might also appear initially contradictory, U.S. companies were becoming less enthusiastic about investing their resources in the Atoms for Peace campaign because the American economy was going into recession.

Still another reason for the decline of Atoms for Peace was the "Sputnik shock" of October 1957. The world's first successful launch of a satellite by the Soviet Union triggered the U.S. government to review the overseas information programs concerning S&T. President Eisenhower and his close aides had been aware of the importance of reconnaissance satellites from early on, and had been quietly developing their own satellites.[1] The Soviet launch of Sputnik I, therefore, did not "shock" the government so much as it did the U.S. public, but the political opponents of Eisenhower seized upon this opportunity to criticize

DOI: 10.4324/9781003243649-8

the government for lagging behind the Soviet Union in S&T. In fact, the Soviet Union had also succeeded in a large-scale thermonuclear test and the launch of ICBM even before the Sputnik, and it further launched Sputnik II in November 1957, with a dog named Laika on board. By contrast, the U.S. postponed the scheduled December 4 launch of its first satellite, Vanguard, and when it was actually launched two days later, the rocket exploded, the endeavor resulting in a miserable failure. This chapter will examine how the Sputnik shock, and the other factors mentioned above, brought about change in the U.S. S&T policies and overseas information programs.

After the Sputnik shock, the U.S. government acted quickly to cultivate more human resources in S&T, passing the National Defense Education Act in 1958 to strengthen S&T education in high schools and universities. The U.S. government also pursued a more centrally controlled system to promote S&T policies. In late 1957, President Eisenhower appointed James Rhyne Killian Jr., president of MIT, as the President's Advisor for S&T. Killian held substantial power within the government, attending the cabinet and NSC meetings, and advised on the activities of the AEC, the Department of Defense (DOD), the CIA, and the Department of State. Under Killian, the President's Science Advisory Committee (PSAC) was re-established. The PSAC had originally been established in 1950 during the Truman administration with the intent of acting as a "counter balance" against the military R&D focus on weapons development. However, its members concurrently served on various military-related committees and as a result, the committee was unable to maintain independence from military influence. By contrast, the newly established PSAC emphasized the importance of basic research not related to military purposes. The 1960 PSAC report, drafted by Glenn T. Seaborg, a former Manhattan Project scientist who was then Chancellor of the University of California at Berkeley, played an important role in allocating budgets for basic research, scholarships for graduate students, and infrastructure of universities.[2]

Changes in the federal S&T policies were also reflected in the overseas information programs. In the 1958 State of the Union message to Congress, President Eisenhower proposed "Science for Peace," an international cooperation in the humanitarian use of science guided by the successful experience of the Atoms for Peace initiative. In his address, the President listed "eight items requiring action" to achieve U.S. "security and peace." They were: (1) defense reorganization, (2) acceleration of the defense effort, (3) mutual aid, (4) mutual trade, (5) scientific cooperation with our allies, (6) education and research, (7) spending and saving, and (8) works of peace. It was in the eighth item, "works of peace," that the President proposed "Science for Peace." He called for international cooperation, including with the Soviet Union, for eradicating malaria, campaigning against cancer and heart disease, driving hunger from the earth, and promoting demilitarization. The president's proposal of Science for Peace can be interpreted as a peace offensive in S&T. The USIA immediately took up Science for Peace as a "priority theme" in its overseas information programs. Of course, the concept of Science for Peace was not conceived of by President

Eisenhower alone, but was developed through discussions in the NSC and OCB.[3]

In this context, the relative importance of nuclear energy in the U.S. overseas information programs shrank, although it did not fade out completely. The USIS films on nuclear energy were screened in many countries throughout the 1960s, and the Atoms for Peace exhibitions also continued to be held. Moreover, the information programs targeting specialists, such as technical training and tours to nuclear facilities, were even accelerated, as the U.S. actually began to export power reactors. Nevertheless, the comparative importance given to nuclear energy within the overall U.S. information programs declined. In turn, the overseas information programs diversified into various themes inclusive of medicine, food, and space flight. This chapter will shed light on the transition of the U.S. overseas information programs, with a particular focus on the discussions in various government agencies and advisory groups, in order to clarify what kind of concrete policies resulted.

6.1 Government Reorganization Concerning S&T after the Sputnik Shock

After the Sputnik shock, S&T was more firmly centered in the U.S. overseas information programs through several different governmental agencies. First, within the Department of State, Wallace Brode, former science advisor of the CIA, was appointed as Departmental science advisor on January 13, 1958. The Department of State's science advisor was originally responsible for collecting scientific information overseas or issuing visas for foreign scientists, but the position had been vacant since 1954, when Joseph Koepfli resigned to return to Stanford University to resume his research in chemistry.[4] The revival of the science advisor position in the Department, and the appointment of a former CIA science advisor, indicated that S&T had been firmly established as an important factor of U.S. foreign policy. Brode directly assisted the Secretary of State, and led the departmental "Science Office" (later renamed the Office of International Science), in which the Deputy Science Advisor, Special Assistant to Science Advisor, and five Assistant Science Advisors (in charge of International Organizations, Physical Science, Biological Science, Engineering Science, and Administration, respectively) plus six secretaries were assigned.[5]

On September 2, 1960, Brode drafted a 25-page report titled "The Department of State Science Program," on his activities over the preceding two and a half years. He explained that "governmental support for science and technology expanded at a very rapid rate" and "international scientific activities assumed greater prominence" throughout the 1950s. Against this backdrop, the Science Adviser was assigned the following responsibilities as they pertained to science, "except in the field of atomic energy":

a. Advise the Secretary and other Department officers on broad policy matters;
b. Advise and assist other Department officers in processing and reaching decisions on issues with scientific implications;

   c. Support the science attachés abroad through guidance and instruction;

   d. Represent the Department's interests on appropriate inter-departmental committees;

   e. Maintain liaison with appropriate Governmental, quasi-official, and non-governmental scientific organizations.[6]

In the past, the Department of State had assumed important responsibilities in nuclear diplomacy such as the conclusion of bilateral agreements and technical aid programs, and had not yet given up such responsibilities when Brode submitted this report. The newly established Science Office, however, was assigned responsibilities "except in the field of atomic energy," a change that may symbolize the diminishing role of nuclear energy *in diplomacy*, although it remained important in trade and industry.

Secondly, science attachés in U.S. foreign posts were replenished; there had been no science attachés after the last retired in 1956, largely because they had difficulty continuing their roles as they were often suspected to be "spies" due to their information collection activities. However, in January 1959, new science attachés were appointed in the U.S. embassies in London, Paris, Rome, Bonn, Stockholm, and Tokyo. Moscow, New Delhi, and Rio de Janeiro followed. Subsequently, Buenos Aires, Brussels, and Canberra joined the list. Under each science attaché, there were deputy attaché, secretaries, and local employees. In some countries where there was no science attaché, such as Mexico, Canada, Switzerland, Israel, Taiwan, South Africa, and Poland, part-time employees called "science contacts" were appointed.[7] The science attachés' responsibilities were very extensive, as the following descriptions of their roles indicate:

   a. Serve as adviser to the Ambassador and his staff on scientific and engineering matters and on the relationship between foreign policy and science;

   b. Assist the Ambassador in assuring that scientific representatives from the United States in his area of assignment are thoroughly cognizant of foreign policy implications of scientific and technical activities;

   c. Assist the Ambassador in coordinating U.S. scientific programs and activities in the area of assignment in order to facilitate maximum effectiveness of these scientific representatives and to assure that the programs are in consonance with foreign policy of the United States and are complementary rather than duplicative or competitive in character; and this coordination is to include, where appropriate, the U.S. governmental, intergovernmental, and non-governmental programs;

   d. Promote and initiate the exchange of scientific information between the scientific community within the United States and the scientific community of the area of his assignment;

   e. Report to the Department of State for appropriate dissemination in the United States, significant scientific developments and trends with an evaluation of implications on international relations; and obtain upon request or upon his own initiative scientific reports and information which may not be generally available;

f. Advise scientific groups within the area of his assignment of the scientific policies of governmental and non-governmental organizations within the United States;

g. Provide, on instruction, scientific representation for the United States at scientific meetings;

h. Provide, on instruction, scientific representation for U.S. governmental agencies which may have no scientific representatives in the area of assignment;

i. Maintain contact with and assist, where feasible, scientific representatives of non-governmental organizations from the United States;

j. Maintain and promote personal and Embassy contacts with the scientific community of the area of his assignment.[8]

As Brode's report mentioned that "the coordinating activities are a heavy responsibility in the Japan area," Herbert Pennington, science attaché in Tokyo who also headed the AEC's Tokyo office, as shown in Chapter 3, oversaw a wide variety of activities not only in Japan but also in other Asian countries. The U.S. and Japanese archival records show that his activities included information collection, liaising for technical assistance, negotiations concerning international agreements, and briefing and debriefing U.S. scientists visiting Asian countries. The roles of science attachés were related both to overseas information programs and to actual technical assistance and technical transfer.

Thirdly, the USIA, since its establishment in August 1953, had played the central role in the overseas information programs on S&T. Soon after the Sputnik shock, the USIA established a "Science Committee" to discuss the new direction of overseas information programs on S&T, and based on the committee's recommendation, created a new post of "Science Advisor." The Science Advisor was expected to recommend, for example, "what scientific and technological exhibits should be shown and what lines in scientific and technological progress should be emphasized." The Science Committee also advised that a suitable candidate be recommended by Killian.[9] The recommendation resulted in the appointment of Harold Leland Goodwin. Goodwin was a well-known writer of science fiction for young readers, under the pen name John Blaine, and also worked for NASA and the National Oceanic and Atmospheric Administration (NOAA).[10] His communication abilities as a science fiction writer and his knowledge of government policies made him an ideal Science Advisor in the eyes of the USIA.

6.2 The New Significance of S&T in the U.S. Overseas Information Programs

Under Goodwin, "The Treatment of Science and Technology in the U.S. Information Program" (called the "Basic Guidance Paper" on S&T in the governmental inner circle) was drafted. Since this document played an extremely important role in defining the USIA's overseas information programs during this transitional period, its contents will be examined in detail in this section. The draft was circulated among several USIA and Department of State divisions and revised by their comments, and the "semi-final draft" was examined and accepted by the NSF and OCB. The final version was further checked by USIA's Division Chiefs, the Department of State, and the OCB.[11] One comment, for example, came from the USIA's Motion Pictures Section (IMS), who emphasized that guidance to the filmmakers was extremely important since "motion pictures communicate[d] so much more than the words spoken on the sound-track," and the producer's "grasp of the over-all situation" would influence the films. In this regard, the IMS felt that the draft was somewhat misleading, as if S&T were "at the apex of our scale of values," and "the principal confrontation between the U.S. and the U.S.S.R." was in S&T. Another comment came from Wallace Brode, Department of State Science Advisor, who highlighted his feeling that the paper should not "concentrate so intensely in the discussion on our competitive scientific status with Russia." Rather, Brode recommended that the paper should "stress the cooperative programs which the U.S. maintain[ed] with many countries for the promotion of scientific research." These comments were reflected in the final version.[12]

The final version, completed on October 30, was an 11-page document titled "The Treatment of Science and Technology in the US Information Program" also known as the "Basic Guidance Paper." According to this document, the objectives of the overseas information program concerning S&T were:

1. To establish and strengthen among foreign people's attitudes favorable towards the U.S. ... with special reference to the applications of science and technology in the U.S. for the betterment of the general welfare;
2. To ensure understanding by foreign peoples of ... U.S. and Free World scientific and technological achievement, particularly as such achievement may demonstrate that the U.S. and Free World systems are in harmony with and will advance their aspirations for freedom, progress, and peace;
3. To reduce the psychological impact of Soviet Scientific and technological achievements.[13]

In the sections that followed, the document pointed out that S&T was assuming new importance in the overseas information program, and that Soviet S&T had been "brought to world public attention suddenly and with great impact" by the successful launching of the world's first satellite. By contrast, it continued, there was not sufficient understanding about the U.S. and the "Free World" S&T, and a common misunderstanding was that the U.S. was "a materialistic nation." This should be corrected by treating "science and technology in perspective and to distinguish between the two," the document recommended. "Science [wa]s international in scope," it argued, with U.S. scientists maintaining cooperative relations with foreign scientists and the U.S. government taking the initiative in "intergovernmental cooperation to advance world science." On the other hand, although U.S. technology was spreading all over the world, technology was nevertheless "more national in character." The paper also cautioned that the "treatment of science and technology in the U.S. Information Program should not be based directly on US-USSR competition," as the struggle between the two countries lay "at a level far more fundamental than that of dramatic scientific or technological achievement." "A major strength of the American system" lay in "a great diversity of technological products with constant product improvement," which made possible "not only the American standard of living" but also the elimination of "geographical and occupational differences." The overseas information program, therefore, should put S&T "in perspective," and avoid item-by-item comparison with the Soviet Union, the paper recommended.[14]

In the final section, the paper provided ten concrete perspectives that should be integrated in the overseas information program. They were, in sum:

1. The internationality of science should be recognized;
2. American scientists are cooperating with scientists of other nations, including the Soviet Union;
3. Any effort designed solely to demonstrate U.S. superiority would be unproductive;
4. It is important to treat science subjects for long-term effect;
5. Overstatements should be avoided, and technical accuracy should be maintained as to the ability of scientists in other countries;
6. Informed speculation or predictions about the future have a particular appeal;
7. A typical area of U.S. contribution is in the biological sciences, including the development of new drugs and biologicals, and the efforts to defeat diseases are "productive fields of exploitation";
8. The military origin of a scientific or technical contribution needs no apology, but the U.S. should deplore the applications of S&T to any heightening of world tension;
9. S&T in the U.S. and the Free World are matters of public interest and do not operate under a blanket of secrecy;
10. S&T holds special promise to the peoples of developing nations. The developments in S&T that may be applied directly to national improvement are the most meaningful, although developments "that stimulate man's imagination, like space flight, may have equal or greater appeal."[15]

It was noteworthy that the "Basic Guidance Paper" strongly cautioned against any exaggeration of U.S. superiority and recommended emphasizing international cooperation. It was also remarkable that the paper named concrete areas of S&T that seemed useful for the overseas information program. Biological sciences and medicine were especially important because they were directly applied to the lives of people in developing countries, while space flight was also a promising theme because it stimulated people's imagination. The importance of medicine and space as the themes of overseas information programs will be explored in Chapters 7 and 8, respectively.

The final version of the "Basic Guidance Paper" was cabled to USIS posts all over the world on November 18, 1958. A short section titled the "Summary of Approach," which was not included in the original Guidance Paper, was inserted at the beginning of the telegram, to draw USIS officers' attention to points of particular importance. Those points were: that American S&T would "improve the general welfare, not only in the U.S, but the world over"; that science and technology should be distinguished, with the understanding that science was "apolitical"; and that the U.S. should not try to "demonstrate item-by-item superiority" over the Soviet Union, but "show the world that American science and technology, in common with other aspects of American life, [were] elements of a system that [was] in harmony with and will advance their aspirations."[16] The narrative that science was "apolitical," and that S&T was inherently "international" and served all human beings, surfaced repeatedly in the U.S. overseas information programs from this time on. Paradoxically, the very concept of "apolitical" and "international" science boosted the usefulness of science and scientists for the U.S. overseas information programs; because science was assumed to be pure, good, and detached from political propaganda, the USIA-sponsored exhibitions, publications, films, and radio and television programs wore a veil of "truthfulness." The overseas information programs which *did not look like* political propaganda were the quintessence of the Cultural Cold War strategies.[17]

Thus far, this section has focused on the overseas information policies on S&T in the Department of State and the USIA. During the same period, several other governmental and non-governmental groups were also engaged in discussion on the relationship between U.S. overseas information programs and S&T, and each group submitted a report to the government. For example, Audra Wolfe introduced two contrasting reports submitted to the NSC: one by the Federal Council of Science and Technology, titled "Strengthening the Free World Position in Science and Technology," and the other by PSAC, titled "International Scientific Activities." Wolfe's analysis concluded that the Federal Council report was based on the "unspoken assumption" that scientific strength was "hard power" yielding "new weapons, a strong economy, and uncontested leadership among allies," although by contrast, the PSAC report held that the objective and apolitical nature of science made it a perfect "tool for political and psychological warfare." The PSAC report was drafted by Detlev Bronk, notable biophysicist and President of the National Academy of Science (NAS). While the NSC adopted the PSAC report, the State Department Science Advisor, Wallace Brode, was critical of its content, as he felt that "insistence on apolitical, international science was naïve and counterproductive." Brode argued that science should be treated as an integral part of national policy alongside economic, social, and political issues. He insisted on this belief, and advocated for the establishment of the Department of Science. However, his opinion was not accepted, and he left the Department in September of the same year.[18] Brode, with his background as the CIA's Science Advisor and his wartime scientific intelligence activities, was unable to accept the PSAC report's assertion that science was objective and apolitical. However, as already discussed in this chapter, the very idea of the objectiveness and apoliticality of science made it an effective theme of the overseas information program precisely because foreign audiences did not see science as political propaganda. In this sense, both the Federal Council of Science report and the PSAC report regarded S&T as a weapon of the Cultural Cold War, be it in terms of hard or soft power.

Just around the same time, the U.S. President's Committee on Information Activities Abroad (also known as the Sprague Committee, taking the name of its chair, Mansfield Sprague) also submitted a nearly 100-page report to President Eisenhower. The Sprague Committee was established in order to review and update the "Jackson Committee report" of 1953 which had guided the overseas information policy of the Eisenhower administration, and to pass on the findings to the succeeding Kennedy administration. According to Nicholas Cull, author of a comprehensive history of USIA, as "a draft bill to return the USIA to the State Department" was being prepared, President Eisenhower hoped to stop the move to diminish the USIA by emphasizing its accomplishments in the Sprague Committee Report.[19] Mansfield D. Sprague was former Assistant Secretary of Defense for International Affairs, and also president of American Machine and Foundry, a private company which, among other products, manufactured small nuclear reactors. The establishment of the Sprague Committee was first conceived of when C. D. Jackson, long-time aide to Eisenhower since psychological operations during the Second World War, wrote a letter to the President on July 10, 1959, decrying the lack of sufficient understanding among the government's policy-makers of the ongoing psychological and political "warfare." The President responded to C. D. Jackson's concern by inviting over 15 representatives of the major national security agencies to the White House on September 10, 1959, to exchange opinions and raise consciousness about psychological aspects of the Cold War. It was only a few weeks later that the President instructed Mansfield Sprague to inaugurate the Committee on Information Activities Abroad (the Sprague Committee).[20]

The members of the Sprague Committee included C. D. Jackson, President's Special Advisor on psychological warfare; George V. Allen, Director of the USIA; Allen Dulles, Director of the CIA; Philip Reed, Chair of the Advisory Commission on Information; Gordon Gray, Special Assistant to the President for National Security Affairs; John N. Irwin II, Assistant Secretary of Defense; Livingston T. Merchant, Under Secretary of State for Political Affairs; Karl G. Harr, Jr., Special Assistant to the President. The staff consisted of officers from the USIA, CIA, State, Defense, and the White House. The Sprague Committee Report was comprised of 7 chapters and 3 appendices, including Appendix II, a list of 33 "Committee Staff Papers" which probably preceded the final paper and functioned as the preliminary studies.[21]

The report had significance as the guiding document of the U.S. overseas information policy in the period from the late 1950s to the early 1960s, and was maintained by the Kennedy administration with the exception of one point, the advised retention of the OCB, which was abolished by President Kennedy. The report proposed the need for "total diplomacy," including an overseas information program, in resisting Communist propaganda. The report emphasized the importance of S&T in three different parts: Chapter Four, "Psychological and Informational Aspects of Economic Aid, Scientific Research and Military

Programs"; in part of Appendix I, "Supplementary Recommendations"; and a Committee Staff Paper (in Appendix II) titled "The Impact of Achievements in Science and Technology upon the Image Abroad of the United States." Since the report had a defining influence on the U.S. overseas information program, it is worth examining its references to S&T.

Chapter Four pointed out, similarly to the USIA's "Basic Guidance Paper," the "increasing impact of scientific and technological achievement upon world opinion," and the "psychological triumph" of the Soviet Union in its successful launch of the world's first satellite. As a result, "the average man in most countries believe[d] that Soviet capability continue[d] to grow relative to that of the United States" and even led "in certain important aspects of space technology." To re-establish U.S. technological prestige, the Sprague Committee felt two things were "indispensable psychologically": (1) that the U.S. maintain a continuing stream of scientific and technological achievements; and (2) that these achievements be more effectively communicated to the world than has been the case in the past. The Committee made the following seven concrete proposals:

1. "The scale and effectiveness of our overseas information efforts to communicate the facts of U. S. scientific achievements should be increased." To this end, "the recruitment and training of qualified information specialists," and "additional appropriations for special projects such as exhibits" should be made available;
2. Increased efforts should be made to "improve our communications both with scientific elites and with the general public";
3. "Interdepartmental efforts" are necessary to coordinate between the "announcement of scientific achievements" and other governmental activities in diplomatic, military, and other fields;
4. Recent organizational measures such as "the establishment of the offices of President's Assistant for Science and Technology and of Science Adviser to the Secretary of State, and the appointment of science attachés at our principal embassies abroad" should be fully utilized;
5. The USIA "should identify programs with unusual interest and psychological impact" and propose actions. "Special psychological value" rested, for example, in "spectacular feats" which did not require "new fundamental research," or in the "development of new, low-cost products, machines and techniques which could directly affect the daily lives of people abroad." Applied chemistry (plastics, fibers, and antibiotics) and public health were promising fields;
6. "Where particular needs are identified," such as in agricultural technology and medicine, "programs for teaching and transmitting American technical knowledge" should be expanded;
7. "Joint scientific and technological programs with other advanced countries of the Free World should be encouraged," and their psychological results be exploited.[22]

Appendix I recommended "an increase in programs to communicate our scientific and technical knowledge and achievements to Asians, particularly to the Japanese and Indian scientific elites."[23] The appendix also distinguished between overseas information programs for "scientific elites," "non-scientific elites," and "the masses." It advised that "communications with science elite" could be improved by the "dissemination of professional books and journals" and "the establishment of science information centers," while "communication with non-scientific elites" could be improved by "written materials and audio-visual programs on U.S. science specially designed for laymen" and "contacts between U.S. scientists and foreign leaders." By contrast, "effective communication with the masses" depended on "motion pictures, television, radio, the press," and "science exhibits."[24] Distinct, separate programs for scientific elites, non-scientific elites, and the general public had already been practiced, for example, in the Foreign Atoms for Peace programs shown in the previous chapters of this book. Training opportunities and technical films were prime examples of the programs for the scientific elite, tours for politicians and business leaders to U.S. nuclear facilities represented the programs for non-scientific elite, and many of the Atoms for Peace exhibitions and USIS films targeted the general public. Such practice was reiterated, more explicitly than ever, in the Sprague Committee Report.

As for the Committee Staff Paper (in Appendix II) titled "The Impact of Achievements in Science and Technology upon the Image Abroad of the United States," several versions (drafts) are stored in the U.S. National Archives. One version, annotated "Study Number 23" and dated July 6, 1960, was comprised of Sections I through IX, among which Section III, "Potential U.S. Scientific and Technological Achievements Having Great Political-Psychological Importance," included themes such as Project Mercury (manned space flight), Project Pluto (development of nuclear-powered ramjet engines), Project Sherwood (controlled thermonuclear reactions), the Mohole (a project for drilling through the earth's crust), and a cure for cancer. Section IV, "Scientific and Technological Programs or Projects Having Broad Public Appeal," included desalination of water, peaceful uses of nuclear explosions for excavation, widespread use of audio-visual communications powered by solar energy, education in the sciences, and educational television. Section V, "Ways of Improving Knowledge of U.S. Scientific and Technical Capabilities," explored effective ways of communication with "the scientific elite," "non-scientific elite," and the "mass audience."[25] Some of the concrete scientific projects named here were omitted in the final Sprague Committee Report. The reasons for the disappearance of certain projects can be found in the minutes of meetings of the Sprague Committee on July 21 and 22. There was "disagreement and some criticism of some of the specific projects recommended," according to the minutes, and the criticized projects included "an anti-gravity project," "controlled nuclear explosions for peaceful purposes,"[26] and another project which was later redacted from the minutes for security reasons. The redacted project may have been either Project Pluto or Project Sherwood, since they were included in the draft Staff Paper, but not in the final report. The minutes also revealed that "considerable discussion" was held on how to ensure that "the so-called 'P' factor" (psychological factor) be considered in the decisions to promote particular scientific projects, and that the Committee agreed on the important role of OCB in that connection.[27]

The essence of the Committee Staff Paper was integrated in the final report of the Sprague Committee. The report was the culmination of the Eisenhower administration's eight-year experiences in overseas information policy, and important guidance to be passed on to the next administration. The report showed that S&T had emerged as an extremely important factor in the overseas information programs, and that scientific and technological knowledge involved not only tangible benefits such as medicine and electric power, but also intangible images of the government, people, and culture of the countries where such knowledge was produced. The Sprague Committee effectively pushed S&T onto the central stage of the Cultural Cold War.

6.3 Overseas Information Programs in the Transitional Period

The previous sections have shown that the three years from January 1958 to the inauguration of the Kennedy administration in January 1961 were a transitional period in which S&T grew more important in the U.S. overseas information program. Science advisors were appointed in various governmental agencies, and important policy guidelines such as the USIA's "Basic Guidance Paper" and the Sprague Committee Report were drafted. Also in this period, the target groups of the U.S. overseas information programs became clearly divided into three groups: the scientific elite, non-scientific elite, and general public. These changes were reflected in concrete overseas information programs, including Foreign Atoms for Peace. The training opportunities at the ISNSE, tours to nuclear facilities, technical films, and AEC's information centers at the industrial exhibitions covered in the previous chapters were examples of programs targeting the scientific and non-scientific elites. Missing from these accounts was the change in the Foreign Atoms for Peace programs targeting the general public. Most of the early Atoms for Peace exhibitions and USIS films targeted the general public. For example, in the 1955 Atoms for Peace exhibitions held in Tokyo, Hiroshima, and several other cities, the spectacular "magic hands" to handle radioactive materials and other eye-catching exhibits entertained the general public and lowered their psychological guard against nuclear energy. Similarly, the USIS film *A Is for Atom*, which was repeatedly shown at Atoms for Peace exhibitions, featured cute animated characters to explain the mechanism of chain reactions, and stirred people's imagination with images of future cities powered by nuclear energy.[28]

Archival records indicate, however, that after 1958, the Foreign Atoms for Peace initiatives targeting the general public became more domesticated and more directly linked to the everyday life of the local people. The Atoms for Peace exhibitions continued to be held all over the world, but their contents became more localized through greater featuring of local scientists, politicians, and industries. For example, the author's analysis of the 43 Atoms for Peace exhibitions held in various Japanese cities between May 1957 and October 1961 showed that many of the exhibitions held after 1958 were sponsored by Japanese private companies, notably electric utilities. Although the USIS continued to either co-sponsor or cooperate in many of those exhibitions, its role and visibility were diminishing. Naturally, the contents of the exhibits also leaned toward the modern lifestyle and advanced industry made possible by the ample supply of electric power. The U.S.–Japan bilateral agreement on atomic energy was revised in 1958 to enable the import of enriched uranium for power reactors. Although it was in the 1960s that Japan actually began to import U.S. power reactors, Japanese electric companies were already busy in the late 1950s preparing for the construction of nuclear power plants in the near future, with the intent to import U.S. light-water reactors.[29] The U.S. Atoms for Peace in Japan, therefore, localized and domesticated to fit the interests of the country.

The situation was somewhat different in developing countries, but the trend toward localization and domestication was the same. One example was the Atoms for Peace exhibition in Saigon, South Vietnam, held in February 1958. The pamphlets for the exhibition prepared in three languages—English, French, and Vietnamese—introduced the displayed items one by one. The first exhibit was the photo panels of President Eisenhower and President Ngo Dinh Diem, and their words on the peaceful use of atomic energy. Eisenhower's were an excerpt from his U.N. address in December 1953:

> The United States pledges before you ... its determination to help solve the fearful atomic dilemma, to devote its entire heart and mind to find the way by which the miraculous inventiveness of man shall not be dedicated to his death but consecrated to his life.

Ngo Dinh Diem's was quoted as follows:

> Here is ample evidence that atomic energy can be used for peaceful purposes and to the tremendous advantage of all humanity. This new force merits the thoughtful attention of all who are dedicated to the goal of providing a better way of life for our children and our children's children. Posterity demands our best energies to that end.[30]

Ngo Dinh Diem, as already seen in Chapter 3, showed his interest in nuclear energy in the summer of 1958, and in September, Bu Hoi, representative of the scientific elite of Vietnam, suddenly released to the press the plan to import a U.S. nuclear reactor. Thereafter, in spite of the strong opposition expressed by the U.S. embassy and local intellectuals, the plan was put into practice. Ngo Dinh Diem's statement at the February 1958 Atoms for Peace exhibition predicted the ensuing development of the South Vietnamese nuclear research program. The exhibition was co-sponsored by the USIS and the Vietnamese Ministry of Information and Youth, which also indicated the interest of both the U.S. and South Vietnamese governments in the nuclear development of South Vietnam.

Ngo Dinh Diem's emphasis of the blessing of atomic energy, i.e., "a better way of life for our children and our children's children" was further reinforced by another exhibit that followed: a panel showing that atomic energy could bring "NEW POWER," "BETTER HEALTH," and "MORE FOOD." These three key words were also used in the Atoms for Peace exhibition in Manila, the Philippines, held in the same year. The brochure of the Manila exhibition read that atomic energy could materialize:

(1) MORE FOOD ...using radioisotopes, farmers will soon be able to grow more and better crops and raise better livestock.
(2) BETTER HEALTH ... medical application of atomic energy assist doctors and researchers in diagnosing diseases such as cancer, tumors and thyroid disorders.
(3) MORE POWER ...atomic fuels supplement world's existing sources of power—coal, oil, water and gas—thus, turning the wheels of industry at a faster rate.[31]

The three key words were essentially the same as in the case of South Vietnam, although in the reverse order. The Sprague Committee's emphasis of agriculture and medicine as the fields "where particular needs are identified" (see the previous section) also supports the recasting of atomic energy as a technology to bring about food and health within the U.S. overseas information program.

The Philippines was one of the first countries to conclude a bilateral agreement with the U.S. and, by the time of this exhibition, had already introduced a research reactor. For this reason, the brochure featured many Filipino individuals who were involved in the nuclear development program—for example, Filipino scientists being trained at the ISNSE, and a government minister receiving a lecture on reactor technology from an American scientist. Photos of Filipino Atomic Energy Commissioners, politicians, military officers, students, farmers, and women were also shown, implying that all citizens were potential beneficiaries of the introduction of nuclear energy (Figure 6.1). By contrast, South Vietnam in early 1958 had not yet commenced talks on the bilateral agreement, and thus, instead of introducing local individuals, the pamphlet emphasized international cooperation: sections titled "Sharing Information" and "International Cooperation" introduced scientists and engineers from many countries receiving training in the U.S., and U.S. foreign assistance in providing nuclear fuels, technical information, and reactors.

In addition to South Vietnam and the Philippines, the USIA in 1958 was planning to hold Atoms for Peace exhibitions in many other countries; exhibitions were planned for 1960 in India, Egypt, Brazil, Pakistan, and Argentina, and for 1961 in Brazil, Lebanon, and Peru, and further planning for future exhibits in Mexico, Chile, Columbia, Greece, and Thailand was underway.[32] Pamphlets, films, and display items were often reused in different countries after translating and modifying the contents. Therefore, the three key words of "more food," "better health," and "more power" may have appeared in the Atoms for

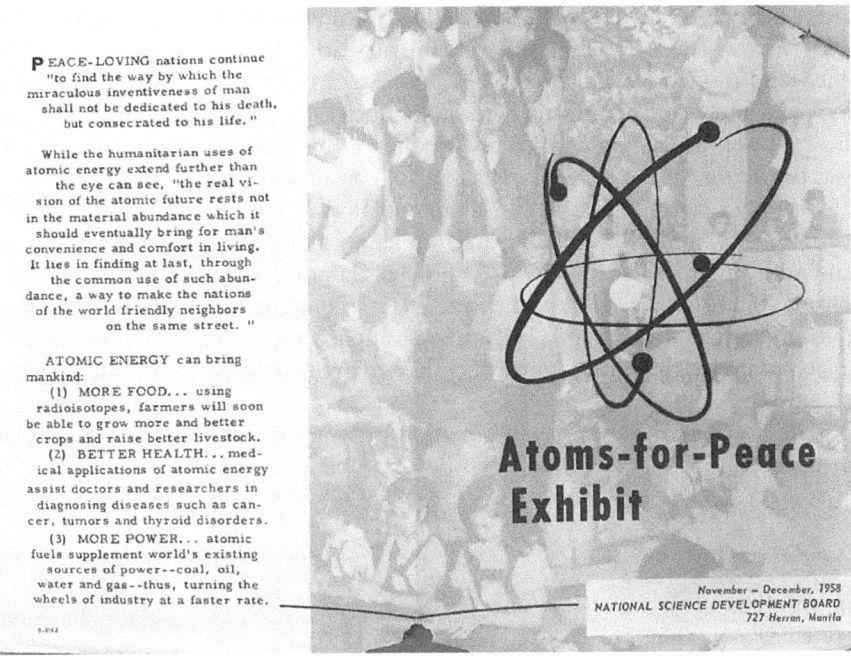

PEACE-LOVING nations continue "to find the way by which the miraculous inventiveness of man shall not be dedicated to his death, but consecrated to his life."

While the humanitarian uses of atomic energy extend further than the eye can see, "the real vision of the atomic future rests not in the material abundance which it should eventually bring for man's convenience and comfort in living. It lies in finding at last, through the common use of such abundance, a way to make the nations of the world friendly neighbors on the same street."

ATOMIC ENERGY can bring mankind:
(1) MORE FOOD... using radioisotopes, farmers will soon be able to grow more and better crops and raise better livestock.
(2) BETTER HEALTH... medical applications of atomic energy assist doctors and researchers in diagnosing diseases such as cancer, tumors and thyroid disorders.
(3) MORE POWER... atomic fuels supplement world's existing sources of power--coal, oil, water and gas--thus, turning the wheels of industry at a faster rate.

Atoms-for-Peace Exhibit

November – December, 1958
NATIONAL SCIENCE DEVELOPMENT BOARD
727 Herran, Manila

Figure 6.1 A brochure for the Atoms for Peace exhibition in Manila, the Philippines (1958). RG306, Entry P46, box 9, NACP.

Peace exhibitions in many developing countries in Asia, Latin America, and the Middle East. Summarizing the Eisenhower-era activities retrospectively, the AEC reported that "U.S. atomic energy exhibits have achieved substantial gains in the international presentation of nuclear energy's peaceful uses on both the popular and technical levels."[33] For those countries already committed to importing U.S. nuclear reactors such as Japan, industrial or technical exhibitions provided specialized knowledge and practical information for the scientific elite. In these countries, exhibitions for the general public rather provided concrete images of electrified lifestyles that seemed achievable in the near future. By contrast, in developing countries where the scientific elite was scarce and scientific education not widespread in the general public, easy-to-understand blessings of atomic energy such as food, health, and power occupied the center of the Atoms for Peace program.

The benefits of S&T for everyday life were emphasized not only in Foreign Atoms for Peace but also in other types of overseas information programs. For example, a 40-page USIA booklet published in 1960, titled "The American Consumer: Key to an Expanding Economy," directly connected the advancement of S&T to the improvement of everyday life; 70,000 copies were printed in Japanese, 4,000 in Nepali, and 2,000 in English. The central thesis of the booklet was that it was consumer-centered American capitalism that had raised the standard of living for everyone in the country. Each page captured an aspect of American society in which technological developments had led to greater efficiency, affluence, and modernity. For example, a photo of a "synthetic textile plant," which looked somewhat like a control room of a nuclear power plant, was accompanied by a caption explaining that "mass production by virtually automatic methods" was providing "an ever-increasing supply of consumer goods" (Figure 6.2).

On other pages, images of washing, cooking, shopping, harvesting, etc., "50 years ago and today" were compared. In laundry, for example, "the wash board and tub," had been replaced by "the electric-powered washing machine that save[d] the housewife time and effort." As for cooking, "the old-fashioned coal-burning range" was replaced by "a stove using gas or, as in this picture, the newest type of electric range" with which housewives could prepare meals quickly[34] (Figures 6.3, 6.4).

The focus of the overseas information program on the high living standard in the U.S. was not anything new. For example, many USIS films since 1946

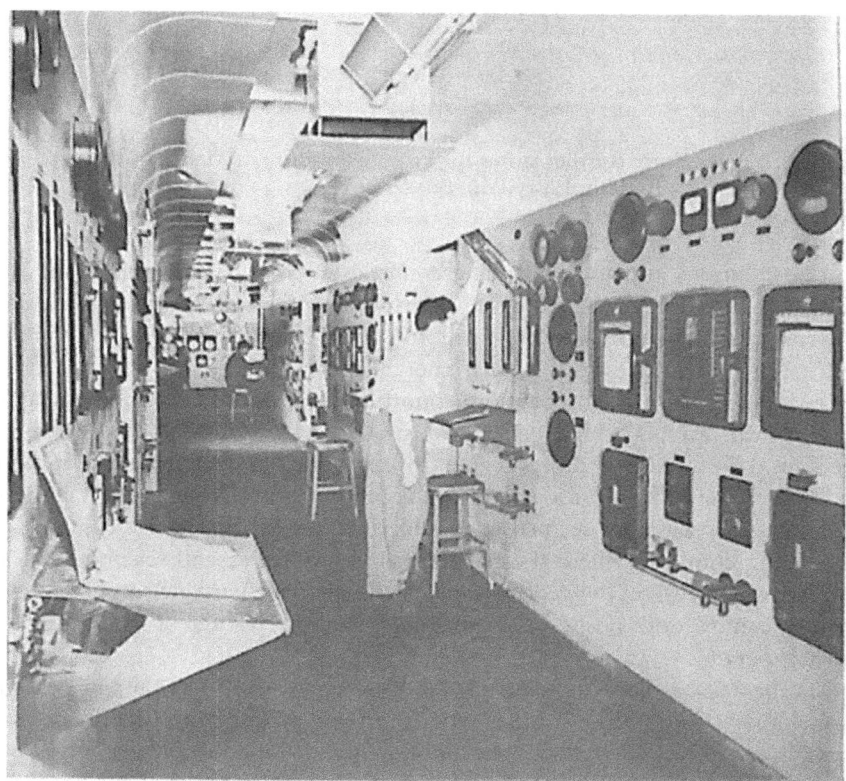

Figure 6.2 A synthetic textile plant with high-tech control panels. RG306, Entry A1 53, box 1, NACP.

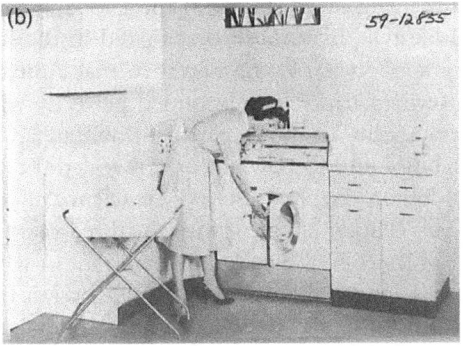

Figure 6.3 Laundry 50 years ago and today. RG306, Entry A1 53, box 1, NACP.

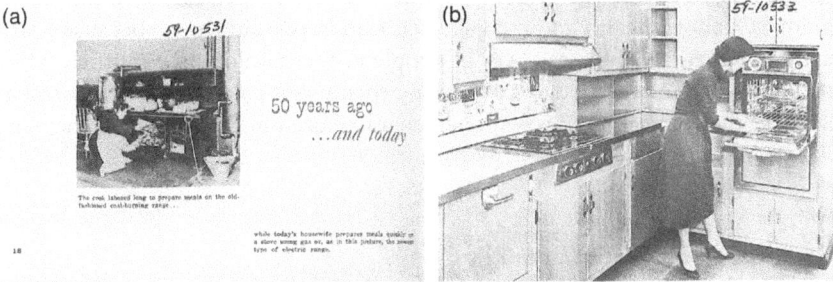

Figure 6.4 Cooking 50 years ago and today. RG306, Entry A1 53, box 1, NACP.

had included images of modern, well-equipped housing, schools, companies, and museums and other cultural institutions. However, images of S&T directly integrated in everyday life were a new trend in the overseas information program after 1958. Electrified kitchens, washing machines, and automated plants might stir foreign readers' emotions of desire but they were also not threatening; in fact, they appeared to be attainable goals, if only sufficient electric power were to become available. This new type of overseas information program on S&T—emphasizing the contribution of American S&T to the improvement of the everyday lives of people all over the world—was a conclusion that U.S. overseas information specialists had reached after reviewing and rebuilding their policies and organizations after the Sputnik shock. The famous "kitchen debate" between Nikita Khrushchev (First Secretary of the Communist Party of the Soviet Union) and Richard Nixon (Vice President) at the American National Exhibition in Moscow in 1959, in which two political leaders had a heated argument on the superiority of household appliances, should be understood in this context. In the model suburban house set up at the exhibition site, Nixon "extolled the virtues of American way of life," and boasted about happy and comfortable American housewives surrounded by "labor-saving devices."[35] Although this episode has generally been understood as a sparring match between the U.S. and the Soviet Union over family, gender, and lifestyle, it can also be situated in the history of U.S. overseas information programs on S&T: the narrative that American S&T reduced labor and made housewives happy corresponded perfectly with the post-Sputnik overseas information policy of the Department of State and the USIA. The American National Exhibition in Moscow not only revealed the U.S. Cold War consensus on gender and domesticity, but also its new science information policy after Sputnik.

This chapter has dealt with the three-year period from the Sputnik shock to the end of the Eisenhower administration, in which the U.S. government perceived the relationship between S&T and foreign relations more clearly than ever, and the roles of S&T in overseas information policy became more important than ever. Interest in S&T increased in the Third World countries, too, and the U.S. government strived to convince them that their technologies, rather than the Communist technologies, would bring about more food, better health, and modern, industrialized society. Out of these efforts emerged new themes of overseas information programs. Of these, health and medicine were an important theme because it was directly related to the everyday lives of people in the developing countries, while space flight was also regarded as a promising theme as it roused people's imagination. These two central themes will be further explored in the following chapters.

Notes

1 Sambaluk, *The Other Space Race.*
2 Wolfe, *Competing with the Soviets*, 37, 47. 50.
3 Cull, *The Cold War*, 135, 148, 150–152. For the full text of President Eisenhower's address, see "Annual Message to the Congress on the State of the Union," January 9, 1958, The American Presidency Project, University of California, Santa Barbara, https://www.presidency.ucsb.edu/documents/annual-message-the-congress-the-state-the-union-10.
4 Wolfe, *Freedom's Laboratory*, 54, 99.
5 "The Department of State Science Program," September 2, 1960, RG306, Entry P243, box 1, NACP.
6 "The Department of State Science Program," September 2, 1960, RG306, Entry P243, box 1, NACP.
7 "The Department of State Science Program," September 2, 1960, RG306, Entry P243, box 1, NACP; Wolfe, *Freedom's Laboratory*, 100.
8 "The Department of State Science Program," September 2, 1960, RG306, Entry P243, box 1, NACP; Ragnar Rollefson, "Science in the Department of State," *Argonne National Laboratory News-Bulletin*, vol. 5, no. 2 (April 1963): 3–5, National Archives at Chicago.
9 From William L. Clark, IAE, to Bradford, IOP, November 12, 1957; From Edmund Schechter to William L. Clark, November 12, 1957, RG306, Entry P243, box 1, NACP. The members of the Science Committee were: Miss Cutter, Messrs. Grossman, West, and Schechter.
10 *New York Times* Obituary, February 23, 1990, http://www.nytimes.com/1990/02/23/obituaries/harold-leland-goodwin-author-75.html.
11 Office Memorandum from Harold L. Goodwin to Kolarek and Halsema, October 30, 1958, RG306, Entry P243, box 3, NACP.

12 Office Memorandum from Anthony Guarco to Saxton Bradford, October 9, 1958; Memorandum from Wallace R. Brode to Philip H. Burris, September 29, 1958, RG306, Entry P243, box 3, NACP.

13 "The Treatment of Science and Technology in the US Information Program," no date, RG306, Entry P243, box 3, NACP. Though this document is not dated, the cover letter signed by Goodwin indicates that it was completed on October 30.

14 "The Treatment of Science and Technology in the US Information Program," no date.

15 "The Treatment of Science and Technology in the US Information Program," no date.

16 Outgoing Message, USIA CA-1367, November 18, 1958, RG306, Entry P243, box 3, NACP.

17 The author has discussed the "apolitical" outlook of science in the VOA radio programs in Yuka Tsuchiya, "VOA *Forum* to kagakugijutsu kohogaiko: reisen radio wa Amerika no kagaku o do tsutaetaka" [VOA *Forum* and S&T Public Diplomacy: Cold War Radio Broadcast American S&T], *Amerika Kenkyu* [*The American Review*], vol. 54 (April 2020): 67–87.

18 Wolfe, *Freedom's Laboratory*, 109–110.

19 Cull, *The Cold War*, 180.

20 "U.S. President's Committee on Information Activities Abroad (Sprague Committee): Records, 1959–61," Dwight D. Eisenhower Presidential Library website, https://www.eisenhowerlibrary.gov/sites/default/files/research/finding-aids/pdf/us-presidents-committee-on-information-activities-abroad.pdf.

21 The Sprague Committee Report's chapters are: 1. The Role of the Psychological Factor in Foreign Policy and the Requirement for an Adequate Information System; 2. Reinforcing the Foundations of the U.S. Information System; 3. The New Importance of Educational, Cultural and Exchange Activities; 4. Psychological and Informational Aspects of Economic Aid, Scientific Research and Military Programs; 5. New Dimensions of Diplomacy; 6. International Activities of Private Persons and Organizations, and of the Mass Media; 7. Organization, Coordination and Review; Appendix I "Supplementary Recommendations"; Appendix II "List of Staff Papers"; Appendix III "Letter of the President to Committee Chairman Sprague Dated December 2, 1959, Concerning the Establishment of the President's Committee on Information Activities Abroad." Yuka Tsuchiya, Shunsuke Okuda, and Shotaro Shindo, "Shiryo shokai: *Sprague iinkai hokokusho* (1960-nen 12-gatsu) shoyaku to kaisetsu" [Primary Source: *The Sprague Committee Report* (December 1960), A Translation of Selected Chapters and Commentary], *Eibungaku Hyoron* [*Review of English Literature*], vol. 91 (February 2019): 1–29.

22 The Sprague Committee Report, 36–38.

23 The Sprague Committee Report, 67.

24 The Sprague Committee Report, 80.

25 The President's Committee on Information Activities Abroad, "The Impact of Achievements in Science and Technology upon the Image Abroad of the United States," June 6, 1960, RG306, Entry P243, box 4, NACP.

26 The idea to use nuclear explosions for excavation, such as in the development of harbors, was suggested by Arthur Larson, Director of the USIA, to Louis Straus, Chairman of the AEC, soon after the Sputnik shock. The plan was embraced by the AEC since "building a harbor in a remote location such as Alaska could meet the need to prove America's technological know-how in the face of the Soviet ICBM test and *Sputnik* while avoiding the danger of domestic or international reaction from fallout." The plan took off as the Project Plowshare, but after many test explosions, the harbor construction never materialized because of public

criticism, local residents' opposition, and other reasons. Scott Kaufman, *Project Plowshare: The Peaceful Use of Nuclear Explosives in Cold War America* (Ithaca, NY: Cornell University Press, 2013), 24–25.

27 "Memorandum for the Record," June 22, 1960, CIA, Freedom of Information Act Electronic Reading Room, https://www.cia.gov/readingroom/docs/CIA -RDP86B00269R001000010022-5.pdf.

28 As for the 1955 Atoms for Peace exhibition, see, for example, Tetsuo Arima, *Genpatsu, Shoriki, CIA: kimitsu bunsho de yomu showa rimenshi* [*Nuclear Reactors, Shoriki, and CIA: The Hidden History of Showa Read through Secret Documents*] (Tokyo: Shinchosha, 2008); and Mitsuo Ikawa, "Genshiryoku hei-wariyo hakurankai to shinbunsha" [Atoms for Peace Exhibitions and Newspaper Companies], in *Sengo nihon no media event: 1945–1960* [*Media Events in Postwar Japan: 1945–1960*], ed. Toshihiro Tsuganezawa (Kyoto: Sekai-shisosha, 2002), 247–265. Concerning USIS films, see Tsuchiya and Yoshimi, eds., *Occupying Eyes, Occupying Voices: CIE/USIS Films and VOA Radio*.

29 Yuka Moriguchi Tsuchiya, "Atoms for Peace Exhibition in Japan: Localization of Nuclear Modernity," Stephan Köhn, Felix Jawinski, Steffi Richter, eds., *Japan's Split Society Between Genbaku and Genpatsu: Media, Propaganda and Science*, forthcoming in 2023.

30 USIS Saigon, "Atoms for Peace Exhibition," Saigon Atoms for Peace Program, RG306, Entry P46, box 48, NACP.

31 "Atoms-for-Peace Exhibit," November–December, 1958, RG306, Entry P46, box 9, NACP.

32 From Larsen to Goodwin, June 5, 1958, RG306, Entry P243, box 3, NACP.

33 "International Scientific Program of the AEC and Future Plans," no date, RG306, Entry P243, box 1.

34 "The American Consumer: Key to an Expanding Economy," RG306, Entry A1 53, box 1, NACP.

35 Elaine Tyler May, *Homeward Bound: American Families in the Cold War Era*, rev. and exp. ed., 4th ed. (New York: Basic Books, 2017), 19–21.

7   Project Hope and Medical
Aid Programs

As we have seen in the previous chapter, the Sputnik shock and the relative decline of the Atoms for Peace transformed the meaning of S&T in the overall U.S. overseas information program. President Eisenhower's Science for Peace address emphasized international cooperation in humanitarian fields such as the fights against disease and hunger. In the USIA's Basic Guidance Paper, the development of new drugs, hygiene, and medicine were named as "productive fields of exploitation," given their direct implications in the lives of foreign citizens: in the midst of the U.S.–Soviet competition over Third World allegiances, medical aid emerged as the center of the U.S. overseas information program in the late 1950s and 1960s.[1]

Malaria, cancer, and heart disease were exemplified as three major health issues in President Eisenhower's Science for Peace address. There were reasons why the president focused on these three diseases; for instance, malaria eradication had been a traditional field of U.S. overseas medical aid since the pre-Second World War years, and it was quickly integrated into the overseas information program in the postwar years. The first section of this chapter focuses on the development of the malaria eradication campaign as a window through which to observe the process of medical aid becoming part of the Cultural Cold War. The second section examines the discussions within the U.S. government on medical aid and the overseas information program triggered by the presidential address on Science of Peace.

In the third section, "Project Hope: Health Opportunity for People Everywhere" (referred to as simply "Project Hope" in U.S. government documents) will be explored as a quintessential example of medical aid programs in the Cultural Cold War. Project Hope was a scheme to convert the *Consolation*, a U.S. Navy medical ship which had seen service in the Second World War and the Korean War, into the *Hope*, a "floating medical center" to visit Southeast Asian countries as a "good-will mission" of the American people. It was part of the People-to-People Program, a "project to enhance international understanding and friendship through educational, cultural and humanitarian activities involving the exchange of ideas and experiences directly among peoples of different countries and diverse cultures," established by President Eisenhower.[2] As part of this project, a non-profit citizens' organization called the People-to-People Health

DOI: 10.4324/9781003243649-9

Foundation, Inc., was established, which became the platform for international medical aid activities.

The People-to-People Program would later (in 1961) become the completely private non-profit People-to-People International (PTPI), and continue activities such as international exchange of youth and the anti-landmines movement. However, even before 1961, Project Hope was publicized as a private voluntary initiative by American medical and health specialists. The *Hope* visited Indonesia, South Vietnam, Peru, Ecuador, Guinea, Nicaragua, Columbia, Ceylon (present-day Sri Lanka), Tunisia, Jamaica, and Brazil before its retirement in 1974, when Project Hope became a non-profit organization of the same name and continued medical aid programs for developing countries.[3]

This chapter deals with the first two trips of the *Hope* during the late Eisenhower administration and subsequent Kennedy administration. Archival documents of the era reveal that the apparently private voluntary activities were in fact closely monitored and supported by the U.S. government. The ambiguous borderline between the governmental and the private, and the contradiction inherent in government-supported voluntarism, characterized the U.S. overseas information program through medical aid: this chapter also illuminates the multi-layered structure of the Cultural Cold War involving not only government policy-makers but also private companies, experts, and the general public. Finally, this chapter examines whether the new goal of U.S. overseas information programs in the post-Sputnik era, i.e., persuading the world that American S&T improved the everyday lives of people, was achieved through the medical aid program.

7.1 Medical Aid as an Arena of the Cultural Cold War:[4] The Example of Malaria Eradication

President Eisenhower deliberately chose malaria, cancer, and heart diseases as examples of health issues to be tackled in his State of the Union address of 1958. There were chosen because, first of all, U.S. national laboratories were developing so-called "nuclear medicine," such as the technology to locate cancers with radioisotopes and cobalt irradiation therapy for the treatment of cancer. Secondly, President Eisenhower was especially interested in the treatment of heart diseases because he had suffered a heart attack himself in September 1955 and nearly lost his life, but had been saved by notable heart disease specialist Paul Dudley White.[5] Thirdly, people in many developing countries suffered from malaria, and the U.S. contribution in this field was therefore expected to have a great impact. In May 1958, Milton S. Eisenhower, the president's brother who had held university presidencies at Kansas State University, Pennsylvania State University, and Johns Hopkins University, gave a speech at the meeting of the World Health Organization (WHO) in Minneapolis. He announced the determination of the U.S. to cooperate with WHO in the development of remedies for cancer and heart diseases, and in the campaign to eradicate malaria, also encouraging the Soviet Union to join their efforts.[6] Milton Eisenhower's WHO speech corresponded with the president's Science for Peace address seven months earlier.

The U.S. medical aid program, however, was not a new initiative of the postwar years but the extension of a long tradition from the prewar era; for instance, the malaria eradication program can be traced back to the years between 1913 and 1939, when the Rockefeller Foundation's International Health Division (IHD) tackled hookworm, yellow fever, and malaria. Some scholars have pointed out that the Rockefeller Foundation's medical aid promoted corporate capitalism and American imperialism. Others have drawn attention to the Rockefeller Foundation's philanthropic activities in the U.S. South which were later applied to overseas medical aid programs.[7] In Latin America, existing studies point out that a major purpose of the Rockefeller's medical aid was to protect the interests of private American firms such as United Fruits. Other private efforts included those of Christian philanthropic organizations which had carried out overseas medical aid for a long time. In addition to these private initiatives, the U.S. Army Medical Corps had carried out anti-malaria activities since the prewar era. Among the various prewar medical aid programs, the IHD in particular became the model for the League of Nations Health Organization (LNHO), to whom it sent experts. In 1943, LNHO was merged into the United Nations Relief and Rehabilitation Administration (UNRRA), a wartime medical aid organization accommodating 1,400 experts. Both LNHO and UNRRA became the precursors to the postwar WHO. When the WHO was inaugurated in New York in July 1946, many of the staff members were former Rockefeller Foundation medical experts, and almost all of the early programs of WHO, including those targeting malaria, tuberculosis, and venereal diseases, were originally the IHD's programs.[8]

During the Second World War, malaria programs developed rapidly out of the need to protect U.S. soldiers in foreign countries, and specialized journals such as *Mosquito News* and *The Journal of the National Malaria Society* were published. The "DDT revolution," brought about by the invention of dichlorodiphenyl-trichloroethane (DDT), dramatically decreased malaria infection.[9] A large quantity of DDT had been used to kill malaria- and typhus-carrying insects during the war, and was also applied to agricultural and domestic use in the postwar years. As Rachel Carson has pointed out in *Silent Spring* (1962), DDT is easily dissolved in fat, and therefore stored in fatty organs in human and animal bodies.[10] Even though DDT's toxicity had been known to scientists, it was exported by the U.S. government and industry, and also used as a foreign aid supply. The Office of Inter-American Affairs (OIAA), a government agency for inter-American cooperation established during the Second World War (later transferred to the Department of State) and directed by Nelson Rockefeller, spent 30 million dollars on medical aid in Latin America, including DDT spraying, before the agency's abolition in 1951.[11]

After the Second World War, Secretary of State John Foster Dulles (1953–1959), in his efforts to cooperate with the WHO, established the Bureau of International Organization Affairs within the Department of State. In the meantime, after Joseph Stalin's death, new Soviet leader Nikita Khrushchev had embarked on the "peace offensive," and medical aid had thus become a battlefront of the Cultural Cold War. When the Kennedy administration was inaugurated in

1961, the U.S. foreign medical aid program took off, with Dean Rusk, former Director of the Rockefeller Foundation, as Secretary of State, and Walt Rostow, an economist who embraced modernization theory, as National Security Advisor. With the cooperation of Warren Weaver, head of the Rockefeller Foundation's Division of Natural Sciences and Agriculture, the Foundation's medical aid programs became closely intertwined with U.S. foreign policies.[12]

The center of the U.S. malaria eradication program, in cooperation with the WHO and UNICEF, was the International Cooperation Agency (ICA), reorganized into the U.S. Agency for International Department (AID) in 1961. Eugene P. Campbell, who headed the ICA's Public Health Division from 1955, had worked in Guatemala as a staff member of the Institute of Inter-American Affairs during the Second World War. Campbell promoted malaria eradication as a high-priority project, and his efforts contributed to the budgetary allocation of millions of dollars for pesticides, spraying devices, and medicines used in developing countries.[13] Since these materials were products of U.S. private companies, the Cold War medical aid was intended not only to win the allegiance of Third World countries but also to contribute to the U.S. market economy.

During the 1950s, the Rockefeller Foundation gradually moved away from medical aid programs. In 1951, the IHD was absorbed into the newly established Division of Medicine and Public Health (DMPH), which turned greater attention to education, policy-making, population control, and agriculture than to medical aid. However, many of the medical aid experts fostered by the Rockefeller Foundation took positions in the WHO. For example, Fred Soper, former chief of the IHD's malaria and yellow fever programs, became director of the WHO's malaria eradication campaign, while he concurrently directed the Pan American Sanitary Bureau (renamed the Pan American Health Organization in 1958), a medical aid organization for Latin America. Many other former IHD staff worked for the WHO, including Paul Russel, John Grant, and Alen Gregg, former Vice Director of the Rockefeller Foundation. Most notably, Marcolino Candau, a Brazilian doctor who had worked for the IHD's malaria program, became Director General of the WHO in May 1953. These examples indicate the degree to which the WHO's malaria eradication campaign was sustained by former Rockefeller Foundation malaria experts.[14]

Although the WHO absorbed human resources from the Rockefeller Foundation, funding aid from the Foundation rapidly shrank, which in turn led the WHO to become increasingly dependent on U.S. government funding. Three months after the inauguration of the WHO, the U.S. congress, after some heated debate, finally voted to join the organization, and U.S. funds poured in after 1956–1957. The U.S. government came to be keenly aware of the "potential of the malaria eradication campaign to combat communism," especially after the Communist countries, once withdrawn from the WHO in protest against U.S. influence, rejoined the organization in 1956, and after influenza pandemic hit the world in 1957.[15]

The history of malaria eradication programs thus far indicates that the U.S. government was a relative newcomer in that field, in contrast to private organizations such as the Rockefeller Foundation. Even so, the U.S. government ventured into the field because of the potential for malaria eradication to be exploited as an overseas information program—many developing countries in Asia, Africa, and Latin America suffered from malaria, and these countries, the U.S. policy-makers expected, would welcome and appreciate U.S. efforts to eradicate the disease. According to a confidential document of USIA prepared immediately after President Eisenhower's State of the Union address, the USIA "should not highlight" the Soviet and Eastern European countries' failure to contribute to the WHO's malaria eradication fund. Instead, the document said, "consistent with the positive spirit" of the president's Science for Peace message, the USIA should treat the malaria eradication campaign "as an example of the sharing of the fruits of science, and the application of modern technology for human betterment."[16] The USIA expected to make the malaria eradication program a symbol of the open and humanitarian S&T that the U.S. was promoting.

The U.S. Congress also found importance in international contribution in the field of medicine, but emphasized international cooperation in medicine on a broad spectrum rather than narrowly focusing on malaria. The Senate voted to designate 1959 as the International Health and Medical Research Year, in which the U.S. would cooperate with other countries in the "discovery and exchange of answers" in the treatment of diseases. The USIA's Science Advisor, Harold Goodwin, objected to the Senate proposal; according to Goodwin, even the international cooperation in malaria eradication had yielded "comparatively little propaganda mileage," and therefore, it was "unlikely that an international program on less widespread diseases would do much more." "Exchange of answers," he asserted, was not important because medical journals all over the world were already publishing research results. Rather than "not yet curable" diseases such as cancer and heart disease, the U.S. should focus on the diseases already eradicated in the U.S., such as "smallpox, diphtheria, typhoid and cholera," which continued to plague "the part of the world we are most anxious to influence." Furthermore, he argued that working through international organizations such as the WHO and UNESCO would lead to "loss of U.S. identity" in the recipient countries. Goodwin believed that a program that would carry U.S. public health techniques "into various countries on a large scale, over a period of years" in such a way that "American interest and aid would be apparent to the recipients" would have the strongest psychological effect.[17]

The different approaches of Congress, the USIA, and its Science Advisor illuminated the conflicting dimensions of the U.S. medical aid program: on the one hand, the U.S. advocated scientific internationalism and cooperation with the WHO, while on the other hand, it preferred bilateral programs showcasing American products and technologies. Project Hope belonged to the latter type, was as did the USIS film *Winged Scourge*, an educational animation film on malaria originally produced during the Second World War for Latin America with the cooperation of the Walt Disney Company, and which was reused as a USIS film in the postwar period and shown in many countries not limited to Latin America. The film provided an easy-to-understand explanation of the protection against malaria, with the well-known Seven Dwarfs of the Disney animation *Snow White* spraying pesticide to kill mosquitoes.[18] The combination of Disney animation and malaria eradication was a quintessential expression of what Goodwin described as "American identity" in the overseas information program.

It is doubtful, however, whether the malaria eradication campaign ultimately won respect and allegiance from foreign countries as expected by the U.S. government. Existing studies have pointed out that the massive spraying of DDT resulted in the emergence of DDT-resistant mosquitoes, and that culturally insensitive intrusions into the lives of local residents (in the process of spraying) rather played a part in alienating them. Records of the ICA also include evidence of an unsuccessful operation of the malaria program. On March 6, 1959, the ICA headquarters in Washington DC telegrammed its Jakarta office expressing concern about the Indonesian government's "failure (to) implement malaria eradication project" in accordance with the plans agreed with the U.S. The planned "National Malaria Eradication Service" had not been established, and moreover, incidents of theft of ICA-provided commodities such as insecticide had occurred; it seemed that, at least in this case, the U.S. aid had gone down the drain.[19]

The emergence of malaria eradication as a theme of the U.S. overseas information program, alongside cancer and heart diseases, illuminated the fact that the U.S. postwar medical aid programs in the late 1950s and the 1960s were built on the long tradition of prewar philanthropic activities, while at the same time, they reflected the Cold War battle for hearts and minds, and furthermore, integrated the business interests of private companies. The next section will show how U.S. policy-makers discussed the relationship between medical aid and the overseas information program, and out of which, how Project Hope emerged as a government-sponsored "private" project.

7.2 Discussion of Medical Aid, Overseas Information Program, and Private Volunteer Activities

In May 1958, four months after President Eisenhower first mentioned Science for Peace, the U.S. and the Soviet governments exchanged notes on mutual cooperation in medical programs. Concretely, the two governments agreed on "exchanges of medical delegations, reciprocal trips of medical specialists, and the exchange of medical films and medical journals." In fact, the Soviet Union had already sent a delegation of six female doctors led by Ekaterina Vasyukova some months prior, and the group had toured the U.S. for one month, visiting various hospitals and laboratories. In return, the U.S. organized a corresponding group of six female doctors led by Helen Taussig and sent them to the Soviet Union in May, in cooperation with the Rockefeller Foundation. The American female doctors visited various medical facilities in Moscow, Leningrad, Kyiv, Sochi, and so on.[20] The Soviet delegation of an all-female medical team was likely intended to emphasize gender equality in Communist society. To compete with the Soviet Union, the U.S. also formed an all-female team, but American society in the 1950s was characterized by the conservative "consensus culture," in which women were first and foremost expected to become housewives and mothers, just as the "kitchen debate" introduced in the previous chapter symbolized.[21]

Helen B. Taussig, leader of the U.S. medical team, was a notable physician specializing in pediatrics and heart diseases, and an exceptional female medical scientist in the 1950s. Born in 1898, Taussig had tried to apply Harvard University's medical school, but the university did not admit women. After some twists and turns, she received a doctoral degree in medicine from Johns Hopkins University in 1927. In the prewar years, she was already noted for her creative methodologies in using X-rays to diagnose heart and lung abnormalities, and in the 1950s, she was a representative U.S. physician in the treatment of children with congenital heart diseases. Her name is also remembered as the first person to sound the alarm, in 1962, about the teratogenicity of thalidomide, a medicine prescribed for morning sickness.[22] Given the unwelcoming environment for women in medical science at that time, the U.S. government must have had to choose female doctors from a scarce pool in order to counter the Soviet cultural offensive, although Taussig's background was perfect for this purpose because she was an authority on heart diseases, and an expert of X-ray ("nuclear medicine"), both of which were the favored themes of the U.S. government.

As medical aid rose to the center stage of the Cultural Cold War, an argument for using medical missions as a Cold War weapon emerged both in and out of the government circle. For example, in October 1958, Navy doctor William V. Healey proposed to the USIA a plan to send medical missions under the auspices of the Department of State to "serve the people of those countries most in need of being shown how our capabilities compare with the Reds." Under the "Selective Service System," Healey explained, young physicians served two years in the Armed Forces, though the great majority of them, after finishing their duty, "contribute[d] little or nothing to the Cold War." In order to "counteract a serious Soviet threat," Healey recommended sending young physicians to developing countries to fulfill the "two-fold" purpose of "propaganda for the West" and the "improvement of health." Director Allen relayed Healey's proposal to Saxton Bradford, head of the USIA's Policy Planning Division (IOP) to ascertain his opinion on Healey's paper. Bradford, in turn, solicited opinions from the National Science Foundation, the ICA, the Department of Defense, the Department of State, and the OCB. Eugene P. Campbell (head of the ICA's Public Health Division, introduced earlier in this chapter) gave a reply in the negative. In the past, various voluntary and church missionary groups had proposed sending medical teams overseas, but Campbell felt that this type of program could "only be implemented under very special conditions" because it was "very expensive," "short range," and its principal effect was "impact." Moreover, foreign governments would view the U.S. efforts "in the direction of complete government control of health matters," which would undermine the image of the American "free enterprise system."[23] Bradford could not ignore Campbell's opinion, given his experience as a medical aid expert, and decided to carry out an evaluation of various existing medical aid programs in terms of "propaganda value." Incidentally, several government agencies, including the Department of

Health, Education, and Welfare (HEW), the ICA, the Department of State, and the Department of Defense, were at the time reviewing various medical aid programs. Bradford thought it would be useful to add "propaganda value" as a perspective for evaluation.[24]

Bradford's idea led to the Interdepartmental Committee on International Health Policy, established in September 1959 with the Director of the HEW as chairman and participated in by representatives of the ICA, USIA, Department of State, and the Public Health Service (PHS). The Committee drafted a report titled "Objectives of United States International Health Policy" in December 1960. The report warned that the Soviet bloc was taking advantage of the global "disparity" in health "to further its own influence and domination" over other countries, and that medical aid was being "used as a tool to accomplish such objectives." The report argued that the U.S. was therefore compelled to demonstrate that its medical aid programs were "preferable to those of the communists." However, the report continued, even ignoring the Cold War aspect, the achievement of better health in various foreign countries would serve the "basic foreign policy objective" of the U.S. to achieve a peaceful and prosperous international order. The report cited concrete examples of medical aid that the U.S. should promote, including "malaria eradication," "health and safety aspects of radiation," and "other matters" including Project Hope.[25]

It seems somewhat strange that Project Hope, a private volunteer service established as a People-to-People Program, was mentioned in the Interdepartmental Committee report. One explanation for this unusual reference can be found in the Sprague Committee report (see Chapter 6); although the Sprague Committee and the Interdepartmental Committee on International Health Policy were established for different purposes, both committees were active during the same period, and both submitted their final reports in December 1960. The Department of State and the USIA participated in both committees, and therefore, it is quite possible that the two committees shared information. Chapter 6 of the Sprague Committee report ("International Activities of Private Persons and Organizations, and of the Mass Media") pointed out the importance of private international activities for the U.S. image overseas, and argued that private activities were an effective countermeasure against the Soviet "front organizations." Although the report endorsed government support for private activities, it discouraged "too much governmental initiative," as any attempt to make "every American an amateur diplomat" would undermine "an attractive image of our voluntary and pluralistic society." The Sprague Committee report also mentioned that the People-to-People Program had succeeded in cultivating friendly contacts abroad with "governmental support through USIA." However, the report recommended that the programs should be continued only on a "selective basis," and that government support should be limited to the programs targeting "audiences of special significance politically," or "strategic geographical areas."[26] In short, the Sprague Committee viewed private activities such as People-to-People Programs as an important part of the overseas information program and endorsed a certain degree of government involvement in such activities, while avoiding the appearance of "too much" government involvement and limiting support to only a select few. Project Hope was among the select few People-to-People Programs for which government support would be retained, and that was the reason why it was cited in the Interdepartmental Committee on International Health Policy: as the next section shows, the *Hope* visited "strategic geographical areas" and offered medical aid to people with "special significance politically."

In sum, reports of the Interdepartmental Committee on International Health Policy and the Sprague Committee showed intricate relations between medical aid, the overseas information program, and private volunteer activities. As the Rockefeller Foundation example shows, there had been a long history of private medical aid programs in the U.S., and Project Hope could be just another example. However, in contrast to other People-to-People Programs, the U.S. government continued to support Project Hope because of its seeming usefulness as an overseas information program. The fact that C. D. Jackson, Eisenhower's advisor of psychological warfare and member of the Sprague Committee, held a seat on Project Hope's Board of Directors also reveals the degree to which the U.S. government viewed it as an important weapon of the Cultural Cold War.

7.3 Contents of Project Hope

President Eisenhower was a fiscal conservative, and proposed the People-to-People Program partly to curtail expenses for the overseas information program by mobilizing private citizens. In June 1956, he invited prominent leaders of business, education, fine arts, motion pictures, religion, and other fields to the White House to explore the possibilities of "better People-to-People contacts and partnerships throughout the world." As a result, 40 committees were established, ranging from "a Cartoonists Committee to a Sports Committee to a Medical Committee" which were "charged with coordinating programs and events in their areas of expertise."[27] In his study of the Eisenhower administration's propaganda policy, Kenneth Osgood has pointed out that the People-to-People Program "served the purposes of domestic as much as foreign propaganda." Among the U.S. government's efforts to make "the maximum use of private groups and nongovernmental organizations" in the overseas information program, Project Hope was the "most dramatic initiative" undertaken by the People-to-People Program.[28] In other words, the People-to-People Program was both an overseas information program and a domestic device to mobilize the entire citizenry into voluntary activities to support the overseas information program.

The People-to-People Program also matched the sensibility shared by middlebrow Americans in the immediate post-Second World War years. Christina Klein aptly coined the term "Cold War Orientalism" to describe the interest of the U.S. in Asia in the immediate post-Second World War years, in which "both middlebrow intellectuals and Washington policymakers produced a sentimental discourse of integration that imagined the forging of bonds between Asians and Americans both at home and abroad." The newly intensified interest in Asia was related to thousands of Americans' visits to Asia during the 1940s and 1950s, and the decolonization of Asian countries, through which the U.S. wished to establish favorable relationships. Many U.S. films, plays, and novels of the 1950s represented friendly or romantic relationships between Americans and Asians, or the overcoming of racism to establish such relationships. Films that belonged to this genre included *Sayonara* (1957) on a romance between an American soldier and a Japanese woman in Occupied Japan; *South Pacific* (1958), which treated the topic of transnational adoption and the overcoming of racism; and *The King and I* (1956) which presented British woman Anna Leonowens as teacher to the King of Siam's children. In these pieces, mutual understanding, friendship, and love were nurtured in settings where Americans (or other Western individuals) uplifted, helped, or educated the less modern or less cultivated Asians.[29] Project Hope, intended to provide medical knowledge to the less developed areas of the world, was in line with these fictional representations; not only fictions but real foreign aid programs were characterized by what Klein termed "Cold War Orientalism."

Since Project Hope was portrayed as an initiative resulting entirely from the good will of American citizens, the annual running cost of $3.5 million was to be covered by contributions from the Rockefeller Foundation and other private groups and individuals. The ship was operated by the President Line, a private company, and the staff were recruited through the American Medical Association. Prominent heart surgeon, William B. Walsh, headed the People-to-People Health Foundation's Project Hope Committee, and Paul E. Spangler, a Navy doctor from California, was Senior Medical Officer of the ship. Spangler was a war hero who had served at Pearl Harbor on December 7, 1941, and was "credited with performing the first operation in the U.S Armed Forces in World War II." The medical staff onboard the *Hope* consisted of "15 physicians, 2 dentists, 20 nurses and about 20 auxiliary medical personnel." In addition, about 35 physicians were to be "flown to the ship every four months on a rotating basis." The ship operated as "a combination [of] floating hospital center, medical school, training and treatment center, a base for medical, nursing and sanitation teams, and a logistical center for medical aid, health and exchange programs," but its especially important role was to educate local medical staff in developing countries.[30] The March 1959 issue of the USIA's monthly newsletter *People-to-People News* carried a photo of the U.S. Navy's hospital ship *Consolation*, which was soon to be converted to the *Hope*. The article explained that the ship would be sent to developing countries as a good-will mission of the American people, and that President Eisenhower congratulated the project (Figure 7.1).

In reality, however, the U.S. government provided both financial and practical support, and in return, attempted to exploit the project as an overseas information program. When Project Hope was inaugurated toward the end of January 1959, Conger Reynolds of the USIA reported to Director Allen that William B. Walsh had met with the Medical Director of the Rockefeller Foundation and ensured the Foundation's funding. The Medical Director showed "considerable interest and enthusiasm" and directed Walsh to submit promptly a request for

People
to
People

NEWS

VOL. 3 NO. 3 MARCH 1959

PTP HOSPITAL SHIP HAILED---Announcement of the People-to-People Health

Figure 7.1 People-to-People News reporting on Project Hope. RG306, Entry P243, box 3, NACP.

a $10,000 grant, and promised "a much larger grant to be considered at the April or May meeting of the Board." According to the Medical Director, the Rockefeller Foundation had turned down requests from many People-to-People Programs because none of them "had submitted a specific project for support," but the Foundation "could and would aid" Project Hope. Reynolds expected that "a thaw in Rockefeller Foundation's previously icy attitude toward People-to-People" might influence other private foundations, too, and labeled Project Hope as the People-to-People Program's most "hopeful" and "significant project to date."[31]

As the Rockefeller Foundation's funding was ensured, the USIA decided to support Project Hope, but also carried out a study to "estimate the public relations implications" of the project. The study results reveal a tension between the private sector and government approaches to the overseas information program. The report regarded Project Hope highly for several reasons: it was "carried out only in those countries to which it ha[d] been willingly invited"; local physicians were to play an active part; it was non-governmental and all American; and the project's directors trusted the U.S. government agencies, were willing to "cooperate with proper briefing and orientation," and promised to make available "the list of selectees," in accordance with the government request. The report concluded that "the public relations impact of this Project" would serve the U.S. national interest.[32] Project Hope's value as an overseas information program lay not only in private voluntarism but also in the leaders' conformity with government policies. Reynolds wrote there was "every reason to believe that Dr. Walsh w[ould] be responsive to guidance with respect to any political or public relations problems" that might happen in the course of Project Hope's operation.[33]

The U.S. government established close ties with Project Hope not only through Walsh but also C. D. Jackson, who joined the Project's Board of Directors and reported to the OCB on its progress. Director Allen wrote to Jackson that he was "encouraged" by his "decision to join the Board of Directors."[34] Moreover, Project Hope actually depended on the government funds to a substantial degree; in July 1961, after the ship's visits to Indonesia and South Vietnam, the ICA summarized:

> Despite the mutual desire of both HOPE and the U.S. Government to have the project run free of any governmental participation as a People-to-People project, the U.S. Government has always had a major interest in it. Financially, since the spring of 1959 to date, through ICA a total of almost $4 million has been made available for Project HOPE.[35]

The details of the $4 million-support were: $2.7 million for refitting the ship, $1 million as a loan for the ship's operation, and $260,000 for the vessel's stay in Saigon. The People-to-People Health Foundation had also raised about $1.9 million in cash and $2.0 million of "usable donations in kind." In addition to the funding aid, the government had provided "liaison agent[s]" and "administrative support" through the USIA and ICA, and also facilitated the "supply of surplus milk" by the Department of Agriculture. In spite of all this support, the government believed that "maximum political benefits" would be derived if the project continued to be "presented as a non-government operation." In order to "preserve the advantages" of "having the project viewed as a private effort," the ICA cautioned that *government financial aid "should be held in confidence by U.S. Government officials involved"* (italics for emphasis by the author).[36]

It sounds potentially paradoxical to keep the government support "in confidence" in order to derive maximum "political benefits," but this stance symbolized the essence of medical aid as an overseas information program. The brochure of Project Hope (Figure 7.2) showed, in bold characters, "HOPE IS NOT GOVERNMENT PROGRAM" with an explanation that it was a good-will mission by American citizens. This message of private initiative was important both for domestic and overseas audiences. According to Kyungjin Ha, American citizens' suspicion of political propaganda was conceived during the First World War. For Americans, "propaganda" belonged to totalitarian regimes such as those of the Nazis or the Soviets, but not to the "free" countries like the U.S.[37] To gain domestic support for fund-raising, therefore, Project Hope had to be presented as a citizens' good-will project, explicitly not government propaganda.

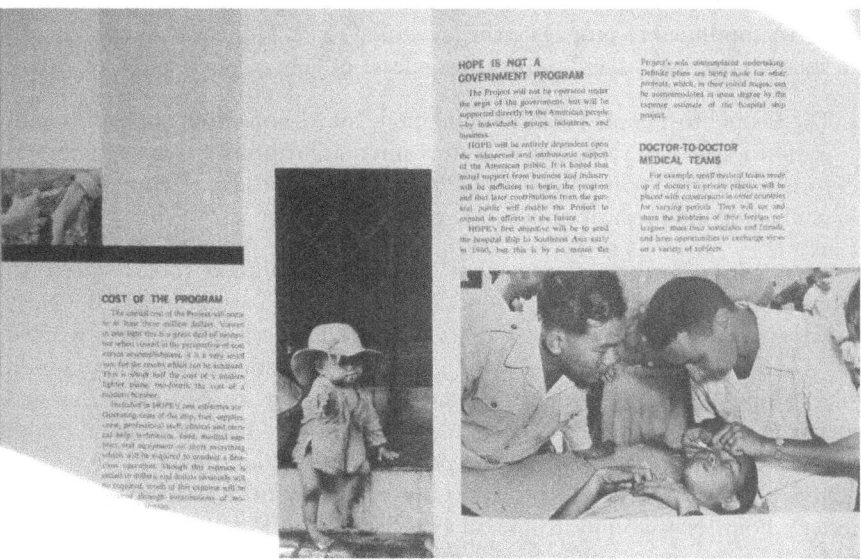

Figure 7.2 Brochure emphasizing that Project Hope was "not a government program." RG306, Entry P243, box 3, NACP.

For overseas audiences, it was also important to emphasize the private initiative and voluntarism because they were representative of the American national identity as distinct from the Communist counterpart.

The idea to keep government support confidential so that Project Hope would not appear to be propaganda was not an officially documented policy, but was nevertheless shared across government divisions and agencies. For example, the Far Eastern Division of the Department of State wrote that Project Hope would contribute to the U.S. foreign policy objectives by projecting the "sincere humanitarian interest of the people of the United States," and the Department should not "become officially involved" but should "maintain informal liaison" and "give assistance" to the project organizers.[38] Director Allen of the USIA instructed all the USIS branches to not "build up exaggerated expectations" or present "excessive USIS-attributed output" to avoid "labeling the trip as a government-sponsored propaganda peg."[39]

The U.S. Congress extolled Project Hope as a prominent volunteer activity by American citizens: senators, including Hubert H. Humphrey, who would later become Vice President of the Johnson administration, proposed that Project Hope should be officially commended in the Senate "as another step forward in increasing good will throughout the world and in bringing the people of all nations together in a bond of mutual trust, friendship, and cooperation." The USIA prepared a list of congressmen who were supportive of Project Hope as a resource useful to the fund-raising drive.[40]

7.4 The *Hope*'s Visit to Indonesia

The USIA continued to propagate the idea that Project Hope was an entirely private volunteer activity. The USIA's press release of February 10, 1959, reported:

> The People-to-People Health Foundation, Inc., a non-profit citizens' organization formed to carry out volunteer international medical aid activities as part of the President's People-to-People Program, today disclosed plans to send a floating medical center on a good-will mission to Southeast Asia.

The press release further reported that President Eisenhower took a "personal interest in the project," called it "a wonderful thing," and assured that "the Navy w[ould] make the ship available" as soon as the financial details were worked out.[41] On September 8, the VOA's Far Eastern broadcasting aired a program on Project Hope, stating that "the American people, in a gesture of friendship and understanding, ha[d] initiated" the project because they knew that good health was "essential to national strength and progress" in all countries.[42]

On September 13, 1960, the launching ceremony of the *Hope* was held at Hunter's Point Naval Shipyard in San Francisco. With Vice President Nixon as a guest of honor, many public figures were invited from industry, and from federal and state governments. A private advertisement company in San Francisco, Whitaker and Baxter, produced the event, but a USIA liaison officer for Project Hope was in contact with media representatives who covered the "commissioning and sailing of the ship."[43]

The *Hope* embarked on its first navigation toward Indonesia on September 22. Indonesia was precisely one of the "strategic geographical areas," that was the terms on which the government would support any People-to-People Program, as explained in the previous section.[44] While President Sukarno had been advocating the unity of non-aligned Third World countries since the Asia Africa Conference held at Bandung, Indonesia, in 1955, the U.S. government was gradually losing tolerance with Sukarno's "brand of neutralism."[45] Sukarno, on his part, strengthened economic nationalism, maintained cooperation with the Indonesian Communist Party (PKI) which had membership of 3 million, and strengthened ties with the Soviet Union and the People's Republic of China. The U.S. government launched a CIA-led clandestine operation to support anti-government insurgency in order to check the Communists. However, in May 1958, an incident involving the shooting down of an insurgency fighter jet, and the capture of an American pilot convinced Sukarno of CIA involvement and deepened his distrust of the U.S. government. Sukarno propelled autocratic governance in the name of "Guided Democracy" and established the "Nasakom," a political system aiming at the balanced union of nationalism, religions, and Communism, under which the PKI further expanded its influence.[46]

However, despite mutual distrust, the U.S. and Indonesia needed each other. Over the territorial conflict with the Netherlands over West Irian, the western half of the island of New Guinea, Sukarno approached President Kennedy to appeal for his mediation. The Kennedy administration, bogged down in the quagmire of the armed conflict in Vietnam, desired stability in the surrounding area, and acquiesced to Sukarno's request.[47] The *Hope* visited Indonesia at precisely a time when the two countries did not necessarily maintain stable and friendly relations but nevertheless needed each other to pursue their own political and diplomatic goals.

However, if we look at ensuing U.S.–Indonesia relations after the *Hope*'s visit, the once-improved relationship turned hostile again over the issue of Malaysia's independence. Sukarno criticized the U.K. harshly as neo-colonialists for the "Malaysia Plan."[48] Sukarno even sent reinforcements to support the anti-British insurgency, and the conflict developed into the burning of the British Embassy in Jakarta.[49] As Sukarno increasingly defected from the Western countries, the U.S. government became disillusioned with him, and by 1964, began to "identify and cultivate potential leaders within Indonesia for the purpose of ensuring an orderly non-Communist succession upon Sukarno's death or removal from office."[50] On September 30, 1965, after an unsuccessful coup d'état by a group of PKI members and pro-Communist military officers, at least 500,000 Communists and leftist political activists were massacred by the National Army led by General Suharto. Although Sukarno survived the incident, his influence gradually diminished, and three years later his political power had completely evaporated.[51] It is yet to be known how deeply the CIA was involved in the process, but in any case, Indonesia thereafter steered the course of pro-American, developmental dictatorship under Suharto. The greatest irony of the U.S. information program in Indonesia, therefore, was that it was through brute force, not through People-to-People friendship, that the pro-American regime was born.

Of course, when the *Hope* was docked at the port of Jakarta in 1960, no one knew of what was to come in the following five years. President Kennedy was inaugurated during the *Hope*'s visit to Indonesia, and was not involved in the preparation of the project at all. However, the USIA dressed up the new president as an earnest supporter of Project Hope. In a USIS film bearing the title of the endeavor, *Project Hope*, and which portrayed American doctors and nurses working for the project, a famous passage of President Kennedy's inaugural address was voiced over toward the end of the film:

To those peoples in the huts and villages across the globe struggling to break the bonds of mass misery, we pledge our best efforts to help them help themselves, for whatever period is required—not because the Communists may be doing it, not because we seek their votes, but because it is right. If a free society cannot help the many who are poor, it cannot save the few who are rich. [52]

The film explained that the President's speech essentially represented the spirit of Project Hope. By inserting President Kennedy's voice, *Project Hope* disseminated the narrative that American volunteer doctors and nurses traveled to Indonesia with the same determination of the president, that was, the U.S. mission to modernize the Third World.[53]

The production process of the film *Project Hope* was somewhat mysterious, and various circumstances suggest the possibility of the USIA's involvement from the early stage of production. The film was sponsored by Ex-Cell-O Corporation, manufacturer of turbine engine parts, produced by notable filmmaker, Frank P. Bibas, and won the 1961 Academy Award for short documentary films. However, the USIA procured the film, thus the film became a USIS film, and its prints were sent to many countries including Japan, Korea, South Vietnam, India, Iraq, and Zambia. Furthermore, the archival records indicate that the film was produced for the purpose of fund-raising for the *Hope*'s trip to South Vietnam. On May 10, 1961, at the ICA in Washington, Walsh, Edward F. Terrar of the People-to-People Health Foundation, and Herbert U. Waters, Special Assistant to the Director of ICA, discussed Project Hope. According to the minutes of the meeting, the South Vietnamese government had requested the *Hope*'s visit to Saigon, but there were insufficient funds to cover the travel. Walsh explained to the others that "around 150 prints of the film made by the Ex-Cello Corporation were in circulation," to help with the fund-raising drive.[54] In the actual film, there is a scene towards the end in which Walsh and Robert Bernard Considine, a journalist who voiced the narration, appeared on screen to appeal for contributions from the audience, explaining that Project Hope was entirely dependent on the good will of American citizens. It is not surprising if the USIA provided uncredited support to the film's production, given the above-mentioned government policy to keep government involvement confidential in order to avoid the appearance of propaganda.

As the author has already analyzed the contents of *Project Hope* in previous publication,[55] this section does not go into the details of the film, but rather focuses on the actual medical aid activities in Indonesia. The *Hope* visited the islands of Sumatra, Borneo, Java, and Bali. In Jakarta, Indonesian and American doctors worked in pairs in treating patients, through which American medical techniques were transmitted to Indonesian doctors. The *Hope* accommodated a limited number of inpatients, and those who necessitated operations were "hospitalized" on board. For example, the film focused on a young boy who underwent an operation to remove a tumor on his face, and recovered under the warm care of American doctors and nurses.[56]

Indonesian doctors, politicians, and even President Sukarno visited the ship. Coinciding with the *Hope*'s being in port, a Soviet industrial exhibition was being held. The USIA quite intentionally invited Sukarno to the ship before he visited the Soviet exhibition so that the media attention would be diverted from the Soviet exhibition. The USIS Jakarta reported that Sukarno's visit "effectively preempted press space from the Soviet exhibit." According to the USIA's report, the Indonesian government "closed the Bandung nurses training school for two weeks during HOPE's stay in Jakarta and sent its students to study on the ship" and pro-Sukarno publications reported the ship's visit favorably. By citing these examples, the USIA expressed the expectation that Project Hope would cultivate "a lasting feeling of good will."[57]

However, a confidential telegram from USIS Jakarta to the USIA on January 31, 1961, pointed out that Project Hope was not yielding sufficient results, and proposed the assignment of a USIA information officer on the vessel to make it a more effective overseas information program. The "USIS hand should not show in HOPE operations," the USIS stated, as this would "give rise additional feeling among Indonesians" that the program was "after all only [a] propaganda device." However, the "risk of leaving this important ... activity unattended" would be greater than the risk from any "criticism which might arise as [a] result of having [a] USIS officer aboard." "As last resort," therefore, the USIS considered asking for Walsh's "consent to having us assign [a] USIS Officer to ship."[58] Although it is unknown whether the USIA officer was actually assigned, this consideration reveals the thin veneer of the "private" outlook of Project Hope, and the limitation of overseas information activities by volunteer citizens.

7.5 The *Hope*'s Visit to South Vietnam

Thanks in part to the USIS film *Project Hope*, Walsh was able to raise sufficient funds to cover the *Hope*'s visit to South Vietnam by May 1961. On May 23, he visited the White House and met with President Kennedy's aides to request the government's permission for the ship's voyage to South Vietnam. Don Wilson of the USIA was also present. The White House staff were initially concerned that Walsh might request extra funding aid, but Walsh "asserted he was not asking for more money" and was rather asking for "lifting the restriction on the SS Hope I, going to Vietnam." After Walsh left, Kennedy's aides discussed whether the *Hope* should visit Saigon, and generally agreed that "it would be in the United States interest" although "no positive action" would be taken until after "review of the May 31, 1961, balance sheet."[59]

Walsh's request was approved, and in early July, the *Hope* arrived at the port of Saigon. The U.S. government created the narrative that the non-governmental Vietnamese Medical Syndicate headed by Dr. Chuong had invited the *Hope*. Although Chuong was aware that the ship was sent by the U.S. government, the U.S. Embassy believed that there was "no reason why he should cause HOPE embarrassment" by revealing that fact.[60] Even before the ship's arrival, local people's expectations had been boosted by the public relations efforts of Walsh and the USIA. Ambassador Frederick Nolting of the U.S. Embassy in Saigon—who replaced Ambassador Durbrow in 1961 and was softer on Ngo Dinh Diem than his predecessor, as shown in Chapter 3—reported to the Secretary of State Rusk that the *Hope*'s "public impact" was "highly favorable," despite the fact that the *Hope* was not "meeting its publicity promises and resultant Vietnamese expectations regarding number of doctors and nurses, range of doctors' specialties, and volume of cases it [could] handle." Nolting recommended that "several internationally known medical personalities and 10 more nurses" be added to the existing staff and the *Hope* remain in Saigon until September 1. While Rusk approved of the *Hope*'s stay until September 1, he did not approve of the dispatch of additional doctors and nurses since Walsh considered "24 doctors now on board as adequate," and he was going to use "Vietnamese student nurses to partially offset [the] nurse shortage."[61]

The Vietnamese Health Minister and physicians had expected that Project Hope would bring to South Vietnam "outstanding medical figures to teach them cardiac and thoracic surgery and demonstrate other highly specialized operations," and when such expectations were not met, Project Hope's propaganda value was undermined. The U.S. Embassy tried in vain to persuade Walsh to recruit more doctors, but Walsh stubbornly insisted such measures were unnecessary. As Walsh himself was busy with fund-raising in the U.S., Richard O. Elliott stayed in Saigon as Senior Medical Officer of the *Hope*. The U.S. Embassy described Elliot as "energetic and level headed," and "devoted to [the] Hope project." Walsh had given Elliott "firm directives on how to run project in his absence," and Elliott "had no knowledge that [the] ICA was providing financial support." Unaware of the government's involvement, Elliott did not lend his ear to the embassy's plea to increase the doctors, saying that "any increase in Hope staff must come through Walsh." To cope with the shortage of medical staff, Elliott, Willard H. Boynton, chief of the USOM's Public Health Division, and South Vietnamese Health Minister Tran Dihn De agreed to persuade "sufferers from minor ailments to go to Vietnamese doctors and hospitals instead of overburdening Hope." The *Hope*'s facilities were "strained by [the] heavy volume [of] applicants," and only 80 beds out of 230 originally expected to be available were usable because of the staff shortage.[62]

Ten days after the arrival of the *Hope*, the ICA in Saigon sent a confidential telegram to the ICA in Washington, in which the serious conditions were reported:

> The Hope Ship is a real problem to us. Walsh promised more than he could deliver and that got him off to a very bad start with the Vietnamese. In ten days he managed to antagonize the Secretary of Health, the Chef de Cabinet, the Surgeon General of the Armed Forces and the President of the medical society ... The chief problems as I see it, are: that Walsh's publicity was too expansive and the ship cannot live up to it; and secondly, there is no proper administration of the project, there is no one who give a definite answer to any questions such as, when will the ship come? who will be aboard? how many people do they need? etc. ... As of this time, the Vietnamese physicians consider HOPE as giving them more trouble than help, and I am inclined to agree with them.[63]

This shocking criticism by the ICA in Saigon alarmed the ICA in Washington, but further problems occurred ten days later. Walsh, after inspecting South Vietnamese hospitals, mentioned in his conversation with ICA officials in Washington that "large amounts [of] medical equipment in various hospitals [in] Vietnam, purchased with U.S. aid" were left unutilized because of the "lack of maintenance and/or repair." The ICA instructed the USOM in Saigon to clarify the fact. As Walsh's statement sounded as if the USOM was mismanaging the medical equipment, the USOM's distrust of Walsh increased further. USOM replied to the ICA in Washington that "USOM [did] not understand [the] basis [of] Walsh['s] remark nor believe it correct." According to the USOM, Walsh had made only one visit to a provincial hospital, where one anesthesia machine was lacking anesthetic gas. However, as the anesthetic gas was supposed to be provided by the government of South Vietnam, the USOM suspected that Walsh was "not distinguishing USOM, WHO, UNICEF, GVN, CARE equipment."[64]

The relationship between Walsh and the embassy/USOM deteriorated. The USOM had been engaged in medical aid well before the *Hope* came to Saigon, and its staff were proud of their tenacious efforts, which they thought were more meaningful than the short-term activities of Project Hope. "USOM doctors and nurses visit as many out of the way places as anyone in the entire U.S. contingent," Ambassador Nolting defended the USOM staff in his telegram to the Department of State. Dr. Boynton, Chief of the USOM's Public Health Division, Nolting remarked, had been "doing an excellent job here for four and one-half years, and ha[d] many good friends in Vietnamese medical circles," and argued that it was Boyton's contribution that had helped the *Hope*'s acceptance in South Vietnam.[65]

In August 1961, the U.S. Embassy in Saigon evaluated the past one month of Project Hope activities from the perspectives of

(1) [the] general public; (2) those seeking treatment or being treated aboard *Hope* or under *Hope* auspices in local hospitals; (3) [the] Vietnamese medical profession; and (4) [the] Ministry of Health, particularly [the] Minister and other officials dealing directly with Project.

According to the evaluation, the reaction of the general public was "rather favorable." The public were extremely impressed with the "sight of [the] imposing white ship in Saigon Harbour and [the] humanitarian nature" of Project Hope. The majority of the Vietnamese public seemed to view Project Hope as a good-will activity by the American people. By contrast, "those seeking treatment or being treated aboard *Hope* or under its auspices in local hospitals" showed "mixed" reactions. Although more than 10,000 people had applied for medical treatment aboard, fewer than 100 applicants were admitted to the ship. However, "nothing but unqualified praise" was heard among those who had been successfully treated aboard.

Generally, "professional health workers of Vietnam" had been disappointed with the Hope Project. Walsh had originally explained to the local Hope Committee that the ship was a "fully equipped hospital, in which 250 beds might be activated," but in actuality, only 100 to 120 beds were in use. "Highly specialized equipment" and "special technical teams of U.S. physicians" that Walsh had promised were not provided. However, "on an individual basis," the *Hope*'s physicians and nurses had established "good working relations" with their Vietnamese counterparts and fostered better Vietnamese-American relations. Especially in instances of "two doctors intimately sharing the care of a patient," their collaboration was the most successful. Up to this point, the evaluation report was written in a fairly moderate tone. However, the atmosphere dramatically changed in the last few paragraphs of the report.

The final few paragraphs of the report included scathing criticisms of Project Hope and Walsh. The first of which was that, "the project's idea of sending teams to provincial towns has been inhibited" not only by the security situation and the short visit, but also by the "apparent misunderstanding between Dr. Walsh and local Hope Committee as to the funding for these trips." The local Hope Committee had no funds for this purpose, while Walsh thought that the committee should cover the domestic travel costs. Secondly, the project suffered from "lack of advance administrative personnel" arriving in Vietnam at least six weeks prior to the ship, and an "administrative head on the ship with full responsibility and authority" to direct the entire operation. The lack of these personnel had "greatly handicapped operations" and "caused considerable confusion during first few weeks." Thirdly, "rude and undisciplined behavior" of some crew members resulted in "numerous encounters with local police." Fourthly, Walsh's "implied and direct criticism of dedicated USOM Public Health Division personnel" had "depressed their morale." This was particularly regrettable since "these individuals devoted many man hours to Hope, and thereby contributed greatly to its success." According to "leading members of [the] Vietnamese medical profession," the USOM's medical education project would have had "more effect in raising the health level of Vietnamese than *Hope* could possibly have during its few months' visit." Some Vietnamese intellectuals pointed out "only approximately 100 hospital beds [were] in operation" in *Hope*, while Vietnam already had 20,000 beds but needed thousands more. They wondered if Project Hope's funds "might not have been put to better use" by using the ship as a vehicle, and by providing more doctors and equipment.[66]

A few days later, Nolting's interview with Samuel Strasburger, Manager of Stanvac Oil Company in Saigon, sustained the various criticisms that Walsh and Project Hope had thus far been targeted with. Strasburger summarized the *Hope*'s activities in Indonesia and South Vietnam as follows: "ESSO ha[s] given $250,000 to the Project Hope, and TEXACO, CALTEX, and DuPont together ha[ve] contributed a million dollars to the project." However, the Hope's visit to Indonesia turned out to be "disappointing on the whole." Although the Indonesians who had "received effective treatment" were happy, "a far greater number of people with ailments who had expected but not received attention" were disappointed, and "this disappointment probably outweighed the satisfaction of the others." Furthermore, the leaders of Project Hope had "alleged that their visit to Indonesia had been curtailed because of lack of funds," and they were "alleging the same thing" again in Vietnam. Strasburger wondered why the project was "chronically out of funds" in spite of the "large sums" Walsh had collected. He also questioned Project Hope's "person-to-person approach" as he thought the USOM's medical aid program "would do much more permanent good, and at a cost in better proportion to the results."[67]

In sum, the *Hope*'s visit to South Vietnam illuminated the conflict between the USOM in Saigon and Walsh, partly because the USOM staff, who were proud of their medical aid program in South Vietnam, viewed Project Hope's short-lived, short-sighted activities as ineffective. Also, Walsh's hardline delivery of Project

Hope, lacking sufficient understanding of the local situation of South Vietnam and the struggle of the USOM thus far, widened the rift between him and the U.S. Embassy members in Saigon. Still another reason lies in the method in which Walsh, a non-governmental physician, worked closely with the USIA and the White House, and kept such a connection secret while carrying out Project Hope. The USOM in Saigon was an ICA branch, and the ICA was part of the Department of State—the medical aid program of the USOM was part of official U.S. diplomatic policy. Yet the USOM staff were requested to assist with another medical aid program, Project Hope, ordered from the White House and the USIA. In other words, their assistance to the project was extra work which did not fit in the normal line of order. Even so, Walsh, while receiving the USOM's assistance and keeping it a secret, was telling the USOM what to do. To the USOM staff, it was likely an intolerable situation: to have to provide assistance to the "private volunteer" activities, and have the private citizen, Walsh, behave as their superior, reporting to Washington on USOM's evaluation. The exploitation of private activities for the government overseas information program without proper recognition resulted in a serious rift among the Americans who were involved in foreign medical aid.

The U.S. military involvement in Vietnam has traditionally been explained as a "proxy war" between the Cold War superpowers. However, some of the recent scholarship emphasizes the aspect of nation-building, as U.S. policymakers believed that building a modern Vietnam would lead to the containment of Communism in Southeast Asia. Jessica Elkind, for example, has framed the Vietnam War as "the failed American and South Vietnamese modernization efforts" in which "civilian aid workers" played a critical role. By the early 1960s, Vietnamese insurgents "routinely targeted Americans, including civilians, for their role in supporting" the authoritarian government of South Vietnam. USOM's medical staff, therefore, were fighting the Cultural Cold War under fire. It was in the midst of this quagmire that Project Hope was an attempt to display a good-will medical mission by American citizens.[68]

The U.S. foreign medical aid programs illuminate the ambiguous borderline between the government and the private, and both cooperative relations and the tension between the sectors. Because medical aid was on the frontline of the U.S.–Soviet competition, Project Hope was assigned an important role as an overseas information program. However, the indispensable human resources for the project, such as doctors, nurses, and technicians, were private citizens, not government personnel. Medical aid programs, in general, could not function without the cooperation of private citizens. Furthermore, as the Sprague Committee had pointed out, having individual Americans play the role of diplomats might harm the image of an autonomous and diverse American society, and therefore the U.S. government had to avoid excessive interference with private activities, despite the financial assistance they provided. The complex relationship between the government and the private sector concerning the overseas information program brought about a certain limitation in government power; when only a few leaders of the private volunteer activities, such as Walsh, were aware of the government's involvement, those who were not had no inherent reason to adhere to government instructions. Herein lies the character and limitations of the overseas information program in the field of S&T, which needed the cooperation of experts. The message of Science for Peace, that American S&T would contribute to the improvement of the lives of ordinary citizens, may have been understood by a few lucky patients and doctors in Indonesia and South Vietnam, and thus the degree of their influence was limited.

Another new theme of the overseas information program was space flight, although it was radically different from medical aid in that foreign citizens could not expect any direct benefit for their everyday lives. It was thus not possible, nor necessary, to create a veneer of "private" activities concerning space, either. The next chapter will examine why this extraordinary, governmental project captured such an important position in the overseas information program.

Notes

1 Concerning the U.S.–Soviet competition over foreign aid, the author consulted: Shigeru Akita, *Teikoku kara kaihatsu enjo e: sengo ajia kokusaichitsujo to kogyoka* [*From Empire to Developmental Aid: The International Order in Postwar Asia and Industrialization*] (Nagoya: University of Nagoya Press, 2017), 9; Shigeru Akita, ed., *Ajia kara mita gurobaru hisutori* [*Global History from Asian Perspectives*] (Kyoto: Minerva Shobo, 2013), 218–219; Shoichi Watanabe, ed., *The Columbo Plan: Formation of the Postwar International Order in Asia*; Shoichi Watanabe, ed., *Reisen henyoki no kokusai kaihatsu enjo to ajia: 1960-nendai o tou* [*International Developmental Aid and Asia in the Cold War in Transition: Inquiry into the 1960s*] (Kyoto: Minerva Shobo, 2017).
2 "People-to-People Program," Dwight E. Eisenhower Presidential Library website, https://www.eisenhower.archives.gov/research/online_documents/people_to_people.html.
3 The non-profit organization Project Hope is active even today. See the organization's website, https://www.projecthope.org/.
4 In this period, the U.S. government often used the term "health aid" or "international health," while this book employs "medical aid program" as it more clearly expresses the contents.
5 Clarence G. Lasby, *Eisenhower's Heart Attack: How Ike Beat Heart Disease and Held on to the Presidency* (Lawrence, KS: University Press of Kansas, 1997); "About Dr. Paul Dudley White," American Heart Association website, https://www.heart.org/en/affiliates/paul-dudley-white-about.
6 Department of State, for the Press, May 29, 1958, RG306, Entry P243, box 3, NACP.
7 E. Richard Brown, "Public Health in Imperialism; Early Rockefeller Programs at Home and Abroad," *American Journal of Public Health*, vol. 66, no, 9 (September 1976): 897–903; Brown, *Rockefeller Medicine Men: Medicine and Capitalism in America* (Berkeley, CA: University of California Press, 1979); John Ettling, *The Germ of Laziness: Rockefeller Philanthropy and Public Health in the New South* (Cambridge, MA: Harvard University Press, 1981); Yumi Hiratai, "Kenkyushi tenbo: Rockefeller zaidan no iryo koshueisei katsudo to bunka gaiko" [State of the Field: The Rockefeller Foundation's Medical and Hygiene Activities and Cultural Diplomacy], *Sapporo Gakuin University Journal of the Society of Humanities*, vol. 92 (October 2012): 111–118.

8 A. E. Birn, "Backstage: The Relationship between the Rockefeller Foundation and the World Health Organization, Part I: 1940s–1960s," *Public Health*, vol. 128, no. 2 (2014): 129–131.

9 Marcos Cueto, "International Health, the Early Cold War and Latin America," *Canadian Bulletin of Medical History*, vol. 25, no. 1 (2008): 20–22.

10 Rachel Carson, *Silent Spring* (London: Penguin Books, 2000; Boston, MA: Houghton Mifflin, 1962), 35–36. Citations refer to the 2000 edition.

11 Cueto, "International Health," 25–26. The OIAA was initially called the Office of the Coordinator of Inter-American Affairs (CIAA or OCIAA). During the Second World War, its primary goal was to counter German propaganda in Latin America.

12 Cueto, 27, 30–31.

13 Cueto, 34–35

14 Birn, "Backstage," 132–134.

15 Birn, 136–137.

16 "Eisenhower's Reference to Malaria Eradication," January 9, 1958, RG306, Entry P243, box 3, NACP.

17 From Goodwin to Bradford, September 29, 1958, RG306, Entry P243, box 3, NACP.

18 *Winged Scourge*, 1943, moving image, RG306, 306.240, NACP.

19 Wolfe, *Competing with the Soviets*, 70–72; From Saccio, ICA to ICA Djakarta, March 4, 1959, RG306, Entry P243, box 3, NACP.

20 Department of State, for the Press, May 29, 1958, RG306, Entry P243, box 3, NACP; "Women to Tour Soviet: Six U.S. Medical Scientists Will Visit for a Month," *New York Times*, March 6, 1958.

21 May, *Homeward Bound*.

22 "Taussig, Helen, Brooke," in Martha J. Bailey, *American Women in Science* (Santa Barbara, CA: ABC-CLIO, 1994), 387.

23 William V. Healey, "The Cold War: A Medical Plan," no date; From Bradford to USIS Morocco, October 6, 1958; Office Memorandum from Campbell to Meagher, October 16, 1958, RG306, Entry P243, box 3, NACP.

24 Saxton Bradford, "Programs of US Medical Aid to Foreign Countries," December 2, 1958, RG306, Entry P243, box 3, NACP.

25 Interdepartmental Committee on International Health Policy, "Report and Recommendations to the President," December 7, 1960, RG306, Entry P243, box 3, NACP.

26 "Conclusions and Recommendation of the President's Committee on Information Activities Abroad," 51–52, CIA, Freedom of Information Act Electronic Reading Room, https://www.cia.gov/library/readingroom/document/cia-rdp86b00269r001400210001-2.

27 "People-to-People Program," Dwight E. Eisenhower Presidential Library website, https://www.eisenhowerlibrary.gov/research/online-documents/people-people-program.

28 Osgood, *Total Cold War*, 215, 240–242.

29 Cristina Klein, *Cold War Orientalism: Asia in the Middlebrow Imagination, 1945–1961* (Berkeley and Los Angeles, CA: University of California Press, 2003).

30 "For Immediate Release," February 10, 1959; "Project Hope Selects Chief Medical Officer," no date; From Reynolds to Allen, January 28, 1959, RG306, Entry P243, box 3, NACP.

31 From Reynolds to Allen, January 28, 1959, RG306, Entry P243, box 3, NACP.

32 From Thoman to Reynolds, November 4, 1959, RG306, Entry P243, box 3, NACP.

33 Office Memorandum from Reynolds to Halsema, November 10, 1959, RG306, Entry P243, box 3, NACP.

34 From Allen to C.D. Jackson, October 13, 1959; "Follow Up on Our Memorandum of August 25 Re USIA Media Coverage of Project Hope," September 2, 1960, RG306, Entry P243, box 3, NACP.
35 From ICA Washington to ICA Saigon, July 17, 1961, RG469, Entry P89, box 7, NACP.
36 From ICA Washington to ICA Saigon, July 17, 1961, RG469, Entry P89, box 7, NACP.
37 Kyungjin Ha, *Public relations no rekishi shakaigaku: Amerika to nihon niokeru <kigyo jiga> no kochiku* [*Historical Sociology of Public Relations: Construction of the Industrial Subjectivity in the U.S. and Japan*] (Tokyo: Iwanami Shoten, Publishers, 2017).
38 "FE's Position Regarding Project HOPE," October 23, 1959, RG306, Entry P243, box 3, NACP.
39 Air Pouch, February 12, 1959, RG306, Entry P243, box 3, NACP.
40 From IOC Conger Reynolds to Mr. Allen, December 8, 1959; Congressional Record, Proceedings and Debates of the 86th Congress, First Session, September 14, 1959, RG306, Entry P243, box 3, NACP.
41 "For Immediate Release," February 10, 1959, RG306, Entry P243, box 4, NACP.
42 "Magazine of the Air #100, Project HOPE," September 8, 1959, RG306, Entry P243, box 3, NACP.
43 "Follow Up on Our Memorandum of August 25 Re USIA Media Coverage of Project Hope," September 2, 1960, RG306, Entry P243, box 3, NACP.
44 As for the Indonesian history from Sukarno to the Suharto era, the author has consulted: Aiko Kurasawa, *9.30: Sekai o shinkan saseta hi* [*September 30: The Day the World Was Shaken*] (Tokyo: Iwanami Shoten, Publishers, 2014); Taizo Miyagi, *Bandung kaigi to nihon no ajia fukki: Amerika to ajia no Hazama de* [*The Bandung Conference and the Comeback of Japan in Asia: Between the U.S. and Asia*] (Tokyo: Soshisha, 2001); Westad, *The Global Cold War*; Yoichi Kibata, "Enjo no hakaba?: 1960-nendai Australia no Indonesia enjo seisaku [Foreign Aid Graveyard?: Australia's Foreign Aid Policies toward Indonesia in the 1960s] Chap. 6 in *International Developmental Aid and Asia in the Cold War in Transition: Inquiry into the 1960s*, ed. Shoichi Watanabe, (Kyoto: Minerva Shobo, 2017); Bradley R. Simpson, *Economists with Guns: Authoritarian Development and U.S.-Indonesian Relations, 1960–1968* (Stanford, CA: Stanford University Press, 2008).
45 Westad, *The Global Cold War*, 129.
46 Miyagi, *The Bandung Conference and the Comeback of Japan in Asia*, 33, 46; Kurasawa, *September 30*, 25, 34.
47 Miyagi, 46; Kurasawa, 9.
48 The plan to establish the Federation of Malay by amalgamating the Federation of Malaya, which had gained independence from the U.K. in 1957, and Singapore, which was under U.K. guardianship.
49 Miyagi, *The Bandung Conference and the Comeback of Japan in Asia*, 39–54; Kurasawa, *September 30*, 15.
50 Westad, *The Global Cold War*, 186.
51 Kurasawa, *September 30*, vi.
52 "Inaugural Address of John F. Kennedy," January 20, 1961, Yale Law School Lillian Goldman Law Library, The Avalon Project, https://avalon.law.yale.edu/20th_century/kennedy.asp.
53 As for President Kennedy's developmental aid for Asia, the author has consulted Watanabe, *International Developmental Aid and Asia*, especially the Introduction.
54 Memorandum of Conversation, May 10, 1961, RG469, Entry P89, box 7, NACP. In this conference, Herbert U. Waters proposed using some Peace Corps to staff Project Hope to Vietnam in order to curtail the budget.

55 Yuka Tsuchiya, "Amerika no seifu koho eiga (USIS eiga) ga egaita reisen sekai: iryo-hoken enjo-sen *Hope*-go o meguru kokusaiseiji" [The Cold War World Portrayed in the U.S. State-Sponsored Films (USIS Films): International Politics Involving the Medical Aid Ship *Hope*], Chap. 9 in *Joho ga tsunagu sekaishi* (*MINERVA sekaishi sosho* vol. 6) [*The World History Connected by Information* (*MINERVA World History Series*, vol. 6)], ed. Shingo Minamizuka, (Kyoto: Minerva Shobo, 2019), 219–241.

56 Frank P. Bibas, *Project Hope* (1961). A U.S. domestic version of the film was available in the past on the non-profit Project Hope website by courtesy of Barbara Bibas Montero, daughter of Frank P. Bibas, but now it is only available on YouTube. *USIS Film Catalog 1966* (Japan), 124.

57 "Project Hope in Indonesia," November 10, 1960, RG306, Entry P243, box 3, NACP.

58 Incoming Telegram from USIS Djakarta to USIA, January 31, 1961, RG306, Entry P243, box 3, NACP.

59 Memorandum for the Files, May 23, 1961, RG469, Entry P89, box 7, NACP.

60 From Nolting to Sterling J. Cottrell, Task Force Vietnam, Department of State, July 27, 1961, RG469, Entry P89, box 7, NACP.

61 From Nolting to Secretary of State, June 27, 1961; From Rusk to Embassy Saigon, July 3, 1961, RG469, Entry P89, box 7, NACP.

62 From Nolting to Secretary of State, July 5, 1961, RG469, Entry P89, box 7, NACP.

63 From Clifford A. Pease to Edward Rawsen, July 10, 1961, RG469, Entry P89, box 7, NACP.

64 From Labouisse to ICA Saigon, July 17, 1961; From Gardiner to ICA Washington, July 20, RG469, Entry P89, box 7, NACP.

65 From Nolting to Sterling J. Cottrell, Task Force Vietnam, Department of State, July 27, 1961, RG469, Entry P89, box 7, NACP.

66 From Frederick Nolting, Embassy Saigon to Secretary of State, August 22, 1961, RG469, Entry P89, box 7, NACP.

67 Memorandum of Conversation, August 25, 1961, RG469, Entry P89, box 7, NACP.

68 Jessica Elkind, *Aid under Fire: Nation Building and the Vietnam War* (Lexington, KY: University Press of Kentucky, 2016), 1–24.

8 Space Flight as a New Overseas Information Program

In 2019, NASA celebrated the 50th anniversary of the Apollo moon landing. TV documentaries, books, posters, and even emblazoned T-shirts with the theme of space flight were produced, and the Smithsonian Air and Space Museum held a spectacular exhibition titled Destination Moon in five museums across the country.[1] The Apollo fever reminded the author of the powerful image of space flight as an icon of national prestige and the victory of S&T in the Cultural Cold War.

The powerful historical memory of Apollo in the U.S. and elsewhere is at least partially attributable to the successful public relations campaigns of NASA and the USIA. On October 1, 1958—almost exactly one year after the Sputnik shock—the U.S. government reorganized the National Advisory Committee for Aeronautics (NACA) into the National Aeronautics and Space Administration (NASA). Space programs hitherto pursued separately by the Army, Navy, and Air Force were unified and placed under the direct jurisdiction of the president.[2] Although NASA was established as a civilian agency and responsible for space programs, excluding those with military purposes, it was in fact difficult to clearly distinguish military and civilian space technologies.[3] For example, satellite technology could be employed for either military reconnaissance or for weather surveillance, and rocketry technologies could be used to haul either humans or nuclear warheads.

Space technologies, therefore, were inherently dual-use technologies, but their use in domestic and overseas information programs imbued them with a "triple-use" quality. NASA astronauts became national and international celebrities, adorning magazine covers and featuring in TV and radio programs, and NASA's activities were introduced through exhibitions and USIS films all over the world. As we have seen in Chapter 6, the USIA's "Basic Guidance Paper" had clearly stated that technologies which would stimulate people's imaginations, such as space flight, had "equal or greater appeal" than more practical technologies in the medical and agricultural fields. The final chapter of this book explores the process by which space technologies, which would not directly benefit the everyday lives of foreign people, as well as being both extraordinary and inaccessible for ordinary citizens, became an important theme of the U.S. overseas information program. Concretely, this chapter will focus on the early manned space flight programs of NASA before the Apollo moon landing project emerged,

DOI: 10.4324/9781003243649-10

roughly between 1958 and 1961. By examining the close cooperation between NASA and the USIA, this chapter will shed light on the role of space flight in the Cultural Cold War.

There is a plentitude of existing studies on NASA's space development, including two edited volumes of interdisciplinary study sponsored by NASA, *Social Impact of Spaceflight* (2007) and *Critical Issues in the History of Spaceflight* (2006). In each of these volumes, science historian John Krige contributed an article on the relationship between NASA's space programs and U.S. foreign policy, which have particular relevance to this chapter.[4] In addition to these volumes, many important scholarly works on U.S. space development during the Cold War have been published in the past two decades. For example, Nicholas Michael Sambaluk discussed President Eisenhower's project to develop reconnaissance satellites and his relations with the Air Force in *The Other Space Race: Eisenhower and the Quest for Aerospace Security* (2015), and Yanek Mieczkowski re-evaluated President Eisenhower's reaction to the Sputnik shock in *Eisenhower's Sputnik Moment: The Race for Space and World Prestige* (2013). Furthermore, Allan A. Needell's *Science, Cold War and the American State: Lloyd V. Berkner and the Balance of Professional Ideals* (2000) showed how S&T and national security strategy became intertwined through the key actors bridging the two fields, although his thesis is not confined to the space program.[5] Japanese-language publications on the U.S. space programs include Yasushi Sato's *People and Technology that Built the NASA: Technological Culture of the Big System Development* (2007), *NASA: 60 Years of Space Development* (2014), Kazuto Suzuki's *Space Development and International Politics* (2011), and Kazutaka Yamamoto's article on the Kennedy administration's space policy (2000).[6]

Krige, in his above-mentioned 2006 article, explained the U.S. collaboration with Western Europe concerning satellite launcher technology in the mid-1960s as "the carrot of technological sharing" to divert European nations "away from national programs which were more difficult to control and which might see the proliferation of weapons delivery systems." Although the plan ultimately collapsed due to opposition within the U.S. government, the argument for the strategic sharing of technology to divert foreign countries away from projects of "which the U.S. might not approve" resurfaced again and again even in later years.[7] NASA's overseas information program may also be explained, at least partly, by the same argument; it is quite reasonable to assume that NASA's movies, exhibitions, and publications were intended to attract foreign countries—albeit on the popular level—to divert them away from technological nationalism and tie them to the international technological framework under U.S. leadership. However, unlike Western Europe in the mid-1960s, most of the world's other countries in the years immediately after NASA's establishment (1959–1962) did not have any space technology at all, but the U.S. nevertheless disseminated information on space technology overseas, including to countries without their own space programs or technologies. What, then, was the purpose of the U.S. overseas information program on manned space flight in this period? Krige's other article (2007) may provide a clue; he argued that the Kennedy administration collaborated with foreign countries in satellite technology to "consolidate the political and cultural solidarity of the free world," and this foreign policy objective was pursued "at a *cultural* level by tangibly demonstrating the values of an open, democratic system over a closed, communist society" (emphasis by the author).[8] NASA's overseas information program also aimed to consolidate political and cultural solidarity with foreign countries, including those with little likelihood of developing space technology in the near future, by stirring up admiration for U.S. space technology. Suzuki also pointed out that manned space flight served as a source of "soft power" which "disseminated positive messages to the world, fulfilled national pride, provoked nationalism, unified domestic society and strengthened the government's legitimacy."[9] In this sense, the overseas information program by NASA and the USIA had been another dimension of the Cultural Cold War since the 1950s, and probably the last phase of the Cultural Cold War, which crumbled in the late 1960s under the weight of the Vietnam War and its domestic and overseas repercussions.

8.1 Space Technology and Information Policy

Space technology emerged as a new theme of the U.S. overseas information program after the Sputnik shock. The Moscow press release proudly announced that the successful launching of the world's first satellite would "pave the way for space travel, and the present generation w[ould] witness how the freed and conscious labor of the people of the new socialist society turns even the most daring of mankind's dreams into reality." Notable Soviet scientist Leonid Sedov told Earnst Stuhlinger, an associate of Wernher von Braun, an émigré scientist from Germany who played a central role in the U.S. Apollo project, "America is very beautiful. The living standard is remarkably high. But it is very obvious that the average American cares only for his car, his home, and his refrigerator. He has no sense at all for the nation."[10] While cars, homes, and refrigerators were of central importance in the U.S. overseas information program, the U.S. needed to avoid the image of being a country which could not manufacture anything beyond consumer goods.

The U.S. failed in launching the Vanguard, the country's first satellite, in December of the same year. Foreign media ridiculed the failure of allegedly the most advanced country in the world: one London newspaper headlined, "OH, WHAT A FLOPNIK!" and a Paris journal quipped, "It seems there is a worm in the grapefruit" (comparing the size and shape of the satellite to a grapefruit). The Soviet delegates to the United Nations asked if the U.S. might accept "foreign aid under Moscow's program of technical assistance to backward nations."[11] To overcome the embarrassment and regain state prestige, the U.S. government entrusted rocketry development to Wernher von Braun, and succeeded in the launching of the first U.S. satellite, Explorer I, in January 1958. In October of the same year, NASA was established. Further, in April 1959, Project Mercury, the first manned space flight program, was announced and this program was succeeded to by the Kennedy administration.

The conventional view of President Eisenhower was that he was not very enthusiastic about space development, and that he established NASA in the wake of the Sputnik shock but tried to contain it to a relatively small scale in consideration of the fiscal balance.[12] However, more recent research in the past ten years has revised such an interpretation and demonstrated that President Eisenhower perceived space as an important field of intelligence, and quietly embarked on the development reconnaissance satellites, and therefore, that he was not "shocked" by the Soviet's launching of Sputnik, and responded with due composure.[13] However, neither the conventional nor the revised image of President Eisenhower fully explains the importance of the overseas information program that was taking place during his administration: the USIA had proposed to use space technology as a theme of overseas information program in the "Basic Guidance Paper," and thus space flight had become a theater of the Cultural Cold War.

President Kennedy turned even more sensitive ears to international public opinion than Eisenhower. According to the USIA's public opinion survey conducted through contracting companies in various countries, the majority of foreigners felt that the U.S. exceeded the Soviet Union in "soft" technologies which would directly influence everyday life, while in regard to "hard" technologies such as space and nuclear weapons, the U.S. lagged behind the Soviet Union.[14] To be sure, such results corresponded with President Eisenhower's Science for Peace message that American S&T contributed to the improvement of everyday life. Thus, the survey results could be seen as evidence that the USIA was successful in disseminating the Science for Peace message all over the world. At the same time, however, such a pervasive image of American S&T might have led to the interpretation that the U.S. could only be successful in "soft" fields. To reverse such an image and demonstrate that the U.S. was strong in both "soft" and "hard" S&T, the Kennedy administration viewed manned space flight as an extremely important overseas information program.

On April 12, 1961, the Soviet cosmonaut Yurii Alekseyevich Gagarin succeeded in the world's first manned space flight. Approximately one month later, on May 5, the U.S. launched Alan B. Shepard, Jr. in Freedom Seven on a 15-minute up and down space flight. The flight length was 486 kilometers, trifling compared to the 40,000 kilometers of the Soviet orbital flight, but NASA and the USIA nevertheless carried out spectacular information programs both domestically and overseas, and the American public were excited. It was three weeks later, on May 25, that President Kennedy gave his famous address in the Joint Congressional Session: "I believe that this nation should commit itself to achieving the goal, before this decade is out, of landing a man on the Moon and returning him safely to Earth." It has generally been believed that Kennedy's moon landing project was influenced both by Gagarin's space flight and the Bay of Pigs incident that happened in the same month (the CIA-supported amphibian landing mission by Cuban exiles to overthrow the Castro administration which ended in disaster). This explanation has merit, especially given that President Kennedy consulted Vice President Johnson on the possibility of a moon landing immediately after the unsuccessful Bay of Pigs invasion.[15] However, the idea of manned space flight was not conceived of by Kennedy, but had already commenced during the Eisenhower administration, as "Project Mercury." Under Project Mercury, the spacecraft Friendship 7, piloted by American astronaut John H. Glenn, Jr., succeeded in the first U.S. orbital flight in February 1962. Project Mercury ultimately carried out six manned space flights from 1961 to 1963, and ten orbital flights from 1964 to 1966.[16] Upon each flight, NASA and the USIA launched elaborate domestic and overseas information dissemination programs.

NASA's public relations program gave birth to the new American hero, that is, the astronaut, and they were put on a pedestal as "icons of Truth, Justice, and the American Way."[17] NASA employed Leo DeOrsey, a lawyer specializing in show business, and through him, entered a $500,000 contract with *Life* magazine which awarded the magazine the exclusive right to carry "personal stories" of the astronauts. On September 14, 1959, *Life* featured photos of the 7 astronauts selected for Project Mercury, and an 18-page feature article titled "The Astronauts: Ready to Make History." In the center of the photo was M. Scott Carpenter, to upper-right was Alan Shepard, the first astronaut to go to space, and from there, counter-clockwise, Donald K. Slayton, Vigil I. Grissom, L. Gordon Cooper, Jr., and Walter M. Schirra, Jr. All smiled confidently, but their clothes and facial expressions represented individuality. The feature article explained that they were the elite selected by very strict screenings, but at the same time, they were typical American citizens who loved individuality, and were good husbands and fathers.

The magazine introduced not only each astronaut's everyday life and his thoughts on space, but also his age, height, weight, color of hair and eyes, birthplace, family members, and pastimes, much as if they were Hollywood movie stars. Other pages of the magazine featured the NASA engineers who developed the re-entry capsule, and various facilities and training machines with accompanying photographs. Not only astronauts but also NASA itself became an object of information dissemination.[18]

Interestingly, the following issue (September 21) of *Life* featured the wives of the seven astronauts in the same layout, under the title: "Seven Brave Women behind the Astronauts: Spacemen's Wives Tell, in Their Own Words, Their Inner Thoughts and Worries." In the feature article, Anna Glenn, for example, disclosed her story of consulting with a Christian minister, when her husband was selected as a candidate for manned space flight, as to whether he should accept the mission, should he be appointed. They discussed "everything from faith in God to faith in the government." The minister responded that there was no reason to object space flight from a religious standpoint, and that "NASA surely would not undertake a program like Project Mercury unless they knew what they were doing." Relieved to know that space technology did not go against Christian teaching, and also with the assurance that the government agency was trustworthy, Anna had ultimately decided to support her husband's space flight. Anna also told *Life* that her confidence in NASA had deepened because her husband told her "everything about the program," and as a result space flight had already become a "family affair."[19] Through the words of an astronaut's wife, the article propagated the ideas which the federal government wished to spread: that space development did not contradict American values such as family togetherness or Christian morality, and that NASA was an open and trustworthy organization. As the 1950s were known to be the "era of consensus," when American citizens subscribed to (or were pressured to observe) conservative family values, gender norms, and loyalty to the government, brand-new concepts such as space flight needed to be carefully coordinated with existing values and norms. At the same time, in the 1950s, expressing individuality through consumption was regarded as a patriotic act in support of American capitalism. Thus, astronauts' wives became the icon of the American way of life, and the cosmetics they wore and cars they drove became hit products. Non-fiction writer Lily Koppel's *The Astronaut Wives Club: A True Story* published in 2013 showed the degree of their influence. For example, when the cover shot was taken, the astronauts' wives wore pink lipstick because pink was "the color of the First Lady, Mamie Eisenhower." However, when the magazine came out, they were shocked to find bright red lips on the photograph: The editors of the magazine had decided that a "bright Patriotic Red" was more suitable for "the Astrowives' lips" in the space age. Soon thereafter, cosmetics brand Revlon launched a bright-red lipstick "Moon Drops," which became a huge hit.[20]

Life expanded the contents of the magazine features to publish *We Seven: By the Astronauts Themselves* in 1962. John Dille, writer/editor of *Life* magazine, explained in the introduction that "there were some striking similarities" among the seven astronauts: they were all "married and had children," grew up in "small towns and cities," had "similar background and technical education," and even had "only small variation in size, shape, and coloring to distinguish them one from the other." At the same time, they had "distinctly original personalities and a rich mixture of private attitudes, personal characteristics and professional ideas," and of course, they were selected through a very strict and competitive screening. Dille portrayed the astronauts as the "ordinary supermen"—the cream of the crop of typical Americans who grew up in small towns, cherished their families, and valued individuality.[21]

This image of American astronauts was employed in domestic propaganda to stimulate national pride and garner support for the space program, but also used in the overseas information program. Photo magazines such as *Life*, as well as NASA pamphlets and films were placed in the 80 some USIS libraries all over the world (Figures 8.1, 8.2). As we have seen in Chapter 3, in the remote towns where there was no library, the USIS's moving libraries—vans loaded with books, magazines, and films—circulated: photo magazines and films were valuable mediums through which American way of life was introduced to foreign citizens who could not read English. The typical image of astronauts as shown in *Life* magazine, therefore, was disseminated both at home and abroad.

English materials on astronauts were also translated into local languages. A good example was a non-fiction account of the seven astronauts, *The Astronauts*, published in 1960 and immediately translated into Japanese. The volume was published in Japan by the Asahi Newspaper Company in December 1960 under

Figure 8.1 USIS library in Saigon, South Vietnam (1956). RG306, No.56-13521, NACP.

the title *Mercury-keikaku* (*Project Mercury*). The book introduced NASA, Project Mercury, and each of the seven astronauts just as *Life* magazine had. The author, Martin Caidin, was a well-known science fiction writer who had published many books on the theme of space, and was also a pilot and Air Force advisor. The fact that the Japanese version came out almost immediately after it was published in the U.S., and the inclusion of numerous photos provided by NASA, strongly indicate that the translation was supported by the U.S. government as part of NASA's overseas information activity.[22]

It is noteworthy that these overseas information activities had already been carried out long before President Kennedy announced the moon landing project. The two issues of *Life* magazine mentioned above were published in September 1959, before the presidential election, and the Japanese translation of *The Astronauts* was released before Kennedy's inauguration. In other words, the overseas information program on manned space flight was put into practice well prior to the Apollo program, in tandem with the domestic information campaign which stirred up excitement among Americans. The early start-up of the overseas information program by NASA and the USIA indicates the value of space flight as the theme of such a program. As will be explored in greater detail later in this chapter, space flight became the standard theme of the USIA's exhibitions and

Figure 8.2 USIS library in Taipei, Taiwan (1959). RG306, No.59-13525, NACP.

films. Before examining such details, however, the next section will first examine how NASA and the USIA established cooperative relations and how they built overseas information programs together.

8.2 Give Mercury Full Play: Cooperation of NASA and the USIA

On May 1, 1961, Laurence P. Dalcher, head of the Policy Planning Section (IOP) of the USIA, circulated the basic guidelines of the overseas information program on space to the International Broadcasting Section (IBS), the International Publication Section (IPS), the International Films Section (IMS), and the International Television Section (ITV). The document included the impressive slogan: "Give Mercury full play." At the same time, the guideline instructed the relevant sections "to keep the project in perspective with other US activities" and "maintain a matter-of-fact tone."[23] One month before this guideline, the Soviet Union had already succeeded in the world's first manned space flight. Lagging behind the Soviet Union, the U.S. could not celebrate the world's second manned space flight as something special. Instead, the U.S. government decided to pose that manned space flight had been prepared step by step in the U.S., and that it had been put into practice matter-of-factly.

NASA and the USIA, however, were establishing a cooperative relationship in order to take maximum advantage of the world's second manned space flight in the overseas information program. The Office of Public Information (OPI) was established soon after NASA came into being, and 30–40 staff were assigned to its Washington headquarters and its 11 Field Centers. In the OPI, the position of USIA Liaison Officer was established on December 15, 1958, to which Creston B. Mullins from the USIA's International Press and Publication Service was appointed.[24] Mullins began his career on *Washington Star*'s foreign desk and joined the Department of State in 1951, placed in Bonn as the Information and Editorial Specialist for the Office of the U.S. High Commissioner for Germany, and later assigned to Paris, also as an information specialist. When the USIA was established in August 1953, he relocated there, and was stationed in USIS posts of several countries before his appointment to the USIA's International Press and Publication Service.[25] In other words, Mullins was a specialist of international public relations, and had engaged in the overseas information program since 1951.

The USIA Liaison Officer's responsibilities were to facilitate "the flow of informational materials from NASA to USIA media services"; to provide "the media services with essential operational information on NASA activities"; to serve "as a ready point of contact between the two Agencies"; and to represent "within NASA/OPI the overseas information program and its responsibility to publicize throughout the world the accomplishments of the US in space exploration." In July 1959, Mullins submitted an activity report of the first half year as a Liaison Officer to the USIA. According to the report, the USIA was now "widely known in NASA," "accepted as a partner in the worldwide information effort of NASA," and "recognized as performing an essential service on behalf of NASA and the national space program." The Liaison Officer enjoyed "free rein to obtain and pass on operational information and information materials as if he were a full-fledged member of NASA/OPI," and was allowed "ready access" to the Directors of the OPI and the Office of International Programs. The Liaison Officer regularly attended OPI staff meetings, lectured in NASA conferences, wrote articles for NASA bulletins, and briefed directors of NASA's overseas stations in Peru, Chile, Ecuador, and Cuba when they visited Washington. The Liaison Officer further cooperated with film production, assisted in the preparation of a pamphlet for the American National Exhibition in Moscow, and set up the channel for all NASA responses to inquiries originating in foreign countries. "Press release, pamphlets, pictures, etc." were sent to USIS officers of the "countries from which the inquires c[a]me," and this system turned out to be useful in that it provided opportunities for the USIS officers to get acquainted with those who were "interested in American scientific developments," especially journalists and writers.[26]

Although Mullins' report gives an impression that close cooperation was established between the USIA and NASA in the first six months, in fact such cooperative relations actually took off in early 1961, when the Kennedy was inaugurated and took over Project Mercury from the Eisenhower administration. Bill

Lloyd was appointed as the head of OPI, and Harry Kendall as a new Liaison Officer (the former USIA Liaison Officer was renamed the USIA-NASA Liaison Officer). The USIA welcomed Lloyd's appointment as he appeared to be "most sympathetic" to overseas information activities: USIA Science Advisor Goodwin instructed Kendall to get in touch with Lloyd immediately to establish "improved NASA support of USIA."[27] The Lloyd–Kendall collaboration turned out to be extremely important for the NASA–USIA partnership thereafter.

According to the oral history of Kendall stored at Georgetown University, and Kendall's autobiography, *A Farm Boy in the Foreign Service: Telling America's Story to the World*, his assignment as the USIA-NASA Liaison Officer had critical significance in his long career in the USIA and the Department of State. Just like Mullins, Kendall was a former newspaperman and became a public information officer of the Department of State in 1951. He worked as a USIA officer in Venezuela, Japan, and Spain, and in Japan, he headed the Takamatsu Japan-U.S. Cultural Center (the former American Cultural Center) in Kagawa Prefecture for two years from September 1955.[28] From January 1961 to March 1964, Kendall worked in the NASA headquarters in Washington as the Liaison Officer. His responsibilities at NASA, according to Kendall himself, were: "keeping USIA media informed about forthcoming developments within the space program, assisting them in coverage of major events and … adapting NASA's information output" to fit the requirements of the USIA and the USIS. He also tried to "educate the NASA information staff about their tremendous overseas audience and the potential for reaching them through USIA." One responsibility of Kendall's which was not included in Mullins' description was to "educate" the NASA officers on the importance of overseas audiences. While Mullins contributed to the "recognition" of the USIA's importance, Kendall pushed NASA–USIA relations one step further by making NASA spontaneously engage in overseas information activities.[29]

NASA's more spontaneous participation in the overseas information activities was encouraged in the large-scale "Staff Conference" on NASA's public relations responsibilities held on June 26–27, 1961. The mission of the Staff Conference was to "develop a working paper for the future of NASA OPI" reflecting the President Kennedy's announcement of "a great new American enterprise," the moon landing project.[30] The conference was jointly planned by Lloyd and Kendall, and the entire staff of OPI participated. James E. Webb, Director of NASA, Pierre Salinger, the White House Public Relations Officer, and Thomas C. Sorensen, Deputy Director of the USIA, participated as lecturers. (Figure 8.3). Thomas C. Sorensen was brother of Theodore C. Sorensen, President Kennedy's aide. He had worked for a newspaper and radio before he became a public relations officer for the Department of State in 1951, served for the USIS Beirut and USIS Cairo, and as an officer of the USIA Headquarters for the Middle East Area, before he was appointed to the position of Deputy Director at the age of 34.[31] In 1968 Sorensen published a book on the USIA activities during his tenure as Deputy Director, in which he said that the USIA attempted to capitalize on the "difference between the American and Soviet systems" in the space race. In

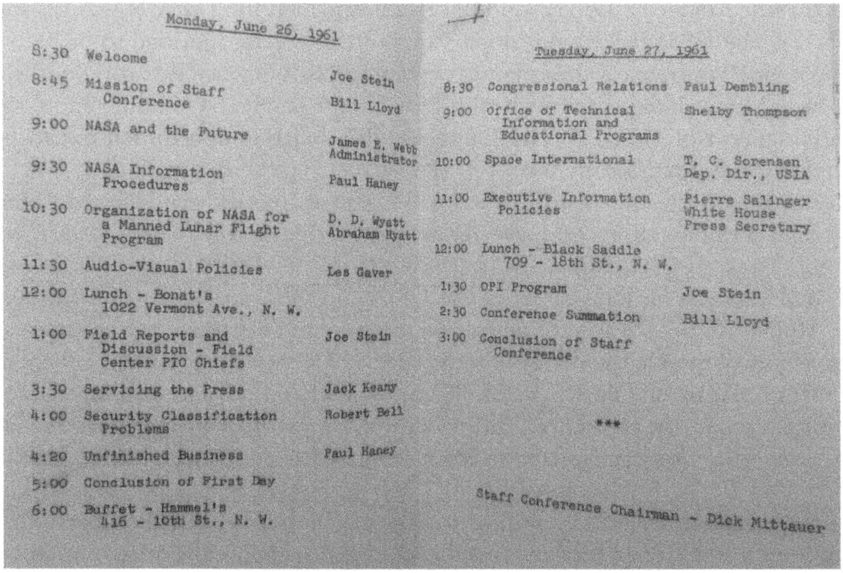

Monday, June 26, 1961

8:30	Welcome	
8:45	Mission of Staff Conference	Joe Stein Bill Lloyd
9:00	NASA and the Future	James E. Webb Administrator
9:30	NASA Information Procedures	Paul Haney
10:30	Organization of NASA for a Manned Lunar Flight Program	D. D. Wyatt Abraham Hyatt
11:30	Audio-Visual Policies	Les Gaver
12:00	Lunch - Bonat's 1022 Vermont Ave., N. W.	
1:00	Field Reports and Discussion - Field Center PIO Chiefs	Joe Stein
3:30	Servicing the Press	Jack Keany
4:00	Security Classification Problems	Robert Bell
4:20	Unfinished Business	Paul Haney
5:00	Conclusion of First Day	
6:00	Buffet - Hammel's 416 - 10th St., N. W.	

Tuesday, June 27, 1961

8:30	Congressional Relations	Paul Dembling
9:00	Office of Technical Information and Educational Programs	Shelby Thompson
10:00	Space International	T. C. Sorensen Dep. Dir., USIA
11:00	Executive Information Policies	Pierre Salinger White House Press Secretary
12:00	Lunch - Black Saddle 709 - 18th St., N. W.	
1:30	OPI Program	Joe Stein
2:30	Conference Summation	Bill Lloyd
3:00	Conclusion of Staff Conference	

***

Staff Conference Chairman - Dick Mittauer

Figure 8.3 Program of the NASA Staff Conference on public relations (June 26–27, 1961). RG306, Entry P243, box 4, NACP.

contrast to "deep Soviet secrecy," the U.S. manned space flight was broadcast on TV simultaneously around the world. Sorensen recollected that the television, radio, and film coverage of the U.S. manned space flight was enthusiastically received in Turkey, Greece, Argentina, Lebanon, and Ghana, among many other countries.[32]

At the conference, Sorensen emphasized the close connection between NASA's and the USIA's tasks, and encouraged even closer cooperation between the two agencies. He stated that Project Mercury especially had contributed to "a tremendous lift" in U.S. prestige overseas by providing the USIA with "a story of vigorous people determined to fulfill their role of world leadership." Since "newly-developing and neutralist countries often lean toward the strongest power," Sputnik and other successes of the Soviet Union were damaging to the U.S. "national posture," he explained. However, he believed that "the President's decision to reach for the moon" would become an incentive to surpass the Soviet Union in the propaganda battle.[33]

Sorensen further introduced the details of overseas information activities concerning the first manned space flight of Alan Shepard. Fully one month before the launch, the USIA sent a "hold-for-release packet" containing photos and articles on Project Mercury to each USIS post. The materials were translated into local languages and, by the day of Shepard's flight, were "on their way to newspapers and radio stations around the world." Thanks to this arrangement, "sixty journalists from 12 countries" covered Shepard's flight, and their reports were backed up by the prepared materials. For example, "the popular Japanese *Shukan Asahi*, which has a circulation of 1,200,000" and the "semi-official *Al Gumhuriya* of Cairo" featured stories. The USIA also produced a 13-minute documentary film on Project Mercury, *Shadow of Infinity*, which was "telecast" in 49 countries, and "released theatrically" in others. Another 10-minute film was produced "within 72 hours after launch" of Friendship 7, and Secretary of State Rusk took a copy to the NATO conference in Oslo to screen it there. Vice President Johnson also took the film on his tour to South Asia "for showing to chiefs of state in the countries he visited." The film had been translated into 28 languages. VOA radio also broadcast the Shepard flight in Arabic, German, Japanese, Russian, and Spanish, as well as in English. In addition, the USIA also used books and exhibitions to introduce Project Mercury to the world; for example, the USIA produced "two full size mock-ups of the Mercury capsule," one of which was to be sent to Europe, and the other to Asia. The Japanese translation of *The Astronauts*, introduced in the previous section, was probably part of these cooperative activities of NASA and the USIA. The aggressive information dissemination through various media yielded the impression, according to Sorensen, that "Shepard really went into space, not Gagarin, and in front of the whole world, too" (underline in the original text).[34]

As space flight increasingly became the central theme of the USIA's overseas information program, the role of the Liaison Officer grew more extensive and more important. According to Kendall, there were experienced "science writers" within the USIA—he cited, for example, the names of Charles Schroth and Walter Froelich—who integrated scientific information originating in NASA into the overseas information program. Also, within the OIP, Lloyd had assigned a public relations officer to each project: for example, Paul Haney was put in charge of manned space flight, Lt. Col. John Powers in charge of Cape Canaveral, Mercury's launch site, and his deputy Jack King, who were all in constant touch with Kendall. Every time a new rocket was launched, these NASA information officers prepared a "press kit," a package of public information materials, and distributed them to domestic news media and science laboratories. Kendall thought the "press kit," which included technical information and profiles of technical staff, would be attractive to overseas citizens, too, and obtained NASA's agreement to have them sent to USIS posts all over the world.[35]

The NASA–USIA collaboration came into full swing upon the first manned orbital flight by John Glenn in February 1962. Following this, out of the seven Mercury astronauts, M. Scott Carpenter and Walter M. Schirra, Jr., took flight, and the final Mercury flight by L. Gordon Cooper, Jr. was carried out on May 15, 1963. Kendall covered all these flights, producing newsreels for television and theatrical showings, sending them overseas, arranging live overseas broadcasting from Cape Canaveral, answering interviews from foreign reporters, and so on. There were differences between the information programs for developed and developing countries; for the developed countries, information was simply transmitted to local science journalists, whereas more detailed explanations were required for developing countries. To address this issue, the USIA recruited two young African American information officers, John Twitty and Elton Stepherson, and sent them to English-speaking and French-speaking African countries, respectively. There they loaded USIS films, 16 mm projectors, and various NASA publications onto vans and drove from one town to another, explaining NASA's space program, and especially the important roles played by NASA's tracking stations on the African continent. NASA, pressed by the need to build more overseas tracking stations, created a second USIA Liaison Officer position, to be in charge of negotiations with foreign countries. Allan Funch, a former newspaperman who had worked in various USIS posts, was appointed to this position.[36]

After his three-year tenure as the USIA-NASA Liaison Officer, Kendall briefly served as the USIA's Acting Science Advisor before Simon Bourgin officially took over Harold Goodwin's position, and then proceeded to the USIS in Panama (1964–1967), Chile (1967–1970), and South Vietnam (1970–1972). Kendall's three years as the USIA-NASA Liaison Officer during Project Mercury proved to be an identity-building experience in the long term. He employed the knowledge and experience he had gained on NASA's space program in each country he served as a USIS officer. In fact, he acted as if he was still a NASA Liaison Officer, continuing to hold slide shows and exhibitions, produce TV programs, exhibitions, and lectures, and sometimes appeared himself on television to lecture on NASA's space program. As he later reflected, people all over the world were eager for access to information on NASA's space program, and therefore, he did not have to "sell" the information: all he had to do was "to get it to them." Since NASA's "openness" provided a "stark contrast with Soviet secretiveness," he "exploited that difference to the hilt." In an age before satellite television or internet, foreign audiences tuned into VOA radio and relied on the USIA press releases for real-time information on the U.S. space program, and watched the USIS films for more in-depth understanding. Of course, Kendall's and other USIS officers' activities were not always welcomed by foreign countries. On occasion, negative rumors about the USIS were spread by local media, and Kendall had been called a "well-known Yankee imperialist." To investigate the source of such anti-American information, Kendall recalled, "the other agency," i.e., the CIA, played an active role.[37] Kendall's memoir illuminates the process in which NASA's space program assumed increasing significance in the U.S. overseas information program after 1958, and became the frontline of the Cultural Cold War.

8.3 USIS Films and Exhibitions

The USIS films were one of the most important overseas information programs promoted cooperatively between NASA and the USIA. In June 1961, the two agencies completed the production of *Project Mercury*, which was shown in many countries around the world.[38] For example, the U.S. embassy in Stockholm reported that the film was shown to members of the scientific and engineering organizations and youth, and the embassy received an audience reaction which remarked, "Many of us knew a little about the American project Mercury, but the production of satellites and technical methods described in the films were complete news." Also, *Explorer in Space* was shown to a group of students at a technical high school, and the embassy received a response stating, "The powerful pictures of the launching of the satellite and the international way the subject has been handled makes a strong impression." From the high school students who viewed *Explorer in Space, Pioneer in Space*, and *Atlas* came the reaction, "The films were much appreciated. It was originally planned to show them once, but the teachers considered them so good that they were shown once more." *Explorer in Space* was also shown at an elementary school.[39]

Every time a new USIS film was released, a USIA circular, a telegram to transmit important information or instructions from the agency to USIS posts around the world simultaneously, was sent. The USIA circular explained the background, purpose, and target groups for each film. For example, a USIA circular on *X Minus 80 Days* was telegrammed on March 13, 1958, explaining that the film had been acquired from the Army "to supplement" *Explorer in Space*, and that it detailed the "preparations, the assembly and finally the count-down and launching" of the first U.S. satellite. The circular also indicated that both *Explorer in Space* and *X Minus 80 Days* emphasized that the launching of the satellites was part of U.S. activities for the International Geophysical Year (IGY), thus it was for the peaceful purpose. In fact, however, *X Minus 80 Days* was originally an Army film, and "a joint presentation of the Jet Propulsion Laboratory at the California Institute of Technology and the U.S Army Ballistic Missile Agency."[40] The circular's (and the movies') emphasis on the peaceful purpose of the satellites seems to be intended to divert the eyes of foreign audiences away from the military aspects of the U.S. space program.

Another USIA circular dated March 27, 1959, introduced *Out Among the Stars*, a film which described "the penetration of the earth's atmosphere by earth satellites" and emphasized U.S. leadership in space research and "the need for friendly international cooperation." The circular explained that the film took "a non-technical approach to a complicated subject" and was therefore suitable for "secondary school students and general audiences."[41] Still another USIA circular, in March, 1961, introduced *Three Years in Space*, which covered the history of U.S. space development from October 1957 to October 1960 (Figure 8.4). The film "described earth satellites and other space probes launched by the United States in the development of its space program," and emphasized the U.S. "contribution to the development of scientific knowledge for the benefit of mankind" and the U.S. release of scientific information "for the use of scientists everywhere." The circular, much like the aforementioned circular of 1959, also explained that the film was "made especially for general program use" and contained "very little technical terminology." The film grouped satellites into four types:

1) Scientific satellites for purely scientific research
2) Service satellites to produce operating systems (e.g., meteorology, navigation, and communications)
3) Developmental satellites to produce research tools for the future

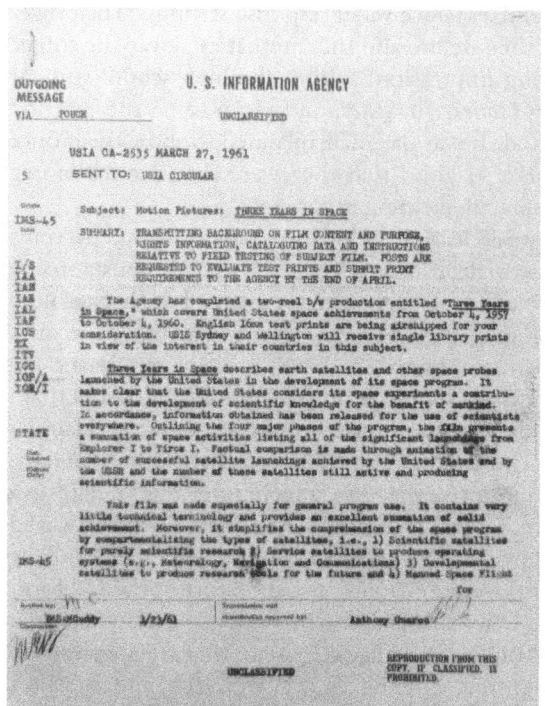

Figure 8.4 USIA Circular on *Three Years in Space* (1961). RG306, Entry P243, box 3, NACP.

4) Manned space flight for scientific research

The circular omitted military reconnaissance satellites, not because NASA was not involved with the development of such satellites, but because the overseas information program intended to portray the U.S. space programs as peaceful initiatives. The circular also listed the following nine relevant USIS films released from October 1957 to October 1960.[42]

> *Vanguard I; Exploring by Satellite; Out Among the Stars; Survey of Astronautics; In Which We Live; Atlas in Orbit; Pioneer in Interplanetary Space; Echo in Space; X Minus 80 Days*

However, the Japanese version of the *USIS Film Catalog* indicates that a far greater number of USIS films on space were produced from the late 1950s to the mid-1960s. While the 1959 *USIS Film Catalog* for Japan included only two titles on space: *Explorer in Space* (released on February 25, 1958) and *The Satellite Explorer* (released on March 25, 1958), the 1966 catalog included as many as

100 films on space (47 Japanese voice-over films, 21 English-language films, and 32 television films). Of the 47 Japanese voice-over films, 31 were of the series entitled the Science Bulletin [*Kagaku jiho*], each of which covered 3 to 4 topics on American science in 14 minutes. Some of the voice-over films were produced in Japan; for instance, *Friendship 7 Comes to Tokyo* dealt with the NASA space exhibition held in a Tokyo department store where the Friendship 7 re-entry capsule was on display.[43] A comparison of the 1959 and 1966 catalogs indicates the dramatic growth in the importance of space technology as a theme of the overseas information program during that period.

Most of the USIS films produced under NASA–USIA collaboration were non-technical, and intended for general audiences. For this reason, some advanced students and young scientists in foreign countries found the films too elementary for them. For example, the USIS in Mexico City reported that Mexican audiences needed "higher level science films particularly on space exploration, astronomy, and closely related subjects," and that "the Soviet or Soviet Bloc Embassies in Mexico" were already providing higher level films for university students and technicians. The USIA responded that the "field of space exploration ha[d] been given a high priority as its importance [wa]s clearly recognized," and therefore, the USIS films should target "lay audiences in view of the considerable publicity given the subject throughout the world."[44] This explanation of the USIA illuminates the role of space technology as a tool of the overseas information program, but it also reveals the limitation of such a tool when the foreign audiences showed a need for more advanced or practical scientific information.

Exhibitions were as important as the USIS films in the NASA–USIA collaborative information programs. The USIA had been preparing for overseas exhibitions on U.S. space technology since the Sputnik shock. When Project Mercury was made public, the USIA's exhibit of the "capsule" in which astronauts returned to the earth attracted the attention of foreign audiences, and requests for the "loan" of the capsule came from both inside and outside of the country. In June 1961, requests for loan of the capsule came from USIS branches in Bonn, Karachi, Athens, London, Kabul, and so on, although criticisms also arose from the U.S. public and Congress of showing the capsule overseas "when not yet seen by American public."[45] The "model capsules" to which Sorensen referred at the above-mentioned NASA conference were produced because of such great demand. The capsule was so popular that USIA Director Murrow recommended to NASA Director Webb to send it "to Paris for display at the U.S. exhibition in the Paris International Air Salon May 26–June 4." Murrow thought that the display of the capsule would "help take press play away from him," referring to Soviet cosmonaut Yuri Gagarin, who had been invited by the French Aircraft Industries Association to attend the Air Salon. "While I understand you have many requests for exhibition of the capsule in this country," Murrow told Webb, "its use in Paris is more important in terms of impact on world opinion."[46] The USIA expected the Mercury capsule to be so effective as to divert foreign media's attention away from the Soviet cosmonaut.

According to Kendall, "many USIS posts, using materials provided by USIA Washington, built their own exhibits of rockets, spacecraft and even lunar landing vehicles." Kendall himself organized such exhibitions in the countries of his overseas posts: Panama, Chile, and South Vietnam. During his tenure in Panama, Kendall also traveled to Argentina and Mexico, where he assisted with NASA space exhibitions. At the exhibitions, the actual Gemini V space vehicle, space suits, space food, photos, and moving images were displayed. The exhibits were extremely popular, according to Kendall, and on one Sunday, 25,000 people visited the exhibition in Argentina. There, he answered many questions of the visitors, and was sometimes overwhelmed by highly technical questions from university engineering students. To the students in group tours, Kendall sometimes lectured using slides. According to Kendall, the materials for lectures sent from Washington assumed the Cold War tone, emphasizing U.S. superiority over the Soviet Union, but Kendall intentionally "toned down the propaganda thrust" based on his lecturing experience in Panama.[47]

Overseas exhibitions, however, were sometimes inevitably affected by the local political climate or international politics. The NASA exhibition held as a part of the "Great Space Exhibition" in Tokyo from June 11 to July 31, 1960, took place in the midst of a massive protest against the renewal of the U.S.-Japan Security Treaty. A document in the USIA records indicates the OCB's surveillance of this exhibition (the OCB was an inter-departmental committee to oversee the psychological aspects of U.S. national security policies as a whole, as already mentioned in the previous chapters). On June 8, 1960, the OCB's "Exhibits Committee" held a meeting on the "Proposal for Outer Space Exhibit in Japan," where members discussed the document entitled "Report on the Status of U.S. Exhibits on Outer Space in Japan" dated May 27, 1960, and concurred that "on the basis of national security interest," the OCB would endorse "NASA's planning and financing a U.S. Exhibit on Outer Space to be held in Tokyo no later than early in 1961." In addition, the OCB concurred that "the establishment of a permanent space-science exhibit program" was desirable.[48] Since the information contained in the OCB minutes is fragmentary and incomplete, no further details are clear. However, the document reveals the OCB's interest in the NASA exhibition in Japan as part of U.S. psychological operations, and also the fact that NASA reported to the OCB on the exhibition. The OCB's interest was likely related to the chaotic situation involving the anti-U.S.-Japan Security Treaty protest. On January 19, 1960, the new U.S.-Japan Security Treaty was signed in Washington, and the ruling Liberal Democratic Party, led by Prime Minister Kishi, forced a vote on the bill through the Diet in May. This angered Japanese citizens opposing the treaty. On June 10, James Haggarty, President Eisenhower's press secretary who had entered Japan ahead of the president's planned visit, had to be rescued by a U.S. Marine helicopter as his car was surrounded by Japanese demonstrators on its way to the airport. Eisenhower's visit was canceled, and on June 15, a violent clash between demonstrators and the Japanese police ensued on the University of Tokyo campus, in which one female student was crushed to death. The Great Space Exhibition was held in the midst of this confusion.

Amidst a storm of anti-U.S. demonstrations, what was to be gained from the holding of the NASA exhibition? Returning to Kendall's memoir for a pertinent comparison, public opinion in Panama was also strikingly anti-American during his service as a USIS officer from 1964 to 1967. Kendall noted, however, that he was "pleasantly surprised to find Panamanian reaction to the U.S. manned space flight was every bit as enthusiastic as the American response." Panama had gained independence from Colombia as a U.S. protectorate in 1903, while the canal region was placed under direct U.S. sovereignty, under which the Panama Canal was completed in 1914. In 1936, Panama was awarded sovereignty of the canal region, although the canal itself remained under U.S. jurisdiction. After the Second World War, stimulated by the Egyptian nationalization of the Suez Canal in 1956, public opinion rose for the total reversion of the canal to Panama. In 1960, the movement to demand display of the Panamanian national flag in the canal region became radicalized, and President Eisenhower permitted the flying of both U.S. and Panamanian flags in that region. While the national flag became an emotional issue, in January 1964, some local high school students flew a U.S. flag at their school, and this incident triggered a nation-wide anti-U.S. demonstration. Some of the demonstrators formed a mob, which tore down the U.S. flag in the canal area and burned it. The mob spread to many places, and their actions included the burning of the USIS library. In the confusion, 20 Panamanians and 4 Americans lost their lives. The U.S. government temporarily severed diplomatic relations with Panama, which were restored on April 3, 1964. Kendall was posted to Panama only two weeks later, imparted with the responsibility of reconstructing the USIS branch and mitigating relations between the two countries through the overseas information program. In these dire circumstances, Kendall took full advantage of his experiences in NASA, lecturing on space technology and inviting American astronauts to the country. According to Kendall, these activities were received by Panamanians with great favor.[49] Kendall's memoir implies that space flight was a theme that stirred audiences' interest regardless of extant political issues in the given society. Just as the ocean was once perceived as having a boundless decontaminating capacity, and therefore radioactive fallout from the thermonuclear tests was regarded as merely a drop of ink, space was perceived as a "new frontier" which could not be polluted or over-developed. Moreover, space was a safe theme of the overseas information program precisely because there were no national borders in space, and therefore it was not easy to connect space with international disputes on the earth, such as the Panama Canal crisis.

The Great Space Exhibition was held in spite of the anti-American feeling among the Japanese public perhaps because—at least partly—for the same reasons Kendall raised regarding Panama. However, there was another, more locally oriented reason; the Great Space Exhibition was sponsored by the *Sankei* and *Chubu-nihon* newspaper companies, and supported by the Science and Technology Agency, the Science Council of Japan, the U.S. Embassy, and the Soviet Embassy. The Japanese sponsors, especially the Science and Technology Agency's newly appointed director Yasuhiro Nakasone (June 18, 1959 to July

19, 1960), were enthusiastic about U.S.–Japan bilateral technological collaboration on space technology. Back in 1954, Nakasone had been a driving force behind the U.S.–Japan bilateral agreement on atomic energy, and he was the first Japanese politician to pass a bill allocating a budget for nuclear power development. Now, Nakasone aimed to build the same kind of momentum concerning the U.S.–Japan bilateral agreement on space technology. Although his scheme was later shipwrecked by the strong opposition of Japanese scientists who preferred the autonomous, domestic development of space science, and the reluctance of the U.S. government to provide sensitive technology, the so-called "Nakasone Space Plan" was still underway when the Great Space Exhibition was held in 1960. The NASA exhibit, therefore, did not represent only the U.S. overseas information program but also Japanese politicians' interest in technological collaboration with the U.S.

The Great Space Exhibition in Tokyo had three exhibition sites, of which a large portion of Site One was occupied by the NASA exhibits. The NASA exhibits included the Redstone engine used in Project Mercury and the capsule in which Alan Shepard returned to the earth.[50] Site One was a 10,000-square-meter dome, in the center of which a 20-meter model of a four-stage rocket towered; a speaker played the count down, followed by the actual sound of the Atlas ICBM test launching accompanied by smoke rising up from the bottom. Visitors were excited at the "Satellite Corner" where the models of Explorer, Pioneer, and Vanguard were displayed, together with the real space suits and the cutting-edge X-15 aircraft for manned space flight. However, the Great Space Exhibition as a whole was not so widely known when compared with the Atoms for Peace exhibitions held a few years prior. The only printed media that reported on the exhibition in detail were the *Weekly Sankei* [*Shukan Sankei*] published by the exhibition's sponsor, and a few specialized magazines such as *Aviation Information* [*Koku Joho*] and *Aviation Fan* [*Koku Fan*], and a student magazine *Middle School Freshman* [*Chugaku-jidai Ichinensei*]. The photos included in these magazines show groups of elementary and middle school students surrounding the exhibits, evidence of group tours organized by the schools, but there is no information on the size of the general audience. The newspaper *Yomiuri* carried a short article on the opening ceremony held on June 11:

> Following the opening address by the Exhibition's Vice Director Mizuno, Minister of Education Matsuda, Director of Science and Technology Agency Nakasone, and Geophysicist Kiyoo Wadachi, who chaired the Science Council of Japan gave congratulatory remarks, respectively. Director Shibusawa presented a certificate of appreciation to Masaji Miyaji, astronomist and head of the Tokyo National Observatory who supervised the exhibition, and Prince Takamatsu, Chairman of the Exhibition, cut the ribbon.[51]

The low-profile media coverage of the Great Space Exhibition could perhaps be attributed to the anti-U.S.-Japan Security Treaty protest that was currently ongoing, and the resultant reluctance of the media to bring attention to the U.S.

under such circumstances. In addition, at the time, the manned space flights of the mid-1960s (either by the U.S. or the Soviet Union) had not yet happened, and the fever for astronauts in the U.S. had thus not yet extended to Japanese audiences. Moreover, in contrast to the speedy conclusion of the 1955 atomic energy bilateral agreement, there was a long way until a similar bilateral agreement would be discussed on space technology; after the Second World War, Japan was prohibited from engaging in research and development of air and space technology. However, after the San Francisco Peace Treaty was enacted and Japan regained sovereignty, Japanese aerospace technology developed rapidly, and in 1955, Hideo Itokawa of University of Tokyo succeeded in the development of the so-called "pencil rocket." Japanese aerospace engineers had been both greatly shocked and stimulated by the success of Sputnik and the U.S. reactions to that event. In 1958, Japanese engineers developed a rocket which reached a height of 60 kilometers and participated in the IGY to carry out high-altitude atmospheric surveys. The majority of Japanese scientists and engineers, including Itokawa, pursued "a system to avoid American influence" in rocketry technology because "dependence on the U.S. meant that Japan was denied free access to the space."[52]

By contrast, the Science and Technology Agency and its charismatic director Yasuhiro Nakasone aimed to conclude a bilateral space technological cooperation agreement with the U.S., following the example of atomic energy. Newspapers reported that the essence of the Nakasone Space Plan was the "conclusion of U.S.-Japan research bilateral agreement for the peaceful uses of the space development." In July 1959, Nakasone established the Preparation Committee for the Promotion of Outer Space S&T [*Uchu Kagakugijutsu Shinko Junbi Iinkai*], and recruited roughly a dozen scientists and engineers, including Hideo Itokawa (University of Tokyo), Masaji Miyaji (head of the University of Tokyo Observatory), and Kiyoo Wadachi (Director of the Meteorological Agency). In February of the following year, Nakasone sent a delegation of three scientists, including Itokawa, to NASA to hold a preliminary meeting on the bilateral cooperation. Nakasone further successfully passed a bill to allocate 235 million yen in the 1960 fiscal year budget for space technology. In 1954, Nakasone had surprised both scientists and the public by suddenly submitting a bill for 235 million yen for atomic energy development, which passed, and thus he employed the same method, and even exactly the same amount, for his bill concerning space technology. Nakasone thereafter held repeated meetings with Ambassador MacArthur of the U.S. embassy to discuss U.S.–Japan collaboration in space technology.[53] It was against this backdrop that the Science and Technology Agency sponsored the Great Space Exhibition, and Nakasone himself participated in the opening ceremony.

However, the Nakasone Space Plan provoked strong criticism in academic circles. For example, astronomist Naosuke Sekiguchi expressed his concern that the politically neutral Japanese approach toward space development might be broken should Japan conclude a bilateral agreement with the U.S.; the capitalist-versus-Communist confrontation had already surfaced in international society, for example, in the Committee on Space Research (COSPAR) and in the

United Nations Committee on the Peaceful Uses of Outer Space (COPUOS). In the COPUOS, the Third World countries expressed strong concerns over the military use of outer space by the superpowers. In the midst of East-West and North-South conflicts, Sekiguchi cautioned that Japan should refrain from aligning with the U.S. Sekiguchi further criticized the Science and Technology Agency for "falling in line with the Preparation Committee for the Promotion of Outer Space S&T, virtually Nakasone's personal advisory board," without listening to scientists' opinions.[54] The *Asahi* newspaper also criticized the Nakasone Plan, pointing out the "possibility that space technology might be applied to the guidance missile development of the Self Defense Agency," the "heavy focus on rocketry development," and the "hasty movement toward the bilateral agreement without clear goals or a research framework."[55] Takashi Mukaibo, who had contributed to the conclusion of the U.S.–Japan atomic energy bilateral agreement as Science Attaché to the Japanese Embassy in Washington, also warned that "although a bilateral agreement was necessary in the case of atomic energy for the receipt of fissional materials," outer space technology was a qualitatively different matter, and should not depend so heavily on the U.S.[56] In the end, the Nakasone Space Plan did not proceed as smoothly as the atomic energy agreement, not only because of domestic opposition, but also because of U.S. caution against Japanese rocketry research and development. It was only toward the end of the 1960s, during the Sato administration, that the U.S. provided the liquid fuel rocketry technology which Japan had long desired.[57]

The goals of the NASA–USIA information program in Japan and elsewhere were to extend a message that the U.S. bettered the Soviet Union not only in "soft" technologies such as electric appliances but also in "hard" areas such as the development of rockets and satellites, and simultaneously, that American S&T was "open" in contrast to the secrecy of their Soviet counterparts. Another long-term goal of the U.S. was to keep foreign countries within the fold of a U.S.-centered outer space technological network, so that those countries would not join with the Soviet Union in missile development. On the Japanese side, however, there was a different reason for pursuing cooperation with the U.S.: the Great Space Exhibition was held in conjunction with the Science and Technology Agency's interest in introducing U.S. outer space technology under the Nakasone Plan. Furthermore, it is important to note that the Great Space Exhibition did not only display U.S. technologies. As already mentioned, NASA's exhibits occupied only a portion of the whole exhibition sites; Soviet and German technologies, as well as Japanese Kappa-7 rockets, were also on display. The *Aerospace Information* magazine also introduced not only the NASA exhibits but also a model of a German V-2 rocket, a model Sputnik II and III, and the Japanese reflecting telescope that had been displayed in the exhibition. Many of the Japanese scientists had a strong interest in autonomous development of outer space technology, independent of reliance on the U.S., and the U.S. technology was merely one of many examples of technologies to refer to. Visitors to the exhibition also viewed the NASA exhibits as part of a larger picture, among various other alternatives. In contrast to Atoms for Peace of the mid-1950s, which demonstrated the overwhelming technological power of the U.S., the NASA exhibition in the 1960s was taken into perspective. This difference might indicate that the U.S. overseas information program was approaching the limits of its approach to demonstrate the overwhelming technological and cultural hegemony of the U.S., one that was becoming outdated as 15 years had passed since the end of the Second World War, and amidst the decolonization and striving for autonomous economic independence of many countries.

Science historian Audra Wolfe has stated that the Apollo moon landing was the "last hurrah of the U.S. military industrial complex" during the Cold War.[58] The U.S. manned space flight program culminated in the Apollo moon landing of 1969 and thereafter the style of state-promoted big science declined. The Cold War itself was also transforming: the dichotomy of East-West confrontation destabilized, demonstrated by the China-Soviet estrangement and the U.S.-China approach, and alliances were also shaken over the issues of natural resources and national security. At the same time, unity within the non-aligned nations loosened, and some developing countries pursued a path toward developmental dictatorship. Furthermore, environmental pollution and nuclear proliferation emerged as new global issues to be tackled, and modernization and developmentalism as propelled by the military-industrial complex came under criticism. All of these changes transformed the very platform of the Cultural Cold War: It was no longer meaningful to win the hearts and minds of the Third World people by providing nuclear or medical technological aid, emphasizing the "free" and "democratic" characters of American S&T, or showing off the modern lifestyle of the U.S. in films, exhibits, and printed media. The outer space technology explored in this chapter was, thus, perhaps the last chapter of the Cultural Cold War.

Notes

1 Smithsonian National Air and Space Museum website, https://airandspace.si.edu /exhibitions/destination-moon.
2 NASA website, https://www.nasa.gov/content/nasa-history-overview.
3 Shohei Yonemoto, "Shizen-kagaku" [Natural Science], in *Jiten gendai no Amerika* [*Encyclopedia on Modern America*], eds. Takahiro Oda, et al. (Tokyo: Taishukan Publishing, 2004), 415.
4 John Krige, "NASA as an Instrument of U.S. Foreign Policy," in *Social Impact of Spaceflight*, eds. Steven J. Dick and Roger D. Launius (Washington, DC: NASA SP-2007-4801, 2007), 207–218; John Krige, "Technology, Foreign Policy and International Collaboration in Space," in *Critical Issues in the History of Spaceflight*, eds. Steven Dick and Roger Launius (Washington, DC: NASA-2006-4702, 2006), 239–260.
5 Nicholas Michael Sambaluk, *The Other Space Race: Eisenhower and the Quest for Aerospace Security* (Annapolis, MD: Naval Institute Press, 2015); Yanek Mieczkowski, *Eisenhower's Sputnik Moment: The Race for Space and World Prestige* (Ithaca, NY: Cornell University Press, 2013); Allan A. Needell, *Science, Cold War and the American State: Lloyd V. Berkner and the Balance of Professional Ideals* (New York and London: Routledge, 2001).
6 Yasushi Sato, *NASA o kizuita hito to gijutsu: kyodai shisutemu kaihatsu no gijutsu bunka* [*People and Technology that Built the NASA: Technological Culture of the*

Big System Development] (Tokyo: University of Tokyo Press, 2007); *NASA: uchu kainatsu no 60-nen* [*NASA: 60 Years of Space Development*] (Tokyo: Chuokoron-Shinsha, 2014); Kazuto Suzuki, *Uchu-kaihatsu to kokusai seiji* [*Space Development and International Politics*] (Tokyo: Iwanami Shoten, Publishers, 2011); Kazutaka Yamamoto, "Kennedy to 'uchu-kainatsu' seisaku" [Kennedy and the 'Space Development' Policies], in *Kennedy to Amerika seiji* [*Kennedy and American Politics*], ed. Kazumi Fujimoto (Tokyo: Tsunan Shuppan, 2004), 151–186. First published 2000.

7 Krige, "Technology, Foreign Policy and International Collaboration in Space," 250.

8 Krige, "NASA as an Instrument of U.S. Foreign Policy," 208, 212.

9 Suzuki, *Space Development and International Politics*, 11–13.

10 T. A. Heppenheimer, *Countdown: A History of Space Flight* (New York: John Wiley & Sons, 1997), 125.

11 Heppenheimer, 127–128.

12 For example, these prevailing views are introduced in Yamamoto, "Kennedy and the 'Space Development' Policies," 163–165.

13 For example, Sambaluk, *The Other Space Race*; Mieczkowski, *Eisenhower's Sputnik Moment*.

14 Mark Haefele, "John F. Kennedy, USIA, and World Public Opinion," *Diplomatic History*, vol. 25, no. 1 (Winter 2001): 63–84.

15 For example, this prevailing interpretation is introduced in Yamamoto, "Kennedy and the 'Space Development' Policies," 166–169. Wolfe, Competing with the Soviets, 94.

16 Ryuichi Kaneko, "Uchu-kaihatsu" [Space Development]; Soichi Kaji, "Kouku-uchu-sangyo" [The Aerospace Industry] in *Encyclopedia on Modern America*, 647–676, 776; Yamamoto, "Kennedy and the 'Space Development' Policies," 66–67.

17 Heppenheimer, *Countdown*, 159–160.

18 "The Astronauts: Ready to Make History," *Life*, vol. 47, no. 11 (September 14, 1959): 26–43.

19 "Seven Brave Women behind the Astronauts: Spacemen's Wives Tell, in Their Own Words, Their Inner Thoughts and Worries," *Life*, vol. 47, no. 12 (September 21, 1959): 142–163.

20 Lily Koppel, *The Astronaut Wives Club: A True Story* (New York: Grand Central Publishing, 2013), 35–40.

21 M. Scott Carpenter, Gordon L. Cooper, John H. Glenn, Virgil I. Grissom, Walter M. Schirra, Alan B. Shepard, and Donald K. Slayton. *We Seven: By the Astronauts Themselves* (New York, London, Sydney, and Toronto: Simon and Schuster Paperbacks, 1990). First published 1962.

22 Martin Caidin, *The Astronauts: The Story of Project Mercury, America's Man-in-Space Program*, 1st ed. (New York: E.P. Dutton, 1960); Martin Caidin, Obituaries, *The New York Times*, March 28, 1997.

23 From IOP Laurence P. Dalcher to IBS Siemer, IPS Mann, IMS Fisher, ITV Stephens, May 1, 1961, RG306, Entry P243, box 4, NACP.

24 "Mr. Mullins' Assignment at the National Aeronautics and Space Administration," December 19, 1958, RG306, Entry P243, box 3, NACP.

25 U.S. Department of State, *News Letter*, 72 (April 1967): 52.

26 "Report on USIA Liaison Office at NASA," July 14, 1959, RG306, Entry P243, box 3, NACP.

27 Memorandum from Donald M. Wilson to Halsema, February 16, 1961; From James J. Halsema to Wilson, February 18, 1961, RG306, Entry P243, box 4, NACP.

28 In 1957, due to the shrinking budget of the USIA, the agency stopped assigning an information officer to Takamatsu Japan-U.S. Cultural Center, and Kendall was transferred to Madrid, accordingly. The jurisdiction over Takamatsu Japan-U.S. Cultural Center was transferred from the U.S. government to the Kagawa Prefectural Library jurisdiction already in 1953, but the Governor of Kagawa had requested the USIA to send an American information officer, Kendall said in his oral history interview.

29 From Thomas C. Sorensen to Goodwin, April 22, 1961, RG306, Entry P243, box 4, NACP; The Association for Diplomatic Studies and Training Foreign Affairs Oral History Project Information Series, Harry Haven Kendall, Interviewed by G. Lewis Schmidt, December 27, 1988, Georgetown University Foreign Affairs Oral History, Box 1, Folder 248; Kagawa Prefectural Library, *Yoran* [*The Bulletin*] (2018), 2.

30 "NASA Staff Conference," June 16, 1961, RG306, Entry P243, box 4, NACP.

31 From O. B. Lloyd, Jr. Director, Public Information, NASA to Edward R. Murrow, June 19, 1961, RG306, Entry P243, box 4, NACP; "Murrow Gets Aide: Thomas Sorensen to Become USIA Deputy Director," *The New York Times*, February 23, 1961.

32 Thomas C. Sorensen, *The Word War: The Story of American Propaganda* (New York: Harper & Row, 1968), 179–183.

33 "Space International," Remarks of Thomas C. Sorensen, Deputy Director (Policy & Plans) of the U.S. Information Agency, Before the Office of Public Information Staff Conference, NASA, June 27, 1961, RG306, Entry P243, box 4, NACP.

34 "Space International."

35 Harry H. Kendall, *A Farm Boy in the Foreign Service: Telling America's Story to the World* (Bloomington, IN: AuthorHouse, 2003), 100–104.

36 Kendall, 106–110.

37 Kendall, 111–124, 166–167.

38 From Marvin W. Robinson, Deputy Director, Office of International Programs, NASA, to Harry Kendall, Code AP, June 30, 1961 RG306, Entry P243, box 4, NACP.

39 From USIS Stockholm to USIA, February 3, 1961, RG306, Entry P243, box 3, NACP.

40 USIA Circular from Washburn, (Acting Secretary of State), March 13, 1958, RG306, Entry P243, box 3, NACP.

41 USIA Circular from Edward Murrow, March 27, 1959, RG306, Entry P243, box 3, NACP.

42 USIA Circular from Edward Murrow, March 27, 1961, RG306, Entry P243, box 3, NACP.

43 *USIS Film Catalog 1959* (Japan), 28, 58; *USIS Film Catalog 1966* (Japan); RG306, Entry P46, box 312, NACP.

44 From Beverly M. Jones, IMS to USIS Mexico City, April 23, 1959, RG306, Entry P243, box 3, NACP.

45 From Glaude Hawley to Goodwin, April 23, 1958; From Harry Kendall to Sorensen, June 15, 1961, RG306, Entry P243, box 4, NACP.

46 From Edward R. Murrow to James Webb, May 16, 1961, RG306, Entry P243, box 4, NACP.

47 Kendall, *A Farm Boy*, 113–115.

48 Memorandum for Members; OCB Exhibits Committee, June 17, 1960, RG306, Entry P243, box 4, NACP.

49 Kendall, *A Farm Boy*, 112, 135–138.

50 "NASA Space Exhibition," RG306, Entry P243, box 4, NACP.

51 Takeshi Sakaguchi, "Uchu daihakurankai manga rupo" [The Great Space Exhibition Manga Report], *Shukan Sankei*, vol. 9, no. 29 (issue 445) (June 13, 1960): 72–73; "Shijo kengaku, uchu-haku" [Visit in the Magazine: Space Exhibition], *Koku Joho*, no. 120 (August 1960): 129–131; "Rocket to uchu-ryoko ten" [Rockets and Space Travel Exhibition], *Koku Fan*, vol. 9, no.8 (August 1960): 40–41; "Uchu hakurankai kenbunki" [Space Exhibition Visiting Report], *Chugaku-jidai, Ichinensei*, vol. 5, no. 5 (August 1960): 166–167, these are all from National Diet Library Digital Collections; "Uchu daihakurankai harumi de hiraku" [The Great Space Exhibition Held in Harumi], *Yomiuri Shinbun*, June 11, 1960, Evening Edition, 7.

52 Suzuki, *Space Development and International Politics*, 176; Institute of Space and Astronautical Science (Japan) website, http://www.isas.jaxa.jp/about/history/.

53 "Uchu-kaihatsu dou susumu Nakasone-koso" [Progress of Space Development: Nakasone Plan], *Yomiuri Shinbun*, July 21, 1959, Evening Edition, 2; "Nihon ni uchu-kaihatsu no yoake, chosadan haken nado hongoshi" [Dawn of Japanese Space Development: Plan to Send a Survey Mission], *Yomiuri Shinbun*, January 6, 1960, Evening Edition, 2; "Kancho-gai: Kagaku gijutsucho" [Government Agencies: Science and Technology Agency], Yomiuri Shinbun, February 7, 1960, Evening Edition, 2; Kagaku Gijutsucho Sosetsu 10-shunen Kinengyoji Jikko Junbi Iinkai, ed. *Kagakugijutsucho 10-shunenshi* [*The 10-Year History of the Science and Technology Agency*] (Tokyo: Kagakugijutsucho sosetsu 10-shunen kinengyoji kyosankai, 1966), 238, National Diet Library Digital Collections. Concerning Nakasone and the atomic energy budget, see Tetsuro Kato, *Nihon no shakaishugi: genbaku hantai, genpatsu suishin no ronri* [*Japanese Socialism: The Logic of Anti-Nuclear Weapons and Pro-Nuclear Power*] (Tokyo: Iwanami Shoten, Publishers, 2013); Kato and Ikawa, *Atomic Energy and the Cold War: Introduction of Nuclear Power Stations in Japan and Asia*, 15–53.

54 Naosuke Sekiguchi, "Nakasone uchu-kaihatsu keikaku: keii to mondai-ten" [The Nakasone Space Development Plan: Details and Problems], *Shizen=Nature*, vol. 15, no. 5 (issue 165) (May 1960): 62–65, National Diet Library Digital Collections.

55 "Nakasone-koso ni hihanteki kuki" [Criticism against the Nakasone Space Plan], *Asahi Shinbun*, August 16, 1959, Evening Edition, 1; "Wagakuni uchu-kagaku no arikata (Shasetsu)" [The Direction of the National Space Science (Editorial)], *Asahi Shinbun*, March 13, 1960, Morning Edition, 2.

56 Takashi Mukaibo, "Nakasone chokan ni chumon: uchu-kaihatsu kyotei kokoro saretashi" [A Request for Director Nakasone: Be Cautious about Bilateral Agreement on Space Program], July 20, 1959, *Asahi Shinbun*, Tokyo Moning Edtition, 3. National Diet Library Digital Collections.

57 Concerning U.S.–Japan agreement on space technological cooperation under the Sato administration, see Suzuki, *Space Development and International Politics*, 176–183.

58 Wolfe, *Competing with the Soviets*, 90.

9 Epilogue

My earliest memory of American science and technology goes back to the World Exhibition held in Osaka, Japan, in March–September 1970, to which I was taken by my late father on two occasions. The Osaka Expo was a huge national event, visited by a cumulative total number of 60 million people—approximately 60% of the whole Japanese population. The U.S. Pavilion was one of the most popular: visitors stood in line for hours to see the "moon rock" brought back by the Apollo astronauts, although my father gave up on seeing it, having decided it would be too much of a wait for a little girl. Beverly Gray, one of "56 young Japanese-speaking American guides who'd been hired to staff the U.S. Pavilion," recollected later that she answered, hundreds of times a day, the same question from enthusiastic Japanese visitors: "Where is the moon rock?" The American guides explained, in accord with the USIA policy, that the moon landing vehicle in the U.S. Pavilion was "not a model, as in the USSR Pavilion, but a *honmono* ('real thing')."[1] I missed the opportunity to encounter the U.S. information program on S&T because I was too young at that time.

I recall the huge, unique, flat-shaped dome of the U.S. pavilion which was, according to the Osaka Expo '70 Commemorative website, the "essence of space engineering," an oval dome of "air film structure" which "spanned 142 meters long and 83.5 meters wide." Inside this innovative air-supported dome were seven separate corners, including American Sports, American Paintings, American Photographers, and so on. The exhibit of "Space Development" occupied the largest area, and attracted the most attention. In addition to the moon rock, the exhibits included the actual Apollo 8 command module, a simulated lunar surface depicting the astronauts' landing, the Lunar Module, the Mercury Capsule, and the Gemini 2 spacecraft.[2] The Osaka Expo was a culmination of a decade of the U.S. overseas information program on the theme of space flight. Apollo had succeeded in its moon landing in July of the previous year, with the whole world watching on live television. In the countries where television sets were not yet widely diffused, USIS posts invited citizens to watch the broadcast in their libraries.[3] The Apollo moon landing and the U.S. scientific spectacle relating to it were still fresh in the Japanese memory when the Osaka Expo was started.

One more thing about Osaka Expo that stuck in my memory was the bright lighting up of the exhibition site after sunset. It was a thrilling experience to roam

DOI: 10.4324/9781003243649-11

freely after dark, with my parents' complete permission, and I was fascinated by the brilliant illumination of various pavilions. It was only after I became a university researcher that I came to know that the electricity that powered the Expo was provided by the Tsuruga Nuclear Power Plant No. 1 Reactor, which commenced operation on the very first day of the Expo.[4] The Tsuruga No. 1 Reactor was a Pressurized Water Reactor manufactured by General Electric Company, imported from the U.S., and it was decommissioned in 2015. Following the Tsuruga Plant, Kansai Electric Company's Mihama Nuclear Power Plant began to send electricity to the Osaka Expo. The Mihama nuclear reactor was a product of Westinghouse Corporation. In other words, the Osaka Expo of 1970 was tangible evidence of the successful U.S. Foreign Atoms for Peace project targeting Japan. While the U.S. Pavilion did not have on display an Atoms for Peace exhibit, that was because "peaceful" use of atomic energy had already taken root firmly in Japan. The American S&T exported during the Cultural Cold War had permeated into Japanese society, and lit up the first World's Fair held in the country, under the slogan: Progress and Harmony for Mankind.

Both the "moon rock" fever and the unquestioned acceptance of nuclear energy in everyday life symbolized how S&T filters into culture and society, and how the U.S. overseas information program played a part in such a process. Science and culture are often perceived as mutually exclusive concepts, just as the Cultural Cold War is often understood to be all about music, art, films, and literature. However, as the many examples in this book have shown, S&T forms a culture in which young scientists and engineers are trained, build international networks, establish their careers, and influence the long-term policies of their countries. S&T also moves people's emotions, just as music, art, films, and literature do. It evokes feelings of respect, awe, fear, or desire, and those emotions sometimes destine the future of countries by influencing the choices their policy-makers and citizens make. The chapters of this book have demonstrated the central role that S&T played in the Cultural Cold War.

Also, as the author discussed in the Introduction, diplomatic history and cultural history are sometimes perceived as oil and water, i.e., separate and unmixing. However, S&T, placed on the central stage of the Cultural Cold War, illuminates just how intricately intertwined with each other diplomatic history and cultural history are. States sometimes employ S&T in traditional hard power diplomacy, in which technological aid is provided in exchange for short-term benefits such as access to markets or natural resources, or in securing political allegiance. At other times, however, states exploit S&T to cultivate seedbeds for nurturing a long-lasting culture: for example, a laboratory or school where people share knowledge and values, an alumni organization of those who share study-abroad experiences, or a scientific device such as a nuclear power reactor through which people modernize and electrify their lives. Many of the examples introduced in this book can be understood only through the combined lenses of diplomatic and cultural analysis, rather than just one or the other. Moreover, this marriage of diplomacy and culture also symbolizes how state power involves S&T. As chapters of this book have shown, government agencies such as the USIA, the Department of

State, the AEC, and NASA supported exhibitions, film shows, study-abroad programs, tours, conferences, and so on. These "cultural" programs on S&T were funded, publicized, and monitored by government authorities. Although there is a general image of S&T as being autonomous and apolitical, it was this very image of S&T that made it a convenient tool of the Cultural Cold War, and thus S&T was closely tied with state power and politics.

S&T as a weapon of the Cultural Cold War succeeded to a certain degree in the countries where American scientific knowledge and technology became absorbed into the fabric of everyday life. In other countries, the U.S. was not able to reap the crop of its substantial investment in the form of various information programs. Especially in politically unstable developing countries, U.S. investment often went down the drain. Even so, scientists and engineers trained in the U.S., and their knowledge and networks, contributed to a certain hegemonic structure on which various scientific research and education were founded, not dissimilar to the operating system on which various software runs.

However, even while the "moon rock" and the Osaka Expo attracted people's attention in Japan, the reputation of U.S. S&T was on the wane as the Vietnam War entered a quagmire and American science launched deadly weapons such as Agent Orange and napalm bombs. People around the world could no longer be swayed by the U.S. "Science for Peace" appeal that the American S&T improves the everyday lives of people everywhere. According to a USIA public opinion survey conducted a few months after the Apollo moon landing in the U.K., France, Japan, Venezuela, India, and the Philippines, 44% to 60% of those surveyed responded that the U.S. should "devote more time and money to troubles on earth" than space flight, even though they were impressed by the Apollo program to some degree.[5] Within the U.S., racial violence and the assassination of the non-violent leader Martin Luther King (1968) undermined the American image that it was a land of freedom. Furthermore, the East-West dichotomy on which the Cultural Cold War stood was destabilizing, as the U.S. sought reconciliation with Mao Zedong's China, while China and the Soviet Union broke up their unified front. Just a year after the Osaka Expo, the U.S. government surprised the world by announcing the prospective visit of President Nixon to China, as the U.S was in dire straits in Vietnam and needed China's help to withdraw from the quagmire. The Cold War in Asia as a whole was undergoing a major transformation as decolonized countries were no longer unified in non-aligned neutralism, and some embarked on "developmental dictatorship." The good-old message of the Cultural Cold War to emphasize the superiority of the "free world" was no longer relevant.

The USIA, which was born to fight the Cultural Cold War, had to look for a new raison d'être in this situation. On the one hand, the overseas information program was renamed "public diplomacy" and, as the name suggests, became a part of authentic diplomacy.[6] Accordingly, former USIA officers, who had been regarded as one rank lower than regular foreign service officers, were upgraded to the rank of ordinary foreign service officers. On the other hand, it was not easy for the information program to make the switch from Cold War mode to détente mode. Frank Shakespeare, Director of the USIA under the Nixon administration, inaugurated in 1969, persisted in the Cold War-type information program, emphasizing the superiority of the U.S. over Communist countries, and clashed with the Department of State. His successor, James Keogh, carefully considered the new role of the USIA, and created new policies such as "dialogue and mutuality" rather than one-way communication, emphasis on "common interests," and use of "local institutions."[7] Although an examination of the details of the USIA's transition is beyond the scope of this book, it suffices here to point out that it was disbanded in 1999, and its public diplomacy role absorbed in the Department of State. Secretary of State Madeline Albright, upon the closure of the USIA, spoke of "agency staff bringing their perspective and expertise into the mainstream of U.S. foreign policy and thereby making all U.S. diplomacy at some level public diplomacy."[8]

One important caution needs to be pointed out before closing this book. In its writing, the author primarily relied on U.S. and Japanese primary sources, although venturing into the cases of other Asian countries such as Vietnam and Burma, and occasionally making reference to China, the Soviet Union, and others. The author's limited language capability has inhibited her from using sources in other languages, and the reliance on U.S. and Japanese sources has its own shortcomings. When the Japanese version of this book was published, one thoughtful undergraduate student of Kyoto University came to me and expressed his doubt about the successfulness of the U.S. overseas information program for Japan; "Actually, more Japanese people visited the Soviet Union Pavilion than the U.S. Pavilion at the Osaka Expo. Why do people (including myself) only write about the moon rock and the U.S. Pavilion?" His question is valid. When one consults only U.S. and Japanese sources, a picture of the U.S. as by far the most effective of the countries in the exhibits at the Expo emerges, and that the Japanese people were by far more enthusiastic about the U.S. exhibits than the others. This picture will surely be revised if one considers other language sources, including Russian, Chinese, Vietnamese, and so on. A scholar of history needs to remain humble and to admit that his or her view is only ever partial, just as the Cultural Cold Warriors had only a partial view of the world in bringing their values and technologies overseas.

Notes

1 Beverly Gray, "When Apollo Went to Japan," Smithsonian Institution, *Air & Space Magazine*, April 2020, https://www.airspacemag.com/space/when-apollo-went-japan-180974469/.
2 Expo '70 Commemorative Park website, https://www.expo70-park.jp/cause/expo/america/. As for Osaka Expo, also see Shunya Yoshimi, *Hakurankai no Seijigaku* [*Politics of Exhibitions*] (Tokyo: Chuo-koron Shinsha, 2000, first published 1992), Chapter 6.
3 Cull, *The Cold War and the United States Information Agency*, 305.
4 Yoshimi, *Atoms for Dream*, 15.
5 Cull, 305.

6 The term "public diplomacy" was proposed by Edmund Gullion, a former foreign service officer who was appointed to Dean of Fletcher School, Tufts University. Gullion had a deep understanding of the overseas information program, and established the Edward R. Murrow Public Diplomacy Center.
7 Cull, 255–261, 293, 320, 335.
8 Cull, 502.

References

Primary Sources

U.S. Archival Records

Atomic Scientists of Chicago. Records, Hanna Holborn Gray Special Collections Research Center, University of Chicago Library.

Bulletin of the Atomic Scientists. Records, Hanna Holborn Gray Special Collections Research Center, University of Chicago Library.

Dwight D. Eisenhower Presidential Library.

White House Central Files.

White House Office, NSC Staff Papers, OCB Secretariat Series.

International Association for Cultural Freedom. Records, Hanna Holborn Gray Special Collections Research Center, University of Chicago Library.

Michigan Memorial Phoenix Project records, 1947–2003 (MMPP), Bentley Historical Library, University of Michigan.

Office of the Historian, *Foreign Relations of the United States* (*FRUS*).

1952–1954, National Security Affairs, Volume II, Part 2, https://history.state.gov/historicaldocuments/frus1952-54v02p2.

1955–1957, Foreign Economic Policy; Foreign Information Program, Vol. IX, https://history.state.gov/historicaldocuments/frus1955-57v09.

1955–1957, Regulation of Armaments; Atomic Energy, Vol. XX, https://history.state.gov/historicaldocuments/frus1955-57v20.

Rabinowitch, Eugene I. Papers, Hanna Holborn Gray Special Collections Research Center, University of Chicago Library.

U.S. National Archives

RG59, General Records of the Department of State, U.S. National Archives at College Park, Maryland.

RG84, Records of the Foreign Service Posts of the Department of State, U.S. National Archives at College Park, Maryland.

RG225, Records of the National Aeronautics and Space Administration, U.S. National Archives at College Park, Maryland.

RG306, Records of the U.S. Information Agency, U.S. National Archives at College Park, Maryland.

RG326, Records of the Argonne National Laboratory, U.S. National Archives at Chicago, Illinois.

RG326, Records of the Atomic Energy Commission, U.S. National Archives at College Park, Maryland.

RG469, Records of U.S. Foreign Assistance Agencies, 1948–1961. U.S. National Archives at College Park, Maryland.

Diplomatic Archives of the Ministry of Foreign Affairs of Japan

"Gensuibaku jikken kankei, beikoku kankei, Eniwetok kansho jikken kankei (1956) vol. 1" [Nuclear Tests, United States, Eniwetok Atoll Nuclear Tests (1956) vol. 1], C' .4.2.1.1-1-1 (microfilm C'-0005).

"Gensuibaku jikken kankei, beikoku kankei, Eniwetok, Johnston tou jikken kankei (1958)" [Nuclear Tests, United States, Eniwetok & Johnston Island Nuclear Tests (1958)], C'.4.2.1.1-1-3 (microfilm C'-0006).

"Gensuibaku jikken kankei, beikoku kankei, Eniwetok, Johnston tou jikken kankei (1958), Kansokusen Takuyo, Satsuma hisai jiken" [Nuclear Tests, United States, Eniwetok & Johnston Island Nuclear Tests (1958), Accident of the Survey Ship Takuyo & Satsuma], C'.4.2.1.1-1-3-1 (microfilm C'-0006).

"Honpou genshiryoku seisaku narabini katsudou kankei, Honpou genshiryoku kagakusha no kyouiku kunren kankei vol. 4" [Japanese Atomic Policies and Activities, Education and Training of Japanese Atomic Scientists, vol. 4], C'.4.1.1.1-4 (paper).

"Kakubakuhatsu jikken ni taisuru honpou no taido" [Japan's Attitude toward Nuclear Tests], C'.4.2.1.2 (microfilm C'-0009).

Others

Argonne National Laboratory News-Bulletin, vol. 5, no. 2 (April 1963), National Archives at Chicago.

Argonne National Laboratory News-Bulletin International, vol. 1, no. 1 (January 1959); vol. 1, no. 2 (April 1959); vol. 1, no. 3 (July 1959); vol. 1, no. 4 (October 1959); vol. 2, no. 1 (January 1960); vol. 3, no. 3 (July 1961); vol. 4, no. 1 (January 1962); vol. 4, no. 2 (April 1962); vol. 4, no. 3 (July 1962), National Archives at Chicago.

Association for Diplomatic Studies and Training Foreign Affairs Oral History Project Information Series, Harry Haven Kendall, Interviewed by: G. Lewis Schmidt, December 27, 1988, Georgetown University Foreign Affairs Oral History, Box 1, Folder 248.

The Bulletin of Atomic Scientists, vol. ix, no. 2 (March 1953); vol. ix, no. 8 (October 1953); vol. ix, no. 9 (November 1953); vol. x, no. 5 (May 1954): vol. x, no. 6 (June 1954); vol. xi, no. 1 (January 1955); vol. xi, no. 4 (April 1955); vol. xi, no. 9 (November 1955); vol. xii, no. 6 (June 1956); vol. xiii, no. 6 (June 1957); vol. xiii, no. 8 (October 1957); vol. xiv, no. 1 (January 1958), Special Collections Research Center, University of Chicago Library.

Genshiryoku sangyo Shinbun [Atomic Industrial News], the National Diet Library, Tokyo, Japan.

Kagawa Prefectural Library, *Yoran* [*The Bulletin*] (2018).

Nichibei genshiryoku sangyo godo kaigi gijiroku [*Minutes of Japan-U.S. Atomic Industrial Forum Joint Conference*] (1957).

Science and Technology Agency (Japan), Department of Atomic Energy, *Genshiryoku iinkai geppo* [*The Monthly Bulletin of the Atomic Energy Commission*], vol. 2, no. 9 (November 1957); vol. 3, no. 9 (September 1958); vol. 10, no. 8 (August 1965); vol. 18, no. 9 (September 1973).

USIS Film Catalog 1959 (Japan).
USIS Film Catalog 1966 (Japan).

Secondary Sources

English Language Books and Articles

Bailey, Martha J. *American Women in Science*. Santa Barbara, CA: ABC-CLIO, 1994.

Benedict, Ruth. *The Chrysanthemum and the Sword: Patterns of Japanese Culture*. London: Routledge, 2020. First published 1946 by Houghton Mifflin Co. (New York).

———. *Patterns of Culture*. London: Routledge, 2020. First published 1934 by Houghton Mifflin Co. (New York).

Birn, A. E. (Anne-Emanuelle). "Backstage: The Relationship between the Rockefeller Foundation and the World Health Organization, Part I: 1940s–1960s." *Public Health*, vol. 128, no. 2 (2014): 129–140.

Brown, E. Richard. "Public Health in Imperialism; Early Rockefeller Programs at Home and Abroad." *American Journal of Public Health*, vol. 66, no. 9 (September 1976): 897–903.

———. *Rockefeller Medicine Men: Medicine and Capitalism in America*. Berkeley, CA: University of California Press, 1979.

Burke, Peter. *What Is Cultural History?* 3rd ed. Cambridge: Policy Press, 2008. First published 2004.

Caidin, Martin. *The Astronauts; the Story of Project Mercury, America's Man-in-Space Program*. 1st ed. New York: E.P. Dutton, 1960.

Carpenter, M. Scott, Cooper, Gordon L., Glenn, John H., Grissom, Virgil I., Schirra, Walter M., Shepard, Alan B., and Slayton, Donald K. *We Seven: By the Astronauts Themselves*. New York, London, Sydney, NSW, and Toronto, ON: Simon and Schuster Paperbacks, 1990. First published 1962.

Carson, Rachel. *Silent Spring*. London: Penguin Books, 2000. First published 1962 by Houghton Mifflin (Boston).

Chapman, Jessica M. *Cauldron of Resistance: Ngo Dinh Diem, the United States, and 1950s Southern Vietnam*. Ithaca, NY: Cornell University Press, 2013. Kindle.

Clifford, James, and George E. Marcus, eds. *Writing Culture*. 1st ed. Berkeley, CA: University California Press, 1986.

Coleman, Peter. *The Liberal Conspiracy: The Congress for Cultural Freedom and the Struggle for the Mind of Postwar Europe*. New York: Free Press, 1989.

Cueto, Marcos. "International Health, the Early Cold War and Latin America." *Canadian Bulletin of Medical History*, vol. 25, no. 1 (2008): 17–41.

Cull, Nicholas. *The Cold War and the United States Information Agency*. Cambridge: Cambridge University Press, 2008.

DiMoia, John. "Atoms for Power?: The Atomic Energy Research Institute (AERI) and South Korean Electrification, 1948–1965." *Historia Scientiarum*, vol. 19, no. 2 (2009): 170–183.

Drogan, Mara. "The Nuclear Imperative: Atoms for Peace and the Development of U.S. Policy on Exporting Nuclear Power, 1953–1955." *Diplomatic History*, vol. 40, no. 5 (November 2016): 948–974.

Elkind, Jessica. *Aid Under Fire: National Building and the Vietnam War*. Lexington, KY: University Press of Kentucky, 2016.

Eschen, Penny Von. *Satchmo Blows Up the World: Jazz Ambassadors Plays the Cold War.* Boston, MA: Harvard University Press, 2005.

Ettling, John. *The Germ of Laziness: Rockefeller Philanthropy and Public Health in the New South.* Cambridge, MA: Harvard University Press, 1981.

Federal Council for Science and Technology. *Proceedings, First Symposium, Current Problems in the Management of Scientific Personnel, October 17–18, 1963.* Federal Council for Science and Technology, 1964.

Foner, Eric. *The Story of American Freedom.* New York: W. W. Norton & Co., 1999.

Foster, Anne L. "Introduction." *Diplomatic History*, vol. 41, no. 2 (2017): 225–227.

Geertz, Clifford. *The Interpretation of Cultures.* New York: Basic Books, 1973.

Gray, Beverly. "When Apollo Went to Japan." Smithsonian Institution, *Air & Space Magazine*, April 2020, https://www.airspacemag.com/space/when-apollo-went -japan-180974469/.

Haefele, Mark. "John F. Kennedy, USIA, and World Public Opinion." *Diplomatic History*, vol. 25, no. 1 (Winter 2001): 63–84.

Hammer, Ellen J. *A Death in November: America in Vietnam, 1963.* New York: E.P. Dutton, 1987.

Harris, Sarah Miller. *The CIA and the Congress for Cultural Freedom in the Early Cold War: The Limits of Making Common Cause.* London: Routledge, 2016.

Heil, Alan L. *Voice of America: A History.* New York: Columbia University Press, 2003.

Heppenheimer, T. A. *Countdown: A History of Space Flight.* New York: John Wiley & Sons, 1997.

Hewlett, Richard G., and Jack M. Hall. *Atoms for Peace and War, 1953–1961: Eisenhower and the Atomic Energy Commission.* Berkeley, CA: University of California Press, 1989.

Hewlett, R. G., and Holl, J. M. "A History of the United States Atomic Energy Commission, 1952–1960: Volume 3." United States. doi: 10.2172/6150636. U.S. Department of Energy, Office of Scientific and Technical Information, https://www.osti.gov/servlets/purl/6150636.

Higuchi, Toshihiro. "'Clean' Bombs: Nuclear Technology and Nuclear Strategy in the 1950s." *The Journal of Strategic Studies*, vol. 29, no. 1 (February 2006): 83–116.

———. "An Environmental Origin of Antinuclear Activism in Japan, 1954–1963: The Government, the Grassroots Movement, and the Politics of Risk." *Peace & Change*, vol. 33, no. 3 (July 2008): 333–367.

———. *Political Fallout: Nuclear Weapons Testing and the Making of a Global Environmental Crisis.* Stanford, CA: Stanford University Press, 2020.

Holl, Jack M. *Argonne National Laboratory 1946–96.* Chicago, IL: University of Illinois Press, 1997.

Ichikawa, Hiroshi. *Soviet Science and Engineering in the Shadow of the Cold War.* London and New York: Routledge, 2018.

Iriye, Akira. *Cultural Internationalism and World Order.* Baltimore, MD: Johns Hopkins University Press, 1997.

———. *Power and Culture: The Japanese-American War, 1941–1945.* Cambridge, MA and London: Harvard University Press, 1981.

Katzenstein, Peter J., ed. *The Culture of National Security: Norms and Identity in World Politics.* New York: Columbia University Press, 1996.

Kaufman, Scott. *Project Plowshare: The Peaceful Use of Nuclear Explosives in Cold War America.* Ithaca, NY: Cornell University Press, 2013.

Kendall, Harry H. *A Farm Boy in the Foreign Service: Telling America's Story to the World.* Bloomington, IN: AuthorHouse, 2003.

Kim, Sonjun. "Formation and Transition of the Korean Atomic Power System, from 1953–1980." PhD diss., Seoul National University, 2012.

Klein, Cristina. *Cold War Orientalism: Asia in the Middlebrow Imagination, 1945–1961.* Berkeley and Los Angeles, CA: University of California Press, 2003.

Koppel, Lily. *The Astronaut Wives Club: A True Story.* New York: Grand Central Publishing, 2013.

Kramer, Paul A. "Is the World Our Campus? International Students and U.S. Global Power in the Long Twentieth Century." *Diplomatic History*, vol. 33, no. 5 (November 2009): 775–806.

Krige, John. "NASA as an Instrument of U.S. Foreign Policy." In *Social Impact of Spaceflight*, edited by Steven J. Dick and Roger D. Launius, 207–218. Washington, DC: NASA SP-2007-4801, 2007.

———. "Technology, Foreign Policy and International Collaboration in Space." In *Critical Issues in History of Spaceflight*, edited by Steven Dick and Roger Launius, 239–260, Washington, DC: NASA-2006-4702, 2006.

———. "Techno-Utopian Dreams, Techno-Political Realities: The Education of Desire for the Peaceful Atom." In *Utopia/Dystopia: Conditions of Historical Possibility*, edited by Michael D. Gordin, et al., 151–155. Princeton, NJ: Princeton University Press, 2010.

Lasby, Clarence G. *Eisenhower's Heart Attack: How Ike Beat Heart Disease and Held on to the Presidency.* Lawrence, KS: University Press of Kansas, 1997.

Leslie, Stuart W. *The Cold War and American Science: The Military-Industrial-Academic Complex at MIT and Stanford.* New York: Columbia University Press, 1993.

MacLeod, Roy. "Consensus, Civility, Community: Minerva and the Vision of Edward Shils." In *Campaigning Culture and the Global Cold War: The Journals of the Congress for Cultural Freedom*, edited by Scott-Smith, 45–68. London: Palgrave MacMillan, 2017.

Maekawa, Reiko. "Rockefeller Foundation and Refugee Scholar during the Early Years of the Cold War." *Eibungaku Hyoron* [*Review of English Literature*], vol. 88 (February 2016): 85–113.

Masuda, Hajimu. *Cold War Crucible: The Korean Conflict and Postwar World.* Cambridge, MA: Harvard University Press, 2015.

Matsuda, Takeshi. *Soft Power and Its Perils: U.S. Cultural Policy in Early Postwar Japan and Permanent Dependency.* Stanford, CA: Stanford University Press, 2007.

Marshall, W. *Nuclear Power Technology: Volume 1: Reactor Technology.* Oxford: Oxford University Press, 1984.

May, Elaine Tyler. *Homeward Bound: American Family in the Cold War Era.* rev. and exp. ed., 4th ed. New York: Basic Books, 2017. First published 1988.

Mieczkowski, Yanek. *Eisenhower's Sputnik Moment: The Race for Space and World Prestige.* Ithaca, NY: Cornell University Press, 2013.

Minami, Kazushi. "Oil for the Lamps of America? Sino-American Oil Diplomacy, 1973–1979." *Diplomatic History*, vol. 41, no. 5 (November 2017): 959–984.

Mizuno, Hiromi, Aaron S. Moore, and John DiMoia, eds. *Engineering Asia: Technology, Colonial Development and the Cold War Order*. London and New York: Bloomsbury Academic, 2018.

Latour, Bruno. *Pandora's Hope: Essays on the Reality of Science Studies*. Cambridge, MA: Harvard University Press, 1999.

Needell, Allan A. *Science, Cold War and the American State: Lloyd V. Berkner and the Balance of Professional Ideals*. New York and London: Routledge, 2001.

Nye, Mary Jo. *Michael Polanyi and His Generation: Origins of the Social Construction of Science*. Chicago, IL: University of Chicago Press, 2011. Kindle.

Oldenziel, Ruth, and Karin Zachmann, eds. *Cold War Kitchen: Americanization, Technology, and European Users*. Cambridge, MA and London: The MIT Press, 2009.

Oreskes, Naomi, and John Krige, eds. *Science and Technology in the Global Cold War*. Cambridge, MA and London: The MIT Press, 2014.

Osgood, Kenneth. *Total Cold War: Eisenhower's Secret Propaganda Battle at Home and Abroad*. Lawrence, KS: University Press of Kansas, 2006.

Sambaluk, Nicholas Michael. *The Other Space Race: Eisenhower and the Quest for Aerospace Security*. Annapolis: Naval Institute Press, 2015.

Saunders, Frances Stonor. *Cultural Cold War: The CIA and the World of Arts and Letters*. New York: The New Press, 1999.

———. *Who Paid the Piper? The CIA and the Cultural Cold War*. London: Granta Books, 1999.

Scott-Smith, Giles. *The Politics of Apolitical Culture: The Congress for Cultural Freedom, the CIA, and Postwar American Hegemony*. London: Routledge, 2002.

Scott-Smith, Giles, and Charlotte Lerg, eds. *Campaigning Culture and the Global Cold War: The Journals of the Congress for Cultural Freedom*. London: Palgrave MacMillan, 2017.

Shanahan, Mark. *Eisenhower at the Dawn of the Space Age: Sputnik, Rockets and Helping Hands*. Lanham, MD: Lexington Books, 2017.

Simpson, Bradley R. *Economists with Guns: Authoritarian Development and U.S.-Indonesian Relations, 1960–1968*. Stanford, CA: Stanford University Press, 2008.

Slaney, Patrick David. "Eugene Rabinowitch, the *Bulletin of the Atomic Scientists*, and the Nature of Scientific Internationalism in the Early Cold War." *Historical Studies in the Natural Sciences*, vol. 42, no. 2 (2012): 114–142.

Smith, Alice K. *A Peril and a Hope: The Scientists' Movement in America, 1945–47*. Cambridge, MA: MIT Press, 1971. First published 1965.

Sorensen, Thomas C. *The Word War: The Story of American Propaganda*. New York: Harper & Row, 1968.

Strange, Susan. *States and Markets: An Introduction to International Political Economy*. 2nd ed. London: Printer Publishers, 1994.

Suri, Jeremi. *Power and Protest Global Revolution and the Rise of Détente*. Cambridge, MA: Harvard University Press, 2005.

Tomotsugu, Shinsuke. "The Bandung Conference and the Origins of Japan's Atoms for Peace Aid Program for Asian Countries." In *The Age of Hiroshima*, edited by Michael D.Gordin and John Ikenberry, 109–128. Princeton, NJ: Princeton University Press, 2020.

Tsuchiya, Yuka Moriguchi. "Atoms for Peace Exhibition in Japan: Localization of Nuclear Modernity." In *Japan's Split Society Between Genbaku and Genpatsu: Media, Propaganda and Science*, edited by Stephan Köhn, Felix Jawinski, Steffi Richter. forthcoming in 2023.

Wang, Jessica. *American Science in an Age of Anxiety: Scientists, Anticommunism & the Cold War*. Chapel Hill, NC and London: The University of North Carolina Press, 1999.

Wendt, Gerald. *You and the Atom*. New York: Whiteside/William Morrow & Company, 1956.

Westad, Odd Arne. *The Global Cold War: Third World Interventions and the Making of Our Times*. New Edition. Cambridge: Cambridge University Press, 2011.

Wilford, Hugh. *The Mighty Wurlitzer: How the CIA Played America*. Cambridge, MA: Harvard University Press, 2009.

Williams, Raymond. *Key Words: A Vocabulary of Culture and Society*, rev ed. New York: Oxford University Press, 1983.

Winner, Langdon. *The Whale and the Reactor: A Search for Limits in an Age of High Technology*. Chicago, IL: The University of Chicago Press, 1986.

Wittner, Lawrence S. *Confronting the Bomb: A Short History of the World Nuclear Disarmament Movement*. Stanford, CA: Stanford University Press, 2009.

Wolfe, Audra J. *Competing with the Soviets: Science, Technology, and the State in Cold War*. Baltimore, MD: The Johns Hopkins University, 2013.

———. *Freedom's Laboratory: The Cold War Struggle for the Soul of Science*. Baltimore, MD: Johns Hopkins University Press, 2018.

———. "Science and Freedom: The Forgotten Bulletin." In *Campaigning Culture and the Global Cold War: The Journals of the Congress for Cultural Freedom*, edited by Scott-Smith, 27–44. London: Palgrave MacMillan, 2017.

Zwigenberg, Ran. *Hiroshima: The Origins of Global Memory Culture*. Cambridge: Cambridge University Press, 2014.

Japanese Language Books and Articles

Akita, Shigeru, ed. *Ajia kara mita gurobaru hisutori* [*Global History from Asian Perspectives*]. Kyoto: Minerva Shobo, 2013.

———. *Teikoku kara kaihatsu enjo e: sengo ajia kokusaichitsujo to kogyoka* [*From Empire to Developmental Aid: The International Order in Postwar Asia and Industrialization*]. Nagoya: The University of Nagoya Press, 2017.

Aochi, Shin. *Gendai no eiyu: jinbutsu raibaru monogatari* [*The Modern Heroes: Rival Stories*]. Tokyo: Heibonsha, 1957. National Diet Library Digital Collections.

Aono, Toshihiko. *"Kiki no toshi" no reisen to domei: Berlin, Cuba, Détente, 1961–63* [*Cold War and Alliance in the Year of Crisis: Berlin, Cuba, Détente, 1961–63*]. Tokyo: Yuhikaku, 2012.

Arai, Tatsuo. *Amerika no taigai enjo: ICA no kino to unei* [*The U.S. Foreign Aid: Function and Administration of the ICA*]. *Keidanren Pamphlet*, vol. 33, 1956.

Arima, Tetsuo. *Genpatsu, Shoriki, CIA: kimitsu bunsho de yomu showa rimenshi* [*Nuclear Reactors, Shoriki, and CIA: The Hidden History of Showa Read through Secret Documents*]. Tokyo: Shinchosha, 2008.

———. *Genpatsu to genbaku: 'nichi bei ei' kakubuso no anto* [*Nuclear Power Plant and Nuclear Bombs: Japan, U.S. U.K.' Hidden Battles for Nuclear Armament*]. Tokyo: Bungeishunju, 2012.

Ayabe, Tsuneo, and Yoneo Ishii, eds. *Motto shiritai Myanmar* [*Learning More about Myanmar*]. 2nd ed. Tokyo: Kobundo, 1994.

Center for Research and Development Strategy, Japan Science and Technology Agency. *ASEAN shokoku no kagakugijutsu josei* [*The Current Conditions of S&T in ASEAN Countries*]. Takamatsu: Bikosha, 2015.

Daigo Fukuryumaru Heiwa Kyokai, ed. *Shinsoban: Bikini suibaku hisai shiryoshu* [*New Edition: Collection of Materials on the Bikini Thermonuclear Disasters*]. With the supervision by Yasuo Miyake, et al. Tokyo: University of Tokyo Press, 2014.

Fujioka, Masaki. *Amerika no daigaku niokeru soren kenkyu no hensei katei* [*The Formation Process of the Soviet Studies in U.S. Universities*]. Kyoto: Horitsu Bunka Sha, 2017.

Fujita, Fumiko. *Amerika bunka to nihon: reisenki no bunka to hito no koryu* [*The U.S. Cultural Diplomacy and Japan: Cultural and Personal Exchanges in the Cold War Era*]. Tokyo: University of Tokyo Press, 2015.

Gaimusho Hyakunenshi Hensan Iinkai, ed. *Gaimusho no hyakunen* [*One Hundred Years of the Ministry of Foreign Affairs*]. Tokyo: Hara Shobo, 1969.

Genshiryoku Gijutsushi Kenkyukai, ed. *Fukushima jikoni itaru genshiryoku kaihatsushi* [*The History of Atomic Energy Development Leading to the Fukushima Accident*]. Tokyo: Chuo University Press, 2015.

Ha, Kyungjin. *Public relations no rekishi shakaigaku: Amerika to nihon niokeru <kigyo jiga> no kochiku* [*Historical Sociology of Public Relations: Construction of the Industrial Subjectivity in the U.S. and Japan*]. Tokyo: Iwanami Shoten, Publishers, 2017.

Hirano, Kenichiro. *Kokusai bunkaron* [*International Culture Theory*]. Tokyo: University of Tokyo Press, 2000.

Hiratai, Yumi. "Kenkyushi tenbo: Rockefeller zaidan no iryo koshueisei katsudo to bunka gaiko" [State of the Field: The Rockefeller Foundation's Medical and Hygiene Activities and Cultural Diplomacy], *Sapporo Gakuin University Journal of the Society of Humanities*, vol. 92 (October 2012): 111–118.

Ichikawa, Hiroshi. "Obninsk 1955-nen: sekai hatsu no genshiryouk hatsudensho to Soviet kagakusha no 'genshiryoku gaiko'" [Obninsk 1955: The World's First Nuclear Power Plant and the Soviet Scientists' 'Nuclear Diplomacy']. In *Kakukaihatsu-jidai no isan: mirai sekinin o tou* [*The Legacies of the Nuclear Development Age: Questioning Responsibilities for the Future*], edited by Yuji Wakao and Eichi Kido, 26–50. Kyoto: Showado, 2017.

Ikawa, Mitsuo. "Genshiryoku heiwariyo hakurankai to shinbunsha" [Atoms for Peace Exhibitions and Newspaper Companies]. In *Sengo nihon no media event: 1945–1960* [*Media Events in Postwar Japan: 1945–1960*], edited by Toshihiro Tsuganezawa, 247–265. Kyoto: Sekai-shisosha, 2002.

Japan Productivity Center. *Seiansei undo 10-nen no ayumi* [*10-Year History of the Productivity Movement*]. Tokyo: Japan Productivity Center, 1965. National Diet Library Digital Collections.

Kagaku Gijutsucho Sosetsu 10-shunen Kinengyoji Jikko Junbi Iinkai, ed. *Kagakugijutsucho 10-shunenshi* [*The 10-Year History of the Science and Technology Agency*]. Tokyo: Kagakugijutsucho sosetsu 10-shunen kinen-gyoji kyosankai, 1966. National Diet Library Digital Collections.

Kamikawa, Ryunoshin. *Denryoku to seiji: nihon no genshiryoku seisaku zenshi* [*Electric Power and Politics: History of the Japanese Nuclear Power Policies*] vol. 1. Tokyo: Keiso Shobo, 2018.

Kamiya, Mitsugi. "Biruma-shiki shakaishgi to nogyo no hatten" [The Burmese Socialism and Agricultural Development]. *Quarterly Journal of Agricultural Economy*, vol. 26, no. 4 (October 1972): 175–198.

Kan, Hideki. *Reisenshi no saikento: henyo suru chitsujo to reisen no shuen* [*Re-examining the Cold War: Transforming Order and the End of the Cold War*]. Tokyo: Hosei University Press, 2010.

———, ed. *Reisen to domei: reisen shuen no shiten kara* [*The Cold War and Alliance: From the Perspective of the End of the Cold War*]. Kyoto: Shoraisha, 2014.

Kan, Hideki, and Ryuhei Hatsuse, eds. *Amerika no kaku gabanansu* [*The U.S. Nuclear Governance*]. Kyoto: Koyo Shobo, 2017.

Karashima, Masato. "Sengo nihon no shakaikagaku to Amerika no philanthropy: 1950–60 nendai niokeru nichibei hankyo liberal no koryu to Rockefeller zaidan" [Postwar Japanese Social Science and U.S. Philanthropy: Association of U.S. and Japanese Liberal Anti-Communists and the Rockefeller Foundation in the 1950s and 60s]. *Nihon Kenkyu*, vol. 45 (March 30, 2012): 155–183.

Kato, Tetsuro. *Nihon no shakaishugi: genbaku hantai, genpatsu suishin no ronri* [*Japanese Socialism: The Logic of Anti-Nuclear Weapons and Pro-Nuclear Power*]. Tokyo: Iwanami Shoten, Publishers, 2013.

Kato, Tetsuro, and Mitsuo Ikawa, eds. *Genshiryoku to reisen: nihon to ajia no genpatsudonyu* [*Atomic Energy and the Cold War: Introduction of Nuclear Power Stations in Japan and Asia*]. Tokyo: Kadensha, 2013.

Kikkawa, Takeo. *Tokyo denryoku, shippai no honshitsu: 'kaitai to saisei' no shinario* [*Tokyo Electric Power, the Essence of Failure: A Scenario toward Dismantling and Rebirth*]. Tokyo: Toyo Keizai Inc., 2011.

Kim, Jiyoung. *Nihon bungaku no <sengo> to henso sareru <Amerika>* [*'Postwar' Japanese Literature and 'America' Played in Variations*]. Kyoto: Minerva Shobo, 2019.

Kishi, Toshihiko, and Yuka Tsuchiya, eds. *Bunka reisen no jidai: Amerika to ajia* [*De-Centering the Cultural Cold War: U.S. and Asia*]. Tokyo: Kokusai Shoin, 2009.

Kobayashi, Somei. "VOA shisetsu iten o meguru kanbei kosho: 1972–73" [The U.S.-Japan Negotiation Concerning the Transfer of VOA Facilities; 1972–73], *Journal of Mass Communication Studies*, vol. 75 (2009): 129–147.

Kume, Shigeru. "Aru tsuihosha: Maeda Hisakichi no baai" [Purged: Maeda Hisakichi's Case]. *Shiso no Kagaku*, vol. 5, no. 53 (August, 1966): 43–54. National Diet Library Digital Collections.

Kurasawa, Aiko. *9.30: Sekai o shinkan saseta hi* [*September 30: The Day the World Was Shaken*]. Tokyo: Iwanami Shoten, Publishers, 2014.

Kurashina, Itsuki. "John Foster Dalles to gunbi kanri: 1958–59 kakujikken kinshi joyaku kosho o chushin ni" [John Foster Dalles and the Arms Control: PTBT Negotiations, 1958–59]. *The Hitotsubashi Journal of Law and International Studies*, vol. 2, no. 3 (November, 2003): 1167–1193.

Kurosaki, Akira. "Amerika no kakusenryaku to nihon no kokunai seiji no kosaku: 1954–60 nen" [The Intersection of U.S. Nuclear Strategy and Japanese Domestic Politics: 1954–60]. In *Chose-hanto to nihon no dojidaishi* [*The Contemporary History of Korea and Japan*], edited by Dojidaishi Gakkai, 189–233. Tokyo: Nihon Keizai Hyoronsha, 2005.

———. *Kakuheiki to nichibei kankei: Amerika no kakufukakusan gaiko to nihon no sentaku, 1960–1976* [*Nuclear Weapons and U.S.-Japan Relations: The U.S. Nuclear Non-Proliferation Diplomacy and Japan's Choice, 1960–1976*]. Tokyo: Yushisha, 2006.

Maekawa, Reiko. *Amerika chishikijin to radical vision no hokai* [*American Intellectuals and the Collapse of Radical Visions*]. Kyoto: Kyoto University Press, 2003.

Masaike, Akira. *Arakatsu Bunsaku to genshikaku butsurigaku no reimei* [*Arakatsu Bunsaku and the Dawn of Nuclear Physics*]. Kyoto: Kyoto University Press, 2018.

Masuda, Minoru, Ikeda, Ryo, Aono, Toshihiko, and Saito, Yoshiomi, eds. *Reisenshi o toinaosu: "reisen" to "hireisen" no kyokai* [*Reconsidering the Cold War: Border of the Cold War and Non-Cold War*]. Kyoto: Minerva Shobo, 2015.

Matsuoka, Hiroshi. *1961 Kennedy no senso: reisen, Vietnam, tonan-ajia* [*1961 Kennedy's War: Cold War, Vietnam, and Southeast Asia*]. Tokyo: Asahi Shimbun Publications, 1999.

———. *Kennedy to Vietnam senso: hanran chinatsu senryaku no zasetsu* [*Kennedy and Vietnam War: Failure of the Counter-Insurgency Strategy*]. Tokyo: Kinseisha, 2013.

———. *Kennedy wa Vietnam ni dou mukiattaka: JFK to Ngo Dinh Diem no anto* [*How Kennedy Confronted Vietnam: The Hidden Struggle between JFK and Ngo Dinh Diem*]. Kyoto: Minerva Shobo, 2015.

Matsuoka, Hiroshi, Yoshikazu Hirose, and Yoshihiko Takenaka, eds. *Reusenshi: sono kigen, tenkai, shuen to nihon* [*The Cold War: Its Origin, Development, Demise and Japan*]. Tokyo: Dobunkan Shuppan, 2003.

Miyagi, Taizo. *Bandung kaigi to nihon no ajia fukki: Amerika to ajia no Hazama de* [*The Bandung Conference and the Comeback of Japan in Asia: Between the U.S. and Asia*]. Tokyo: Soshisha, 2001.

Miyasaka, Naofumi. "Terrorism taisaku ni okeru senryaku bunka: 1990 nendai kohan no nichibei kankei o jirei toshite" [Strategic Culture in Counter-Terrorism: Cases of U.S. and Japan in the Late 1990s]. *Kokusai Seiji [International Relations]*, vol. 129 (February 2002): 61–76.

Nakakita, Koji. *Nihon rodo seiji no kokusai kankei-shi, 1945–1964: shakai minshushugi toiu sentakushi* [*History of International Relations Concerning Japanese Labor Politics: Social Democracy as an Alternative*]. Tokyo: Iwanami Shoten, Publishers, 2008.

Nakazawa, Shiho. "Eisenhower Seiken koki ni oeru kaku-gunshuku kosho: kakujikken teishi o meguru mondai o chushinni" [The Nuclear Disarmament Negotiation in the Latter Half of the Eisenhower Administration: A Focus on the Moratorium Issue]. *Bunka Joshidaigaku Kiyo: Jimbun Shakaikagaku Kenkyu [Bulletin of the Bunka Women's University: Humanities and Social Sciences]*, vol. 13 (January 2005): 41–53.

———. "Leo Szilard to genshi kagakusha undo: genshiryoku no kaihatsu tokanri no shiten kara" [Leo Szilard and the Atomic Scientists' Movement: From the Perspective of Atomic Power Development and Control]. *Kokusai Kankeigaku Kenkyu [Study of International Relations]*, no. 18 (January 1991): 51–60.

———. *Oppenheimer: Genbaku no chichi wa naze suibaku kaihatsu ni hantai shitaka* [*Oppenheimer: Why Did the Father of Atomic Bombs Oppose the Thermonuclear Bombs Development*]. Tokyo: Chuokoron-Shinsha, 1995.

———. "Suibaku kaihatsu hantai kankoku to kagakusha no tachiba" [The Anti-Thermonuclear Bombs Development Recommendation and Scientists' Standpoint]. *Kokusai Kankeigaku Kenkyu [Study of International Relations]*, no. 17 (January 1990): 19–29.

Oda, Takahiro, Hiroshi Kashiwagi, Takayuki Tatsumi, Masako Notoro, Matsuo Kazuyuki, and Shunya Yoshimi, eds. *Jiten gendai no Amerika [Encyclopedia on Modern America]*. Taishukan Publishing, 2004.

Oishi, Matashichi. *Bikini jiken no shinjitsu: inochi no kiro de* [*The Truth about the Bikini Incident: At the Crossroads of Life*]. Tokyo: Misuzu Shobo, 2003.

Okuaki, Satoru. *Umi no hoshano ni tachimukatta nihonjin: Message from Bikini to Fukushima* [*Japanese Scientists Confronting Radioactive Contamination of the Sea: Message from Bikini to Fukushima*]. Tokyo: Junpo-sha, 2017.

Onozawa, Toru. *Maboroshi no domei: reisen shoki Amerika no chuto seisaku* [*The Illusion of Alliance: The U.S. Middle Eastern Policies in the Early Cold War Era*]. Nagoya: The University of Nagoya Press, 2016.

Oyane, Satoshi. "Constructivism no shiza to bunseki: kihan no shototsu, chosei no jissho bunseki e" [The Perspective and Analysis of Constructivism: Toward the Positivist Analysis of the Collision and Coordination of Norms], *Kokusai Seiji* [*International Relations*], vol. 143 (November 2005): 124–140.

Saito, Yoshiomi. *Jazz Ambassadors: 'Amerika' no ongaku gaikoshi* [*The Jazz Ambassadors: History of the U.S. Music Diplomacy*]. Tokyo: Kodansha, 2017.

Sakuma, Hirayoshi. *Burma (Myanmar) gendai seijishi (zohoban)* [*Burma (Myanmar) Modern Political History (Updated)*]. Tokyo: Keiso Shobo, 1993.

Sasaki, Hideki. *Kaku no nanmin: Bikini suibaku jikken 'josen' gono genjitsu* [*Nuclear Refugees: Realities after the 'Decontamination' of the Bikini Thermonuclear Tests*]. Tokyo: NHK Publishing, 2013.

Sato, Yasushi. *NASA o kizuita hito to gijutsu: kyodai shisutemu kaihatsu no gijutsu bunka* [*People and Technology that Built the NASA: Technological Culture of the Big System Development*]. Tokyo: University of Tokyo Press, 2007.

———. *NASA: uchu kainatsu no 60-nen* [*NASA: 60 Years of Space Development*]. Tokyo: Chuokoron-Shinsha, 2014.

Shimada, Go. "Sengo Amerika no seisansei kojo, tainichi enjo ni okeru nihon no hienjokoku toshiteno keikenwa nanika: minshuka, rodo-undo shien, ajia eno tenkai" [Japanese Experience as a Recipient Country in the Postwar U.S. Productivity Improvement and Foreign Aid: Democratization, Support of the Labor Movement, and Advancement to Asia]. JICA Research Institute, Background Paper: A Historical Perspective, no. 2 (October 2018).

Suzuki, Kazuto. *Uchu-kaihatsu to kokusai seiji* [*Space Development and International Politics*]. Tokyo: Iwanami Shoten, Publishers, 2011.

Takahashi, Hiroko. *(Shintei zoho-ban) Fuin sareta Hiroshima, Nagasaki: bei kakujikken to minkan boei keikaku* [*(Updated and Enlarged Edition) Sealed Up Hiroshima and Nagasaki: U.S. Nuclear Tests and Civil Defense Program*]. Tokyo: Gaifusha, 2012.

Takeuchi, Keiji. *Denryoku no shakaishi: naniga Tokyo Denryoku o undanoka* [*Social History of Electric Power: What Gave Birth to Tokyo Electric Power Company*]. Tokyo: Asahi Shimbun Publications, 2013.

Takemine, Seiichiro. *Marshall shoto: owarinaki kakuhigai o ikiru* [*The Marshall Islands: Living the Endless Nuclear Victimization*]. Tokyo: Shinsensha, 2015.

Tanaka, Shingo. "Genshiryoku kaku mondai ni okeru tokushu na nichibei kankei no hoga: Truman seiken no tainichi genshiryoku kenkyu kisei to kanwa, 1945–47" [The Emergence of Special U.S.-Japan Relations in Atomic Power and Nuclear Issues: The Truman Administration's Restriction of Japan's Atomic Power Research and Its Relaxation, 1945–47]. *Kokusai Kokyo Seisaku Kenkyu* [*International Public Policy Studies*], vol. 17, no. 2 (March 2013): 113–126.

———. "'Nichibei genshiryoku kenkyu kyotei' eno dotei, 1951–1955: beikoku ni okeru kakuheiki shiyo no kioku to reisen senryaku" [The Pathway toward the U.S.-Japan Atomic Power Research Agreement, 1951-55: The U.S. Memory of the Use of Nuclear Weapons and the Cold War Strategy]. *Doshisha American Studies*, no. 52 (March 2016): 1–17.

———. "Taigai seisaku ketteiron ni okeru bunka: shuyo model no hyoka to kongo no kadai" [Culture in the Foreign Policy-Making: Evaluation of Major Models and Future Issues], *Kokusai Kokyo Seisaku Kenkyu* [*International Public Policy Studies*], vol. 12, issue 1 (September 2007): 243–257.

Tanikawa, Takeshi. *Amerika eiga to senryu seisaku* [*American Cinema and the Occupation Policy*]. Kyoto: Kyoto University Press, 2002.

Tomotsugu, Shinsuke. "'Ajia genshiryoku center' koso to sono zasetsu: Eisenhower Seiken no tai Ajia gaiko no ichi danmen" [The 'Asia Nuclear Center' Plan and Its Collapse: One Aspect of the Eisenhower Administration's Asia Diplomacy]. *Kokusai Seiji [International Relations]*, vol. 163 (January 2011): 14–27.

Tsuchida, Eiko. "Tekunoroji ga tsukuru kokumin, ethnicity: bunka-teki icon to shiteno kagaku gijutsu to shudan identity" [Technology Creates Nation and Ethunicity: Science and Technology as Cultural Icons and Group Identity]. In *"Hate" no jidai no Amerika-shi: jinshu, minzoku, kokuseki o kangaeru* [*American History in the Age of "Hate": Race, Ethnicity, and Nationality*], edited by Ayumu Kaneko and Yoshiyuki Kido, 95–117. Tokyo: Sairyusha, 2017.

Tsuchiya, Reiko, ed. *Kindai nihon media jinbutsu-shi: sogyosha, keieisha hen* [*Who's Who in Modern Japanese Media: Founders and Owners*]. Kyoto: Minerva Shobo, 2009.

Tsuchiya, Yuka. "Amerika-sei keisuiro no sentaku o meguru joho kyoiku program: 1950- nendai sue no nichibei kankei" [Information and Education Programs Concerning American Light Water Reactors: U.S.-Japan Relations in the Late-1950s]. *Rekishigaku Kenkyu [Historical Studies]*, vol. 976 (October 2018): 129–138.

———. "Amerika no seifu koho eiga (USIS eiga) ga egaita reisen sekai: iryo-hoken enjo-sen *Hope*-go o meguru kokusaiseiji" [The Cold War World Portrayed in the U.S. State-Sponsored Films (USIS Films): International Politics Involving the Medical Aid Ship *Hope*]. Chap. 9 in *Joho ga tsunagu sekaishi (MINERVA sekaishi sosho* vol. 6) [*The World History Connected by Information (MINERVA World History Series*, vol. 6)], edited by Shingo Minamizuka, 219–241. Kyoto: Minerva Shobo, 2019.

———. "'Hankyo' to 'hankaku': 1950 nendai ni okeru kagakuzasshi *Genshikagakusha Kaiho* to Bunka jiyu kaigi" ['Anti-Communist' and 'Anti-Nuclear': *Bulletin of the Atomic Scientists* and the Congress for Cultural Freedom in the 1950s]. *Americashi Kenkyu* [*Japanese Journal of American History*], no. 41 (September 2018): 36–51.

———. "Kagaku-gijutsu kohogaiko to genshiryoku heiwa-riyo: Sputnik shock igo no Atoms for Peace" [Science & Technology Public Diplomacy and the Peaceful Use of Atomic Energy: The Atoms for Peace After the Sputnik Shock]. In *Kaku no seiki: nihon genshiryoku kaihatsu-shi* [*The Nuclear Century: History of Japanese Atomic Power Development*], edited by Yasunao Kojita, et al., 193–223. Tokyo: Tokyodo Shuppan, 2016.

———. "Koho bunka gaiko toshiteno genshiryoku heiwariyo kyanpen to 1950nendai no nichibei kankei" [The Atoms for Peace Campaign as Public and Cultural Diplomacy: the U.S.-Japan Relations of the 1950s]. In *Nihibei domeiron: rekishi, kino, shuhenshokoku no shiten [U.S.-Japan Alliance: Perspectives of History, Function, and Neighboring Countries*], edited by Toshitaka Takeuchi, 180–209. Kyoto: Minerva Shobo, 2011.

———. "Maguro enyo-gyogyo to tuna-kan sangyo o meguru nichibei kankeishi: 1950–60 nendai no boeki-masatsu, suibaku jikken, soshite senzenki karano renzokusei" [U.S.-Japan Relations through Tuna Fisheries and Canneries: The Trade Conflict of the 1950s-60s, Nuclear Tests, and Continuity from the Prewar Era]. *Chushikoku American Studies*, vol. 8 (2017): 111–131.

———. "Senryo-ki no CIE eiga (Natco eiga)" [CIE Films (Natco Films) in Occupied Japan]. In *Fumikoeru Documentary (Nihon eiga wa ikite iru*, vol. 7) [*Documentaries beyond Boundaries (Japanese Films Are Alive*, vol. 7)], edited by Kiyoshi Kurosawa, et al., 155–181. Tokyo: Iwanami Shoten, Publishers, 2010.

———. *Shinbei nihon no kochiku: Amerika no tainichi joho kyoiku seisaku to nihon senryo* [*Constructing a Pro-U.S. Japan: U.S. Information and Education Policy in the Allied Occupation*]. Tokyo: Akashi Shoten, 2009.

———. "VOA *Forum* to kagakugijutsu kohogaiko: reisen radio wa Amerika no kagaku o do tsutaetaka" [VOA *Forum* and S&T Public Diplomacy: Cold War Radio Broadcast American S&T]. *Amerika Kenkyu* [*The American Review*], vol. 54 (April 2020): 67–87.

Tsuchiya, Yuka, and Shunya Yoshimi, eds. *Senryo suru me, senryu suru koe: CIE/USIS eiga to VOA rajio* [*The Occupying Eyes, Occupying Voices: CIE/USIS Films and VOA Radio*]. Tokyo: University of Tokyo Press, 2012.

Tsuchiya, Yuka, Shunsuke Okuda, and Shotaro Shindo. "Shiryo shokai: *Sprague iinkai hokokusho* (1960-nen 12-gatsu) shoyaku to kaisetsu" [Primary Source: *The Sprague Committee Report* (December 1960), A Translation of Selected Chapters and Commentary]. *Eibungaku Hyoron* [*Review of English Literature*], vol. 91 (February 2019): 1–29.

Watanabe, Shoichi, ed. *Columbo Plan: sengo ajia kokusi chitsujo no keisei* [*The Columbo Plan: Formation of the Postwar International Order in Asia*]. Tokyo: Hosei University Press, 2014.

———, ed. *Reisen henyoki no kokusai kaihatsu enjo to ajia: 1960-nendai o tou* [*International Developmental Aid and Asia in the Cold War in Transition: Inquiry into the 1960s*]. Kyoto: Minerva Shobo, 2017.

Yamamoto, Kazutaka. "Kennedy to 'uchu-kainatsu' seisaku" [Kennedy and the 'Space Development' Policies]. In *Kennedy to Amerika seiji* [*Kennedy and American Politics*], edited by Kazumi Fujimoto, 151–186. Tokyo: Tsunan Shuppan, 2004. First published 2000.

Yamashita, Masatoshi. *Kaku no umi no shogen: Bikini jiken wa owaranai* [*Testimony of the Nuclear Sea: The Bikini Incident Has Not Concluded*]. Tokyo: Shin Nihon Shuppan-sha, 2012.

Yamazaki, Masakatsu. "1968-nen no nichibei genshiryoku kyotei kaitei to kaku-fukakusan Taisei, 1955–1970: beikokusei keisuiro yui taisei no seiritsu" [The Revision of the 1968 U.S.-Japan Atomic Power Agreement, 1955–1970: The Establishment of the Supremacy of U.S. Light Water Reactors]. *Gijutsu Bunka Ronshu* [*TITech Studies in Science, Technology and Culture*], no. 15 (2012): 25–37.

———. "Keisuiro no nihon eno donyu to beikoku no kaku-fukakusan seisaku, 1964–1968: chugoku no kakujikken to nihon no kakuhoyu soshisaku to shiteno Atoms for Peace" [Introduction of Light Water Reactors to Japan and the U.S. Non-Proliferation Policy, 1964-1968: China's Nuclear Tests and the Atoms for Peace as Prevention against Japanese Nuclear Armament]. *Kagakushi Kenkyu* [*Journal of History of Science, Japan*], vol. 53, no. 270 (2014): 199–210.

———. *Nihon no kaku kaihatsu, 1939–1955: genbaku kara genshiryoku e* [*The Japanese Nuclear Development, 1939–1955: From Nuclear Bombs to Atomic Power*]. Tokyo: Sekibundo, 2011.

Yamazaki, Masakatsu and Shizue Hinokawa, eds. *Genbaku wa koshite kaihatsu sareta* [*This is How Atomic Bombs were Developed*]. Tokyo: Aoki Shoten, 1990.

Yoshimi, Shunya. *Yume no Genshiryoku: Atoms for Dream* [*The Dream of Atomic Power: Atoms for Dream*]. Tokyo: Chikuma Shobo, 2012.

———. *Hakurankai no Seijigaku* [*Politics of Exhibitions*]. Tokyo: Chuokoron-Shinsha, 2000. First published 1992.

Yoshioka, Hitoshi. *Shinban genshiryoku no shakaishi: sono nihon-teki tenkai* [*New Edition, Social History of Atomic Power: Japanese Development*]. Tokyo: Asahi Shimbun Publications, 2011.

Newspapers

<Japanese>

Asahi Shinbun
Sankei Shinbun
Shikoku Shinbun
Chugoku Shinbun
Nihon Keizai Shinbun
Mainichi Shinbun
Yomiuri Shinbun

<English>

Baltimore Sun
Boston Glove
New York Times

Magazines

"The Astronauts: Ready to Make History." *Life*, vol. 47, no. 11 (September 14, 1959): 26–43.

"Rocket to uchu-ryoko ten" [Rockets and Space Travel Exhibition]. *Koku Fan*, vol. 9, no. 8 (August 1960): 40–41. National Diet Library Digital Collections.

Sekiguchi, Naosuke. "Nakasone uchu-kaihatsu eikaku: keii to mondai-ten" [The Nakasone Space Development Program: Details and Problems]. *Shizen=Nature*, vol. 15, no. 5 (issue 165) (May 1960): 62–65. National Diet Library Digital Collections.

"Seven Brave Women behind the Astronauts: Spacemen's Wives Tell, in Their Own Words, Their Inner Thoughts and Worries." *Life*, vol. 47, no. 12 (September 21, 1959): 142–163.

"Shijo kengaku, uchu-haku" [Visit in the Magazine: Space Exhibition]. *Koku Joho*, no. 120 (August 1960): 129–131. National Diet Library Digital Collections.

"Uchu daihakurankai manga rupo" [The Great Space Exhibition Manga Report]. *Shukan Sankei*, vol. 9, no. 29 (issue 445) (June 13, 1960): 72–73. National Diet Library Digital Collections.

"Uchu hakurankai kenbunki" [Space Exhibition Visiting Report]. *Chugaku-jidai, Ichinensei*, vol. 5, no. 5 (August 1960): 166–167. National Diet Library Digital Collections.

U.S. Department of State, *News Letter*, 72 (April 1967).

TV Programs and Moving Images

Bibas, Frank P., dir. *Project Hope*. 1961.

NHK BS1 Special, "Gokuhi shirei, uran nenryo o kaishu seyo: senka no genshiro 40-nenme no shinjitsu" [Secret Mission, Recover the Uranium Fuel: Nuclear Reactor in Fire, Truth Uncovered after 40 Years]. Aired June 20, 2015, on NHK BS1.

NHK ETV Special, "Amerika kara mita Fukushima genpatsu jiko" [The Fukushima Nuclear Accident from the American Perspective]. Aired September 4, 2011, on NHK ETV.

NHK ETV Special, "Umi no hoshano ni tachimukatta nihonjin: Bikini jiken to Shunkotsu-maru" [Japanese Scientists Confronting Radioactive Contamination of the Sea: The Bikini Incident and Shunkotsu-maru]. Aired September 28, 2013, on NHK ETV

Winged Scourge, 1943. RG306, 306.240, National Archives at College Park.

Websites

American Heart Association website, https://www.heart.org/en/affiliates/paul -dudley-white-about.

Argonne National Laboratory website, http://www.ne.anl.gov/About/hn/ news961012.shtml.

Bulletin of the Atomic Scientists website, https://thebulletin.org/.

CIA, Freedom of Information Act Electronic Reading Room, https://www.cia.gov /library/readingroom/.

Database of Japanese Diplomatic History website, https://drive.google.com/file/d /0B_wk3O1slLl7amdfOUVqSVp5alU/view.

Densho Encyclopedia, http://encyclopedia.densho.org/A.L._Wirin/.

Dwight D. Eisenhower. "Address at the Centennial Comment of Pennsylvania State University." June 11, 1955. Online by Gerhard Peters and John T. Woolley, The American Presidency Project, http://www.presidency.ucsb.edu/documents/ address-the-centennial-commencement-pennsylvania-state-university.

———. "Annual Message to the Congress on the State of the Union." January 9, 1958, The American Presidency Project, University of California, Santa Barbara, https://www.presidency.ucsb.edu/documents/annual-message-the-congress -the-state-the-union-10.

Dwight D. Eisenhower Presidential Library website, https://www.eisenhowerlibrary .gov.

Energy Citations Database by U.S. Department of Energy, https://www.energy.gov /eere/bioenergy/databases.

Expo '70 Commemorative Park website, https://www.expo70-park.jp/cause/expo/america/.

Gray, Beverly. "When Apollo Went to Japan." *Air & Space Magazine*, April 2020, https://www.airspacemag.com/space/when-apollo-went-japan-180974469/.

Hanna Holborn Gray Special Collections Research Center, University of Chicago Library, Guide to the Atomic Scientists of Chicago Records 1943–1955, https://www.lib.uchicago.edu/e/scrc/findingaids/view.php?eadid=ICU.SPCL.ASCHICAGO.

Hanna Holborn Gray Special Collections Research Center, University of Chicago Library, Guide to the Bulletin of the Atomic Scientists Records 1945–1984, https://www.lib.uchicago.edu/e/scrc/findingaids/view.php?eadid=ICU.SPCL.BULLETIN.

Hanna Holborn Gray Special Collections Research Center, University of Chicago Library, Guide to the Eugene I, Rabinowitch Papers, https://www.lib.uchicago.edu/e/scrc/findingaids/view.php?eadid=ICU.SPCL.RABINOWITCH.

IAEA website, https://www.iaea.org/about/history/atoms-for-peace-speech.

"Inaugural Address of John F. Kennedy." January 20, 1961, Yale Law School Lillian Goldman Law Library, The Avalon Project, https://avalon.law.yale.edu/20th_century/kennedy.asp.

Institute of Space and Astronautical Science (Japan) website, http://www.isas.jaxa.jp/about/history/.

The Japanese Ministry of Foreign Affairs website, FAQ, "What is Public Diplomacy? What is Soft Power?" http://www.mofa.go.jp/mofaj/comment/faq/culture/gaiko.html.

Kokkai kaigiroku kensaku shisutemu [Database of the Minutes of Japanese Diet Meetings], https://kokkai.ndl.go.jp/simple/detail?minId=102605077X01419570329&spkNum=37#s37.

Ministry of Foreign Affairs (Japan) website, http://www.mofa.go.jp/mofaj/comment/faq/culture/gaiko.html.

MYANMAR JAPON Online (November 2015), https://myanmarjapon.com/1511interview.html.

NASA website, https://www.nasa.gov/content/nasa-history-overview.

ONR Website, https://www.onr.navy.mil/About-ONR/History/History-Research-Guide.

Pennsylvania State University, College of Engineering website, https://www.rsec.psu.edu/Penn_State_Breazeale_Reactor.aspx.

Project Hope website, https://www.projecthope.org/.

Smithsonian National Air and Space Museum website, https://airandspace.si.edu/exhibitions/destination-moon.

"Underground Lab Capability at WIPP" The U.S. Department of Energy, Waste Isolation Pilot Plant (WIPP) website, http://www.wipp.energy.gov/science/ug_lab/gnome/gnome.htm.

University of Michigan Energy Institute website, http://energy.umich.edu/about-us/phoenix-project.

Index